ISNM
International Series of Numerical Mathematics
Vol. 136

A Variational Inequality Approach to free Boundary Problems with Applications in Mould Filling

Jörg Steinbach

Birkhäuser Verlag
Basel · Boston · Berlin

Author:

Jörg Steinbach
Gärtnerstr. 8
86153 Augsburg
Germany

2000 Mathematics Subject Classification 35R35, 35R45, 49J40, 35D05, 35D10, 65N30, 65N12, 65N15, 76D08, 76M30

A CIP catalogue record for this book is available from the Library of Congress, Washington D.C., USA

Deutsche Bibliothek Cataloging-in-Publication Data
Steinbach, Jörg:
A variational inequality approach to free boundary problems with applications in mould filling / Jörg Steinbach. - Basel ; Boston ; Berlin : Birkhäuser, 2002
(International series of numerical mathematics ; Vol. 136)
ISBN 978-3-0348-7599-8 ISBN 978-3-0348-7597-4 (eBook)
DOI 10.1007/978-3-0348-7597-4

Softcover reprint of the hardcover 1st edition 2002
Member of the BertelsmannSpringer Publishing Group
Printed on acid-free paper produced of chlorine-free pulp. TCF ∞

ISBN 978-3-0348-7599-8

9 8 7 6 5 4 3 2 1

www.birkhauser.ch

Contents

Preface ix

1 Introduction

2 Evolutionary Variational Inequality Approach

2.1 The degenerate free boundary problem 8

2.2 Some application problems 10

2.3 Different fixed domain formulations 18

2.3.1 Front tracking and fixing methods versus fixed domain formulations exemplified by injection and compression moulding 18

2.3.2 Weak formulation 19

2.3.3 The evolutionary variational inequality approach 23

3 Properties of the Variational Inequality Solution

3.1 Problem setting and general notations 32

3.2 Existence and uniqueness result 36

3.3 Monotonicity properties and regularity with respect to time . . . 40

3.3.1 Time-independent convex sets 40

3.3.2 Time-dependent convex sets 46

3.4 Regularity with respect to space variables 55

3.4.1 Dirichlet boundary conditions 56

3.4.2 Boundary conditions of Neumann/Newton type 63

3.5 Some remarks on further regularity results 69

4 Finite Volume Approximations for Elliptic Inequalities

4.1 Finite element and volume approximations for the obstacle problem 74

4.1.1 The elliptic obstacle problem 74

4.1.2 Finite element approximations for the obstacle problem 76

4.1.3 Basics of finite volume approximations 80

4.1.4 Finite volume approximations for the obstacle problem 86

4.2 Comparison of finite volume and finite element approximations . . . 93
4.3 Error estimates for the finite volume solution . . . 103
4.4 Penalization methods for the finite volume obstacle problem . . . 110
4.4.1 Discrete maximum principle . . . 110
4.4.2 Discussion of penalization techniques . . . 111
4.4.3 Iterative solution of the penalization problems . . . 122
4.5 The Signorini problem as a boundary obstacle problem . . . 126
4.6 Results from numerical experiments for elliptic obstacle problems . . . 131
4.6.1 Examples with known exact solution . . . 132
4.6.2 Numerical results for the error between the finite element and the finite volume solution . . . 135
4.6.3 Error behaviour of the finite volume and the penalization solutions . . . 137

5 Numerical Analysis of the Evolutionary Inequalities

5.1 Finite element and volume approximations for the evolutionary problems . . . 144
5.1.1 Formulation of the finite element and finite volume approximations . . . 145
5.1.2 Properties of the discrete inequality problems . . . 149
5.1.3 Time evolution of the finite volume solution . . . 156
5.2 Error estimates for the finite element and finite volume solutions . . . 157
5.2.1 Comparison of the finite element and finite volume approximations . . . 158
5.2.2 A priori estimates for the finite element and finite volume solutions . . . 162
5.2.3 Convergence rate for the finite element and finite volume solutions . . . 170
5.3 Penalization methods for the evolutionary finite volume inequalities . . . 179
5.3.1 Discussion of penalization techniques . . . 179
5.3.2 Iterative solution of the penalization problems . . . 187
5.4 Numerical experiments for evolutionary variational inequalities . . . 191
5.4.1 Two evolutionary variational inequalities and the related free boundary problems . . . 192
5.4.2 Numerical results for the errors between exact, finite element and finite volume solution . . . 194
5.4.3 Error behaviour of the penalization solutions . . . 199

6 Injection and Compression Moulding as Application Problems
6.1 Classical Hele–Shaw flows and related moving boundary problems . . . 204
6.2 Mathematical modelling of injection and compression moulding . . . 206
6.2.1 Injection and compression moulding – Technical background and requirements on simulation . . . 207
6.2.1.1 Technical background . . . 207
6.2.1.2 Short comparison of injection/compression moulding and metal casting . . . 209
6.2.1.3 Some aims of the numerical simulation . . . 210
6.2.2 Balance and state equations . . . 212
6.2.3 Rheological behaviour of polymer melts . . . 213
6.2.4 Temperature-dependent Hele–Shaw flow in the injection and compression moulding process . . . 217
6.2.4.1 The generalized Hele–Shaw flow . . . 217
6.2.4.2 Viscosity models and non-isothermal effects . . . 221
6.2.4.3 The numerical core problems . . . 224
6.2.5 The distance concept – a geometrical approach for injection moulding . . . 225
6.2.6 Recent three-dimensional simulation developments . . . 231
6.3 Simulation results . . . 238
6.3.1 Variation of gate location and thickness, non-isothermal effects, narrow flow region . . . 238
6.3.2 Comparison with the distance model . . . 246
6.3.3 Comparison with three-dimensional simulations . . . 252

7 Concluding Remarks

Bibliography . . . 263
List of Figures . . . 277
List of Tables . . . 279
List of Symbols . . . 281
Index . . . 289

Preface

Since the early 1960s, the mathematical theory of variational inequalities has been under rapid development, based on complex analysis and strongly influenced by 'real-life' application. Many, but of course not all, moving free (i.e., a priori unknown) boundary problems originating from engineering and economic applications can directly, or after a transformation, be formulated as variational inequalities.

In this work we investigate an evolutionary variational inequality with a memory term which is, as a fixed domain formulation, the result of the application of such a transformation to a degenerate moving free boundary problem. This study includes mathematical modelling, existence, uniqueness and regularity results, numerical analysis of finite element and finite volume approximations, as well as numerical simulation results for applications in polymer processing.

Essential parts of these research notes were developed during my work at the Chair of Applied Mathematics (LAM) of the Technical University Munich. I would like to express my sincerest gratitude to K.-H. Hoffmann, the head of this chair and the present scientific director of the Center of Advanced European Studies and Research (caesar), for his encouragement and support. With this work I am following a general concept of Applied Mathematics to which he directed my interest and which, based on application problems, comprises mathematical modelling, mathematical and numerical analysis, computational aspects and visualization of simulation results. In preparation of the final version of this work, especially Chapter 6, I have tried to take into account not only the mathematical point of view but also industrial requirements on flow analysis software. In this context I have benefitted from my present work in the software development group of Magma Gießereitechnologie GmbH. This company develops commercial codes for mould filling analysis (and related processes) for metal casting and injection moulding of plastics and elastomers.

I want to thank my former supervisor W. Weinelt for introducing me to Applied Mathematics and, in particular, to the topics of variational inequalities and finite volume methods in connection with application problems. He has strongly influenced my mathematical and general way of thinking.

Many ideas and various aspects of the present work have been strongly influenced by useful discussions and valuable suggestions from colleagues of the international free boundary community, who are too numerous to mention here. At this point I would like to thank them all. In particular, I wish to mention J.F. Rodrigues of CMAF Lisbon from whom I have learned a lot about obstacle problems in mathematical physics.

Very special thanks are due to G. Aronsson, F. Berntsson and P. Johansson at the Department of Mathematics, Linköping University for fruitful cooperation concerning the comparison of their distance model as a geometrical approach with finite element/volume models for injection moulding. I am grateful to them for performing several computations with the distance model and letting me use some figures in Chapter 6.

Furthermore, I want to extend my thanks to my former colleagues at the LAM for interesting and stimulating discussions on mathematics in general and several aspects of this work, their help in performing computations by means of the finite element toolbox FeliCs (developed at the LAM) and their reading of parts of the manuscript.

I wish to thank my parents and my wife Andrea for their continuous moral support. In particular, I am indebted to my wife for her encouragement and for her active participation in proof reading. Best thanks to Andrea and my daughter Nicole for patiently overlooking my absence of mind during several phases of the preparation of this work.

Finally, I wish to express my sincere appreciation to Birkhäuser-Verlag, especially to Mrs. S. Lotrovsky and Mr. Th. Hintermann, for the competent help in all editorial questions and for their patience during the preparation of the final version of this book.

August 2001 Jörg Steinbach

International Series of Numerical Mathematics
Vol. 136, © 2002 Birkhäuser Verlag Basel/Switzerland

Chapter 1

Introduction

Founded in the 1960s the *variational inequality theory* is a relatively young and fast-growing field of Applied Mathematics. It is a typical example of how application problems stimulate and intensify development of both mathematical analysis and numerical treatment. During the last 30–40 years, variational inequalities have proved their importance in the mathematical modelling and numerical simulation of diverse application problems. Besides the pioneering work by Fichera in 1964 (see [FICH64]) on solution of the Signorini problem in the theory of elasticity, applications of variational inequalities in physics, mechanics, engineering, and also in mathematical programming, control and optimization problems, have been considered in many monographs, see e.g., [DULI76], [KIST80], [GLT81], [BELI82], [BENS82], [ELOC82], [BACA84], [BELI84], [CHIP84], [CRAN84], [GLOW84], [RODR87], [FRIE88], [KIOD88], [FRSP93] and [HANE96].

These application problems show also the close connection of variational inequalities to *free boundary problems (FBP)*, since most of them are characterized by a (moving) free boundary dividing the domain into different phases. This boundary is not known in advance and has to be determined as part of the solution of the problem. We emphasize that the term 'free boundary problem' is commonly used for two types of problems:

- a stationary boundary as part of a steady-state problem and
- time-dependent problems when the position of the *moving* boundary has to be determined as a function of space and time.

Over the last years the wide field of FBP has been under rapid development. Both mathematical aspects of these problems and their numerical solution have been widely studied. This has been indicated, for instance, by the interdisciplinary conference series on 'FBP: Theory and Applications' which started in 1981. For a survey see the several conference proceedings [FAPR83], [BDF85], [HOSP90], [CHRA93], [DHLV95], [NIST96] and [ATMR99].

Let us return to the connection between FBP and variational inequalities. First, we emphasize that not all FBP can be reduced to variational inequalities. However, numerous FBP admit a direct variational inequality formulation. Moreover, a narrow but, owing to its 'real-life' significance important, subclass of FBP can, after applying an appropriate transformation, be formulated as variational inequalities. The problems to be studied here belong to this second subclass which is characterized by a jump discontinuity of the gradient across the free boundary. At the beginning of the 1970s, Baiocchi (in his work on the dam problem) and Duvaut (with respect to the classical Stefan problem) pointed out how such FBP can be converted into variational inequalities by performing an integral transformation to the dependent variable in the original problem. We refer to [RODR87], Section 2.9, 2.12 and [NAUM84], Chapter 4 for more historical comments.

In this monograph an evolutionary variational inequality approach to a degenerate moving free boundary problem is studied with respect to both analytical and numerical methods. The main features of this variational inequality formulation of obstacle type are a second order elliptic differential operator, a memory term, time-dependent convex (constraint) sets and different types of boundary conditions. The study of such variational inequality problems is motivated by their applications, e.g., the generalized Hele–Shaw flow in injection and compression moulding, the electro-chemical machining process with a time-dependent conductivity of the electrolyte or a quasi-stationary Stefan type problem with zero-specific heat. These application problems can be formulated as degenerate moving free boundary problems characterized by space- and, in particular, time-dependent coefficients in both the differential operator and the boundary conditions. The evolutionary inequality problem as a fixed domain formulation is the result of the application of an integral (Baiocchi type) transformation to the free boundary problem.

The starting point of our work is a general (moving) free boundary problem (see Section 2.1) 'lying between' the classical (one-phase) Stefan problem and the classical Hele–Shaw problem. The one-phase Stefan problem describes the melting of a body of ice maintained at zero degree in contact with a region of water. In a two-phase Stefan problem one considers additionally the variation of the temperature in the (second) solid phase. The classical Hele–Shaw problem arises when an incompressible, viscous, isothermal liquid is moving within a (thin) region between two slightly separated plates. From the mathematical point of view the Hele–Shaw problem can be regarded as a degenerate one-phase Stefan problem with zero-specific heat. It is beyond the scope of this book to give a survey of the comprehensive bibliography concerning classical, weak ('enthalpy-type') and variational inequality solutions of these problems and their variants. At this stage we only refer to the monograph [MEIR92] and to the research-expository work [RODR94] concerning the Stefan problem (see also the references cited below in Chapter 2). In Section 6.1 we will give a brief overview of the mathematical study of the classical Hele–Shaw flow problem (see the references cited there). In a general context let us mention the monograph [JERO83] dealing with the global existence of weak solutions and the local classical theory of general nonlinear evolution systems. Furthermore, various aspects of the numerical approximation of parabolic free boundary problems of Stefan type are discussed in the research notes [VERD94] (see also the literature cited therein and the references given in our Chapter 2).

We remark that application of the above mentioned Baiocchi transformation to one- or two-phase Stefan problems leads to parabolic variational inequalities, whereas this integral transformation with respect to time gives a family of elliptic inequalities with time as a parameter for the classical Hele–Shaw problem.

We emphasize that both the weak ('enthalpy type', see Section 2.3.2) and the variational inequality formulation (see Section 2.3.3) represent *fixed domain formulations* for the corresponding FBP. Typically, the free boundary disappears as an explicit unknown in such fixed domain formulations. However, it can be recovered *a posteriori* as a level set of the solution. Compared to classical formulations of FBP, fixed domain formulations have numerous advantages both with respect to the mathematical analysis and to the numerical treatment. More precisely, fixed domain formulations allow us to apply variational methods for the mathematical-analytical investigation, see e.g., [RODR94] for the Stefan problem. Furthermore, their numerical solution does not require us to track the free boundary, which may exhibit singularities such as cusps (see e.g., [VERD94] for the Stefan problem and the discussion in Section 2.3.1 for the problems considered here). These general advantages of fixed domain formulations are the motivation for this work to derive a variational inequality formulation for the moving free boundary problem although we obtain a memory term. At least for the well-posed situation (advancing moving boundary) of the problem studied in this work, such fixed domain formulations can be given (for more details see Chapter 2).

As indicated above, the free boundary problem to be considered here occupies an intermediate position between Stefan and Hele–Shaw type problems. To be more precise, it is characterized by a vanishing time-derivative as well as space- and, in particular, *time-dependent* coefficients in the differential operator and in the different types of boundary conditions. Then, as a result of the application of a generalized Baiocchi type transformation, we obtain evolutionary variational inequalities of obstacle type containing a second order elliptic differential operator, *a memory term* and different types of boundary conditions. The solution has to belong to a *time-dependent* convex set. The time-derivative of the solution does not appear in the inequality formulation. The memory term is caused by the time-dependent coefficients.

Consideration of the time-dependent coefficients is motivated by application problems (injection and compression moulding, electro-chemical machining process, etc.; see Section 2.2 and Chapter 6). Based on these applications the main topic of our work consists in the mathematical and numerical analysis of such evolutionary variational inequalities. At the moment we do not go into further details in the description of the evolutionary inequalities, because this is the topic of Chapter 2. There we will formulate the moving free boundary problem, discuss connections to related problems, derive and compare different fixed domain formulations and introduce some of the underlying application problems.

Let us continue with the outline of the following chapters. We mention that at the beginning of each chapter and section a brief introduction and an overview on the content will be given.

In Chapter 3 we are concerned with the mathematical analysis of evolutionary variational inequalities. Besides the memory term, the time-dependent convex (constraint) sets and the different types (Dirichlet, Neumann, Newton) of boundary conditions occuring in the inequality formulation represent essential difficulties for the mathematical treatment. The general problem of existence and uniqueness of a solution is treated by means of a fixed point argument in connection with the theory of elliptic inequalities. Having proved existence of the solution, we study its regularity properties both with respect to time and space as well as its time evolution. For these subjects we exploit a Lewy–Stampacchia type penalization/regularization (see [RODR87], Section 5.3 and [BELI84], Section 2.5.4) which seems to be more suitable than other penalization methods (in particular, with respect to the different types of boundary conditions). To investigate the spatial regularity we additionally use a semi-discretization in time (Rothe method, see e.g., [ROTH30], [KACU85]). Moreover, having in mind the numerical error analysis (see Section 5.2), we look for the associated complementarity problem and Lewy–Stampacchia type inequalities (see [RODR87], Section 5.3).

We note that penalization methods are essential mathematical techniques for the study of regularity properties of the solutions of variational inequalities (see e.g., [KIST80], [BELI82], [BELI84], [RODR87] and [RODR94]). In this context it is worthwhile to emphasize that the question of regularity for variational inequalities differs from its analog in the theory of boundary value problems. In general, constraints defining the convex sets impede regularity of the solutions. In other words, the solutions cannot surpass a 'limit regularity' regardless of smoothness of the data. Moreover, the convergence rate of discretization methods is also limited (see Remark 4.18 and Figure 4.2).

In Chapters 4 and 5 we deal with the numerical treatment of elliptic and evolutionary variational inequalities. Besides finite element approximations we focus our attention on finite volume approximations which fulfill a discrete maximum principle (at least for a weakly acute triangulation). We refer to Section 4.1.3 for an introduction of finite volume methods (often also called box methods).

In Chapter 4 we study finite volume schemes in a variational framework for the numerical solution of *elliptic* obstacle problems. Stimulated by papers (see e.g.,

[HACK89]) concerning the comparison of finite element and finite volume approximations for second order elliptic equations, we will extend this comparison to *interior and boundary* obstacle problems with mixed boundary conditions in two and three dimensions (see Sections 4.2 and 4.5). However, we will not only compare finite volume schemes with finite element approximations, but we will also derive error estimates between the exact and the finite volume solution. Furthermore, we analyse two penalization techniques for solution of the (discretized) inequality problems. The theoretical results are examined by numerical tests. Let us anticipate that we will show a comparable accuracy of the finite volume and finite element approximations.

Besides being interesting by itself, the reason for the inclusion of Chapter 4 is that many results of these investigations will be exploited for the numerical treatment of the *evolutionary inequalities.* This is the topic of Chapter 5 and comprises the investigation of both finite element and finite volume approximations in two and three space dimensions. After introducing a suitable (efficient) discretization of the memory term, we will study the stability of the discrete problems, compare finite element and volume approximations, derive error estimates based on the regularity gained in Chapter 3 (see, in particular, Section 3.5) and discuss different penalization (regularization) methods. On several occasions we exploit monotonicity properties by means of the above mentioned discrete maximum principle, which is an important reason to consider finite volume approximations. Finally, at the end of Chapter 5 we report on results of numerical tests illustrating the convergence behaviour between the exact, the finite element, the finite volume and the penalization solutions.

Chapter 6 contains an overview of the mathematical modelling, the numerical core problems and simulation results concerning our main application problems. These are the injection and compression moulding processes for polymers which represent relevant manufacturing techniques for industrial mass-produced plastic articles. In injection moulding the polymer melt is injected from gates into a cavity. In compression moulding a premeasured initial charge is placed between the mould halves and then, by closing the mould, the polymer melt is squeezed out to fill the cavity (squeeze flow). Although both manufacturing techniques appear to be simple, there is a complex interaction between operating and geometrical conditions as well as material properties. The physical processes of both techniques to be simulated are characterized by the flow of the plastic melt, the a priori unknown free boundary between melt and escaping air (flow front), the energy balance containing convection, diffusion and viscous heating and, finally, by the rheological behaviour of the melt. The coupled phenomena of fluid flow, heat transfer and rheology during the filling stage of the cavity determine to a large extent the final properties of the moulded plastic parts.
This complexity of interaction motivates mathematical modelling and the development of software based on special finite element and finite volume techniques which supplement and improve commercial software packages. The evolutionary

variational inequality that we consider as a *fixed domain formulation* for the flow process exhibits essential progress in this context.

In Chapter 6 we do more than just describe the application of the evolutionary variational inequality using a generalized Hele–Shaw flow in injection and compression moulding. We also discuss a geometrical (asymptotic) approach (pseudo-circle principle as a distance concept) and some recent developments concerning full three-dimensional simulation methods governed by generalized Stokes and Navier–Stokes flow, respectively.

The simulation results presented for the application problems that we consider are concerned with the study of the influence of basic geometrical and operating conditions as well as with comparisons between different mathematical models.

We emphasize that the type of evolutionary inequalities considered in this work can not only be used advantageously in mathematical modelling and numerical simulation of applications in polymer processing, but also for the electro-chemical machining process and other 'real-life' applications (see Section 2.2).

The final Chapter 7 contains a short discussion of the main results and some recommendations for further study.

Summarizing, the main aim of this work consists in the study of an evolutionary variational inequality approach to a moving free boundary problem, both with respect to the mathematical analysis and to the numerical treatment. In the course of our investigations, stimulated by application problems, we point out connections to mathematical modelling, to computational aspects of numerical algorithms and to simulation results for manufacturing techniques in polymer processing.

International Series of Numerical Mathematics
Vol. 136,

Chapter 2

Derivation of the Evolutionary Variational Inequality Approach

In this chapter we will provide motivation for the mathematical (analytical and numerical) investigation of evolutionary variational inequalities containing a memory term. These inequality problems are the result of a Baiocchi type transformation applied to a general degenerate moving free boundary problem with space- and, in particular, time-dependent coefficients in the differential operator and also in the boundary conditions.

At first we introduce the moving free boundary problem. Then, we shall derive different fixed domain formulations. In particular, we are interested in a variational inequality approach to the general free boundary problem. This inequality of obstacle type is characterized by an elliptic bilinear form, an additional memory term (caused by the time-dependent coefficients) and time-dependent convex (constraint) sets. In the derivation of the inequality approach we follow a generalization of the natural relation between variational inequalities of first and second kind and the enthalpy formulation for one- or two-phase Stefan problems (see e.g., [RODR89] and [RODR94]). Furthermore, we will point out the connections between different fixed domain formulations (weak, variational inequality, complementarity formulation; see, in particular, Remarks 2.8, 2.9). Concerning the distinction of variational inequalities into first and second kind we use the terminology introduced by Duvaut, Lions and Glowinski (see [DULI76], [GLT81] and [GLOW84]).

In a further part (Section 2.2) of this introductory chapter some application problems will be introduced and discussed in connection with the general free boundary problem.

2.1 The degenerate free boundary problem

Let $\Omega \subset \mathbb{R}^n$, $n = 1,2,3$ be a bounded smooth domain. Its boundary $\Gamma = \partial\Omega$ should be partitioned into $\Gamma_D \cup \Gamma_N = \Gamma$ such that $\Gamma_D \cap \Gamma_N = \emptyset$. The subscripts D and N refer to Dirichlet and Neumann/Newton (Robin) boundary conditions, respectively. We denote the space-time domain by

$$Q = \Omega \times (0,T),$$

where $0 < T < \infty$ is fixed. Accordingly, the lateral boundary of Q is given by

$$\Sigma = \partial\Omega \times (0,T) \quad \text{with} \quad \Sigma_D = \Gamma_D \times (0,T) \quad \text{and} \quad \Sigma_N = \Gamma_N \times (0,T).$$

Now, let us consider the following general moving free boundary problem (cf. Figure 2.1).

$$\left.\begin{aligned}
&-\operatorname{div}\big(k(x,t)\,\nabla\theta(x,t)\big) + q(x,t)\,\theta(x,t) = f(x,t) \quad \text{in } \Omega(t),\\
&\theta = \eta(t) \quad \text{and} \quad k\,\nabla\theta\cdot\vec{n} = -\lambda(x,t)\,\vec{w}\cdot\vec{n} \quad \text{on } \Gamma_f(t) = \partial\Omega(t)\cap\Omega,\\
&\theta = \gamma_D(x,t) \quad \text{on } \Gamma_D, \quad k\,\nabla\theta\cdot\vec{n} + \kappa(x)\,\theta = \gamma_N(x,t) \quad \text{on } \Gamma_N,\\
&\Omega(0) \quad \text{is given,} \quad t\in(0,T].
\end{aligned}\right| \tag{2.1}$$

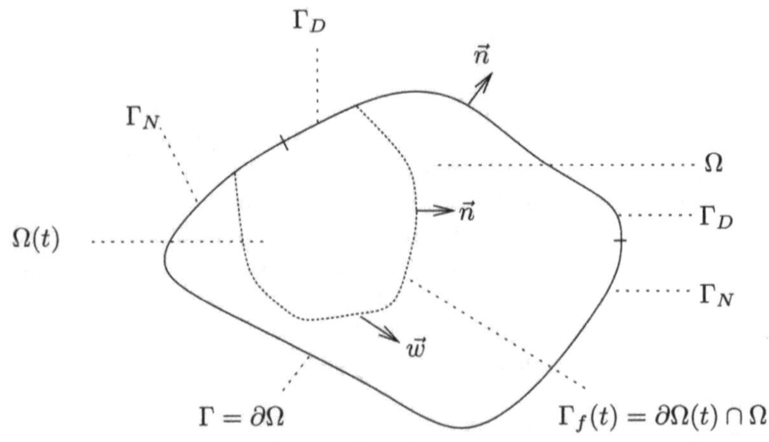

Figure 2.1: *Free boundary value problem: geometrical situation (here spatial dimension $n = 2$) at a fixed time t.*

A second order partial differential equation with space- and time-dependent coefficients is given in the unknown (moving) domain $\Omega(t)$ (the first phase, 'liquid' phase in some of the application problems, see Section 2.2). This region $\Omega(t)$ is separated by an interface Γ_f from the remaining part (the second phase) $\Omega\backslash\Omega(t)$ of the domain Ω and (possibly) partially limited by the external boundary Γ of Ω.

The motion of the interface is described by the second condition on $\Gamma_f(t)$ in (2.1) (the jump condition), where $\vec{n}$ denotes the (spatial) outward unit normal to $\partial\Omega(t)$ and $\partial\Omega$, respectively. Furthermore, $\vec{w}$ is the velocity of the interface $\Gamma_f(t)$ at a fixed time t. The first condition on $\Gamma_f(t)$ in (2.1) expresses the continuity of the solution θ across the interface. Later, θ will be extended by η in the second phase.

Finally, we consider mixed boundary conditions of Dirichlet and Neumann/Newton (Robin) type on the fixed boundary $\Gamma = \partial\Omega$. Due to the just mentioned extension of θ in the second phase, the boundary conditions are formulated on the whole boundary $\partial\Omega$ in problem (2.1).

In a sense, problem (2.1) can be regarded as a *degenerate* one-phase Stefan type problem characterized by a *zero-specific heat.*

It should be emphasized that, due to the underlying application problems (see Section 2.2), we consider coefficients $k,\ q,\ f,\ \lambda$ (the latent heat in terms of the Stefan problem), γ_D and γ_N depending not only on the space variable x, but, in particular, also on the time t. Only for the sake of brevity we assume $\kappa = \kappa(x)$ (zero order coefficient in the Newton boundary conditions).

As usual for a one-phase problem, the solution θ of (2.1) is extended by η in the second phase $\Omega\backslash\Omega(t)$, such that $\eta = \eta(t)$ should hold.

Furthermore, in this introductory part the coefficients are assumed to be smooth and

$$k = k(x,t) \geq k_0 > 0, \quad q(x,t) \geq 0, \quad \kappa(x) \geq 0.$$

More precise conditions will be given later (see Chapter 3).

Due to our aim to construct fixed domain formulations for problem (2.1), we consider the well-posed situation, i.e., an advancing moving boundary Γ_f, which can be concluded by the maximum principle (for the ill-posed (suction) problem see e.g., [DIFR84]).

Lemma 2.1 *Let θ be a solution of problem (2.1) with $f \geq q\,\eta,\ \gamma_N \geq \kappa\,\eta,\ \gamma_D \geq \eta$ and $\lambda > 0$ (function describing the jump at the interface). Then,*

$$\theta \geq \eta \quad \text{in} \quad \Omega(t) \times (0,T] \qquad \text{and} \qquad \Omega(t) \subseteq \Omega(t') \quad \text{for} \quad t < t';\ t,\ t' \in (0,T].$$

Proof. The multiplication of the second order differential equation in (2.1) by a function $v \in H^1(\Omega(t))$ defined by $0 \geq v = (\theta - \eta)^- = \min\{\theta - \eta, 0\}$ and the application of Green's formula lead to

$$C \,\|(\theta-\eta)^-\|^2_{H^1(\Omega(t))} \le \int_{\Omega(t)} \left(k\,\nabla(\theta-\eta)\,\nabla(\theta-\eta)^- + q\,(\theta-\eta)(\theta-\eta)^-\right)\,dx$$
$$+ \int_{\Gamma_N\cap\partial\Omega(t)} \kappa\,(\theta-\eta)\,(\theta-\eta)^-\,ds = \int_{\Omega(t)} (f - q\,\eta)\,(\theta-\eta)^-\,dx$$
$$+ \int_{\Gamma_N\cap\partial\Omega(t)} (\gamma_N - \kappa\,\eta)\,(\theta-\eta)^-\,ds \le 0.$$

Consequently, $(\theta-\eta)^- = 0$ holds, which implies $\theta \ge \eta$.

The second statement of the lemma follows from $w_n = \vec{w}\cdot\vec{n} \ge 0$ as the normal interface velocity on $\Gamma_f(t)$. ■

2.2 Some application problems

In this section we introduce a *selection* of application problems related to the general free boundary problem (2.1). Our aim is to indicate that the free boundary problem (2.1) may model different physical processes.

Remark 2.2 We refer to [RODR87], [RODR94] and [DALI90] for a general derivation of free boundary problems obtained by means of continuum mechanics techniques consisting of a combination of conservation laws and constitutive relations (balance and state equations, see also Section 6.2.2). □

At this point we will not give a detailed mathematical modelling of these processes. The injection and compression moulding processes as important engineering applications in polymer processing will be considered in more detail in Chapter 6.

Application 2.1 *Quasi-steady one-phase Stefan problem.*
The classical (one-phase) Stefan problem (cf. [STEF91]) describing the melting of ice (at zero degree, i.e., $\eta \equiv 0$ in (2.1)) surrounded by water has been investigated intensely in the mathematical literature as a typical free boundary problem. We refer to [RUBI71], [KIST80], [JERO83], [RODR89], [RODR94], [MEIR92] with respect to the mathematical analysis and to [CRAN84], [VERD94] concerning the numerical treatment as well as the extensive bibliography cited in these monographs and survey articles. Further references will be given in the subsequent remarks of this introductory chapter (see, in particular, Remarks 2.4, 2.5 and 2.10).

In the one-phase Stefan problem, the unknowns are the temperature distribution in the water and the free boundary consisting of the ice-water interface. The latent heat $\lambda > 0$ in the Stefan condition of (2.1) describes the jump of the internal energy (enthalpy) at the interface Γ_f (cf. [STEF91]).

Supposing that the product of specific heat and time-derivative of the temperature can be neglected, this parabolic problem degenerates to a moving free boundary problem characterized by an elliptic equation. Considering a space- and, in particular, time-dependent heat conductivity $k = k(x,t)$, we are in the situation of problem (2.1). □

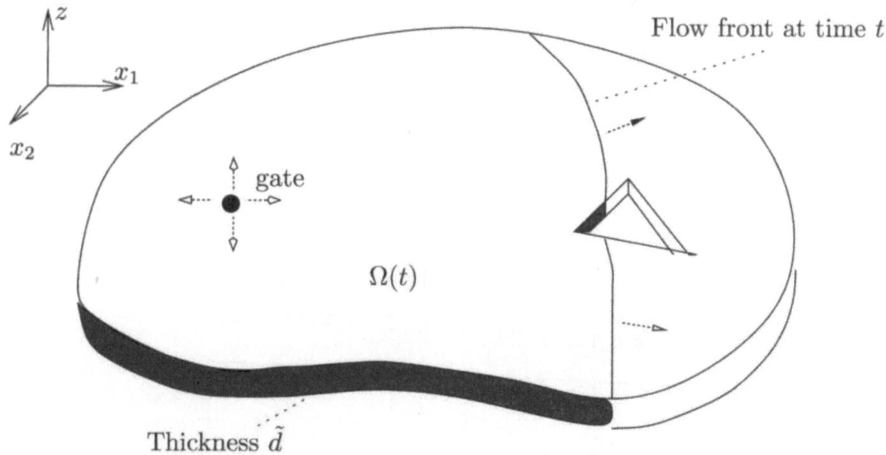

Figure 2.2: *Schematic view of the injection moulding problem – Hele–Shaw flow in a thin cavity with a slowly varying thickness.*

For manufacturing of plastic articles, injection and compression moulding (see Applications 2.2, 2.3 and Chapter 6) are two essential forming methods for polymers (cf. e.g., the monographs [TAGO79], [BERN83], [MANZ87], [TUCK89], [ADVA94] and [KENN95]).

Before we come to the free boundary problems describing the flow process in injection and compression moulding, let us recall the basic types of polymers: *thermoplastics* and *thermosets* as well as *elastomers*. In the technological processes there are known different operating conditions depending essentially on the used polymer. At this point we will not discuss details (see Section 6.2.1.1 and the books [TAGO79] and [WHCG85]). We only mention that for thermoplastics the mould is colder than the melt. In contrast to this, due to activation of the cross-linking (cure) reaction of the compound, the mould is heated for elastomers (rubber) and thermosets. Moreover, by including inhibitors in thermosetting polymers and elastomers, one ensures that the (main) curing process starts after the mould is filled (cf. [LFTU84]).

Application 2.2 *Non-isothermal injection moulding.*
In the injection process of thermoplastics a polymer is forced into a thin cavity

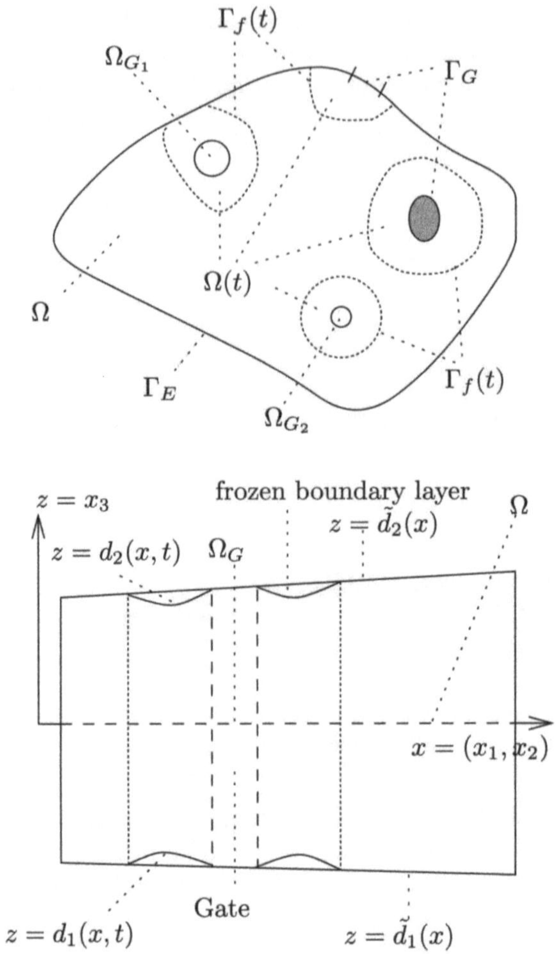

Figure 2.3: *Schematic representation of the ground plan and the side view of a cavity (thickness* $\tilde{d} = \tilde{d}_2 - \tilde{d}_1$ *is exaggerated to show the flow region;* $\tilde{d} \ll \operatorname{diam}\Omega$ *holds).*

surrounded by a cold mould, while the mould is hotter than the compound for elastomers and thermosets.

Let the cavity be described by a two-dimensional region Ω and the gap thickness

$$\tilde{d} = \tilde{d}(x) = \tilde{d}_2(x) - \tilde{d}_1(x), \quad x \in \overline{\Omega}$$

(cf. Figures 2.2, 2.3 and Figure 6.3 (concerning real plastic parts)). It should be emphasized that we do not assume the (geometrical) symmetry of the cavity with respect to the plane $z = 0$.

The domain $\Omega_G = \bigcup_{i=1}^{m} \Omega_{G_i}$ is characterized by the places where the hot polymer is injected into the cavity. At time $t \in (0,T]$, the plastic melt occupies the region $\Omega(t)$, which is limited by the free boundary $\Gamma_f(t)$ (location of the advancing melt front) and possibly by a part $\Gamma \cap \partial\Omega(t)$ of the fixed boundary $\Gamma = \partial\Omega$. At the last part the melt is attached to the lateral container walls. The moving free boundary $\Gamma_f(t)$ separates the melt in $\Omega(t)$ from the empty part (air). Sometimes, due to multiple gates or inserts in the cavity, the melt region can have a rather complicated planar geometry $\Omega(t)$.

The basic principles describing the dynamics of a viscous fluid flow are the conservation of mass and momentum (cf. Section 6.2.2). Due to the geometrical assumption that the thickness $\tilde{d}$ of the cell (cavity) is much smaller than its diameter, application of the lubrication theory is justified. This leads to a Hele–Shaw type flow which is based on Hele–Shaw's experiments made already in 1898 (see [LAMB93], Chapter VI, §330, §330a). However, the corresponding simplifications in the momentum equation can also be verified by an order-of-magnitude analysis. Summarizing (for details see Section 6.2.4), we obtain the following free boundary problem for the pressure $p = p(x,t)$ in the injection moulding process.

$$\left.\begin{aligned}
&-\sum_{i=1}^{2} \frac{\partial}{\partial x_i}\left(k(x,t)\,\frac{\partial p(x,t)}{\partial x_i}\right) + q(x,t)\,p(x,t) = f(x,t) \quad \text{in } \Omega(t),\\
&p = 0 \quad \text{and} \quad k\,\nabla p \cdot \vec{n} = -\tilde{d}(x)\,\vec{w}\cdot\vec{n} \quad \text{on } \Gamma_f(t) = \Omega \cap \partial\Omega(t),\\
&k\,\nabla p \cdot \vec{n} = \gamma_G > 0 \quad \text{on } \Gamma_G \subseteq \Gamma,\\
&k\,\nabla p \cdot \vec{n} + \kappa\,p = 0, \quad 0 \le \kappa \ll 1 \quad \text{on } \Gamma_E \subseteq \Gamma; \qquad t \in (0,T],
\end{aligned}\right| \tag{2.2}$$

where $\Omega(0) \supseteq \bigcup_{i=1}^{m} \Omega_{G_i}$ is given (cf. Figure 2.3). Furthermore, T denotes the filling time, $\vec{n}$ is the exterior unit normal and $\partial\Omega = \Gamma = \Gamma_E \cup \Gamma_G$ is the (fixed) boundary of Ω. The injection of the melt, i.e., the blowing of the blob can be represented by a known (high) pressure or flow rate at the gates Ω_{G_i}, $i = 1, 2, \ldots, m$. In problem (2.2) both situations are expressed by

$$f \ge 0 \quad \text{and} \quad q \ge 0 \quad \text{with} \quad \operatorname{supp} f \subset \Omega(0), \quad \operatorname{supp} q \subset \Omega(0).$$

Another possibility for the injection of melt is given by a boundary condition on Γ_G describing how the melt enters across the contour Γ_G.

Since capillarity is assumed to be negligible, the pressure at the free boundary $\Gamma_f(t)$ is constant and equal to the atmospheric pressure (set equal to zero with no loss of generality). On the lateral container wall Γ_E the flow is assumed to be tangential for $\kappa = 0$ (no leakage), whilst for a small parameter $0 < \kappa \ll 1$ this wall is not completely impermeable, which is justified from the practical point of view. Concerning an open mould we refer to the application problem 2.3 (compression).

Furthermore, $w_n = \vec{w} \cdot \vec{n}$ is the (thickness-averaged) normal interface velocity.

The coefficient $k = k(x,t)$ (flow conductivity) plays an essential role in the model. It depends on the viscosity of the melt and the gap thickness. Since the viscosity depends on the temperature and, in general, the rheological behaviour deviates from that of Newtonian melts, the coefficient function k should be assumed to be space- and *time*-dependent. Finally, we mention that, due to the possible formation of frozen boundary layers on the lower and upper container walls (caused by a cooling regime there), the effective thickness

$$d(x,t) = d_2(x,t) - d_1(x,t)$$

for the flow can be reduced, cf. Figure 2.3. We refer to Section 6.2.4, in particular, to (6.13) and Remark 6.2 for a more detailed discussion and an explicit expression for $k(x,t)$ as a measure of the so-called fluidity or flow conductance.

Let us note, that, in terms of the lubrication theory, the free boundary problem (2.2) has much in common with a generalized Reynold's equation.

With respect to our general problem (2.1) we have $\theta = \tilde{\theta} = p$ and $\eta = 0$. The 'latent heat' λ is now given by $\lambda = \lambda(x) = \tilde{d}(x)$ (thickness of the cavity). The fixed boundary of Ω is split into $\Gamma = \partial\Omega = \Gamma_N = \Gamma_E \cup \Gamma_G$. □

Application 2.3 *Non-isothermal compression moulding.*
In the compression moulding process schematically shown in Figure 2.4 (see also Figure 2.3) a pre-measured initial polymer charge, denoted by $\Omega(0)$, is placed into the cavity between the halves of a mould, for instance by injection. Then, these halves are brought together with the closing speed $\partial_t d(x,t)$ to squeeze the polymer melt (z is the direction of squeezing). Thus, the initial charge is transformed into its final shape or, in other words, the cavity is filled. Here, $\tilde{d}(x) = d(x,0)$ denotes the initial gap thickness and $d(x,t)$ is the distance between the mould halves at the time t. Hence, we have

$$d(x,t) \leq d(x,t') \quad \text{for} \quad t \geq t'.$$

Besides the filling of closed moulds, compression moulding techniques are also frequently used for open moulds, i.e., no lateral walls appear in this case. Sometimes, both mould halves are moved, whereas from time to time one surface remains stationary (non-symmetric situation with respect to $z = 0$). Furthermore, both a cooling regime in the mould and a heated mould (thermosets) are used.
But, in any case, no frozen layer should occur during the mould filling.

Assuming that the cavity width and length are much larger than the part thickness, the justification of applying the Hele–Shaw theory to compression moulding techniques is successfully discussed e.g., in [LETU87], [LFTU84] and the monograph [TUCK89].

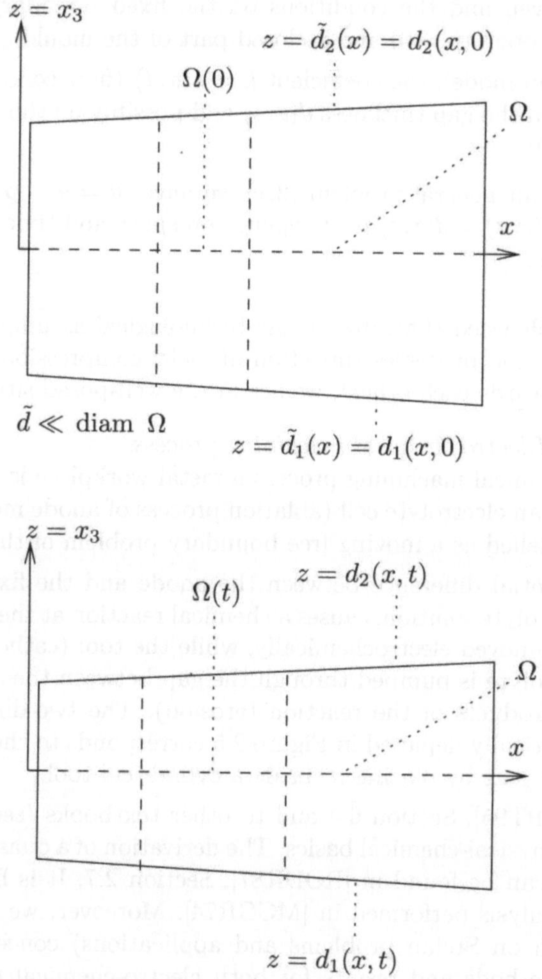

Figure 2.4: *Compression moulding. Top: initial charge. Bottom: charge shape after the time t.*

Similar to injection moulding (see Application 2.2), the classical formulation of the free boundary problem for the flow process in compression moulding can be formulated by

$$
\left.
\begin{aligned}
&-\sum_{i=1}^{2} \frac{\partial}{\partial x_i}\Big(k(x,t)\,\frac{\partial p(x,t)}{\partial x_i}\Big) = f(x,t) = -\frac{\partial d(x,t)}{\partial t} \quad \text{in } \Omega(t),\\
&p = 0 \quad \text{and} \quad k\nabla p\cdot\vec{n} = -d(x,t)\,\vec{w}\cdot\vec{n} \quad \text{on } \Gamma_f(t) = \Omega\cap\partial\Omega(t),\\
&p = 0 \quad \text{or} \quad k\nabla p\cdot\vec{n} + \kappa\, p = 0,\ 0\le\kappa\ll 1 \text{ on } \Gamma_E;\quad t\in(0,T],
\end{aligned}
\right| \qquad (2.3)
$$

where $\Omega(0)$ is given and the conditions on the fixed boundary $\Gamma = \partial\Omega = \Gamma_E$ correspond to an open ($p = 0$) and a closed part of the mould.

As in the injection model, the coefficient $k = k(x,t)$ (flow conductivity, cf. (6.13) below) depends on the gap thickness $d(x,t)$ and possibly on the viscosity (temperature) of the melt.

With respect to our general problem (2.1) we have $\theta = \tilde{\theta} = p$ and $\eta = 0$. The coefficient $\lambda = \lambda(x,t) = d(x,t)$ now depends on space and time. We mention that $f + \partial_t \lambda = 0$ holds. □

It should be emphasized that, due to the technological assumptions of the injection and compression processes (injection of melt; compression: lower and upper surfaces move towards each other), we are in the well-posed situation.

Application 2.4 *Electro-chemical machining process.*
In the electro-chemical machining process a metal workpiece is shaped by placing it as an anode in an electrolyte cell (ablation process of anode metal). This process may also be modelled as a moving free boundary problem of the form (2.1).

An applied potential difference between the anode and the fixed cathode, separated by an electrolyte solution, causes a chemical reaction at the anode. Therefore, anode metal is removed electrochemically, while the tool (cathode) remains unaltered. The electrolyte is pumped through the gap between the electrodes in order to remove the products of the reaction (erosion). The two-dimensional annular situation schematically depicted in Figure 2.5 corresponds to the shaping of a long cylindrical metal part by placing it inside a cylindrical tool.

We refer to [GERT95], Section 6.4 and to other textbooks (see e.g., [MCGE74]) concerning the physical-chemical basics. The derivation of a quasi-stationary mathematical model can be found in [RODR87], Section 2.7. It is based on an order-of-magnitude analysis performed in [MCGR74]. Moreover, we refer to [HOSP90] (chapter: Session on Stefan problems and applications) concerning a survey of mathematical methods and results for both electro-chemical machining and related problems. For more recent studies in this application problem we refer to [WEOS97] (further developments of conformal mapping techniques) and [VOLG97]. The last paper contains investigations and a comparison of different methods, especially from the application point of view, including a lot of references (in English and Russian). The considered methods are fixed domain, front tracking and level set methods.

Based on the assumption that the electric field $\vec{E}$ between the electrodes is approximately irrotational, we get the relation

$$\vec{j} = k\,\vec{E} = -k\,\nabla\Phi$$

for the electric current $\vec{j}$ from Ohm's law. Here, Φ denotes a potential and k is the conductivity of the electrolyte. Furthermore, assuming that the density of

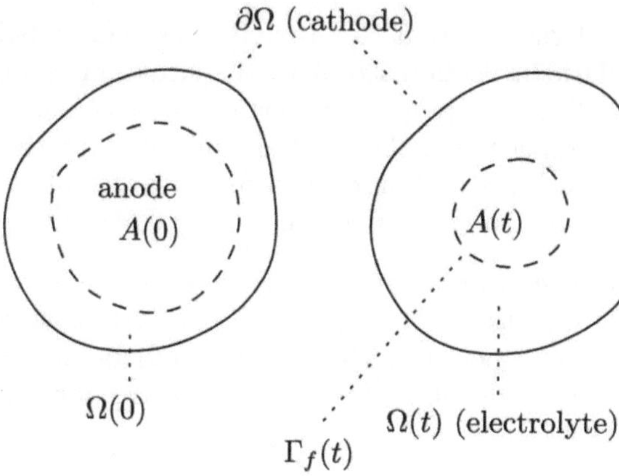

Figure 2.5: *Cross section of an annular electro-chemical machining process.*

the electric charge is constant, the conservation of charge leads to $\operatorname{div}\vec{j} = 0$. Consequently,

$$\operatorname{div}(k\ \nabla\Phi) = 0$$

holds in the electrolyte region $\Omega(t)$. A potential difference $\gamma_A = \gamma_A(t) > 0$ is applied across the electrodes such that we can take $\Phi = \gamma_A > 0$ on the anode and $\Phi = 0$ on $\partial\Omega$ (cathode).

Due to Faraday's law, the dissolution rate of the workpiece (i.e., the velocity of the interface Γ_f (anode surface)) is proportional to the normal magnitude of the local current density. Hence, we have the condition

$$k\ \nabla\Phi \cdot \vec{n} = -\lambda_E\ \vec{w}\ \cdot \vec{n}$$

at the anode surface as moving boundary Γ_f, where $\vec{n}$ is the unit normal directed towards the workpiece. Furthermore, $\vec{w} \cdot \vec{n} = w_n > 0$ represents the normal velocity of the interface and $\lambda_E < 0$ denotes the jump of the electric charge across Γ_f (the so-called electrochemical equivalent).
Summarizing, the classical formulation of the ECM problem can be written as follows.

$$\left.\begin{array}{l} -\operatorname{div}\big(k(x,t)\ \nabla\Phi(x,t)\big) = 0 \quad \text{in } \Omega(t), \quad \Phi = 0 \quad \text{on } \Gamma_D = \partial\Omega, \\ \Phi = \gamma_A(t) > 0 \quad \text{and} \quad k\ \nabla\Phi \cdot \vec{n} = -\lambda_E\ w_n \quad \text{on } \Gamma_f(t) = \partial A(t) \cap \partial\Omega(t) \end{array}\right| \tag{2.4}$$

for $t \in (0,T]$, where $A(0)$, $\Omega(0) = \Omega \backslash A(0)$ are given and $0 < T < \infty$ is the machining time.

By means of the maximum principle we verify $0 \leq \Phi(x,t) \leq \gamma_A(t)$ in $\Omega(t)$. The potential $\Phi(x,t)$ can be extended by continuity, i.e., we set $\Phi(x,t) = \gamma_A(t)$ in $A(t)$ for each $t > 0$.

As a generalization of the constant conductivity (see e.g., [CROW79], [ELLI80], [LORO86] and [RODR87], Section 2.8, 9.5), the situation to be considered here (see also [STEIc97], [STEIb98]) is characterized by a conductivity $k = k(x,t)$ depending both on the space variable x and, in particular, on the time t. From the physical background such a situation occurs, for instance, when the conductivity of the electrolyte depends on the temperature (cf. [GERT95]).

With respect to our general problem (2.1), we have $\theta = -\Phi$, $\eta = \eta(t) = -\gamma_A(t)$ and $\lambda = -\lambda_E > 0$, such that $\tilde{\theta} = \gamma_A - \Phi$ holds. Furthermore, *time-dependent* Dirichlet conditions are imposed at the whole (fixed) boundary (cathode). Due to $\gamma_A(t) > 0$ we have a shrinking anode, which corresponds to the well-posed situation (cf. Lemma 2.1). □

2.3 Different fixed domain formulations

2.3.1 Front tracking and fixing methods versus fixed domain formulations exemplified by injection and compression moulding

In the engineering literature on flow analysis in polymer processing (for a survey see [TUCK89] and [KENN95]) front tracking and front fixing are very popular techniques to simulate such moving boundary problems as (2.2), (2.3).

An introduction and overview of front tracking and front fixing methods from the mathematical point of view can be found in [CRAN84], Chapter 4 and 5. We refer to papers by Meyer (see [MEYE90] and the references cited there) for further developments in this context.

In what follows we restrict ourselves to sketching some techniques from the above mentioned engineering literature. At this point we will not go into the particulars of the application of boundary element methods (see [TUCK89], Chapter 7 and the references cited there) and level set methods (for a general introduction of level set methods see [SETH96]) in polymer processing. We refer to Section 6.2.6 for some remarks concerning further front tracking (capturing) and level set techniques.

Dynamic remeshing techniques exploiting body-fitted coordinates are frequently used. To be more precise, at each time step the physical domain with a curvilinear coordinate system is (often numerically) mapped onto a computational domain having a regular geometry. The transformed boundary problem is solved by a finite difference method. Then the flow front is moved and a new grid with more nodes is generated.

Sometimes, when the initial charge covers a considerable portion of the cavity in compression moulding, an expansion of the mesh (with a fixed number of nodes) is proposed. The nodes are moved according to the calculated velocity and afterwards a mesh equalization routine is applied (see [LFTU84], [LETU87]).

The so-called finite element/control volume approach is also very popular. In this case, the pressure equation with $p = 0$ on $\Gamma_f(t)$ is solved with finite elements and afterwards the melt front is tracked by means of a mass balance. To realize this, the second condition on the moving boundary $\Gamma_f(t)$ is exploited on a secondary mesh built up by joining the barycentres of the elements to the midpoint of the edges (in this context cf. also Section 4.1.3).

In general, the just mentioned front tracking techniques include some drawbacks and difficulties, e.g., the handling of multiply-connected or disconnected flow regions resulting from inserts in the cavity or multiple gates/charges. Another difficulty occurs when the flow front touches the lateral wall. Then the boundary condition changes from $p = 0$ to $\nabla p \cdot \vec{n} = 0$. Moreover, the selection of time increments has to be based on a criterion that at most one mesh zone (control volume layer) is filled at each time step.

Consequently, from the mathematical and, in particular, numerical point of view, the problems (2.2), (2.3) and, more generally, the moving free boundary problem (2.1) should be reformulated by means of fixed domain formulations. This will be done in Section 2.3.2 ('enthalpy type', weak formulation) and in Section 2.3.3 (variational inequality formulation). These reformulations are possible at least in the well-posed situation considered here (cf. Lemma 2.1).

2.3.2 Weak formulation

Similar to the classical Stefan problem a weak ('enthalpy type') formulation for problem (2.1) can be derived using the results of Lemma 2.1 (maximum principle). This weak formulation incorporates the relations of (2.1) into a single nonlinear equation, where the interface Γ_f disappears as an explicit unknown. However, Γ_f can be recovered *a posteriori* as a level set of the solution.

For the derivation of the weak formulation let us introduce the ('liquid') region $\mathcal{L}$ by

$$\mathcal{L} = \Omega(t) \times (0, T).$$

Multiplication of the second order differential equation in the problem (2.1) by a function $\phi \in H^1(Q)$ with $\phi = 0$ on Σ_D and $\Omega \times \{T\}$, application of integration by parts and, finally, consideration of the boundary and jump conditions in (2.1) lead to

$$\int_{\mathcal{L}} (k\ \nabla\theta\ \nabla\phi + q\ \theta\ \phi)\ dxdt + \int_{\Sigma_N \cap \partial\mathcal{L}} (\kappa\ \theta - \gamma_N)\ \phi\ dsdt$$
$$- \int_{\Omega(0)} \lambda\ \phi\ dx - \int_{\mathcal{L}} (\lambda\ \partial_t\phi + \partial_t\lambda\ \phi)\ dxdt\ = \int_{\mathcal{L}} f\ \phi\ dxdt.$$

As already mentioned above, the solution θ of (2.1) is extended continuously by means of $\eta = \eta(t)$ to the whole space-time domain $\overline{Q}$. Inserting the function $\tilde{\theta}$, which is defined by

$$\tilde{\theta} = \theta - \eta \quad \text{in} \quad \overline{Q},$$

we obtain

$$\begin{aligned}
&\int_Q [(f + \partial_t\lambda - q\,\eta)\,\phi - k\,\nabla\eta\,\nabla\phi]\,\chi_{\mathcal{L}}\,dxdt + \int_{\Sigma_N} (\gamma_N - \kappa\,\eta)\,\phi\,dsdt \\
&= \int_Q (k\,\nabla\tilde\theta\,\nabla\phi + q\,\tilde\theta\,\phi)\,dxdt + \int_{\Sigma_N} \kappa\,\tilde\theta\,\phi\,dsdt - \int_Q \lambda\,\partial_t\phi\,\chi_{\mathcal{L}}\,dxdt \\
&- \int_\Omega (\lambda\phi)(0)\,\chi_{\Omega(0)}\,dx,
\end{aligned}$$

where $\chi_{\mathcal{L}}$ and $\chi_{\Omega(0)}$ are the characteristic functions related to $\mathcal{L}$ and $\Omega(0)$, respectively.

Let us introduce the maximal monotone graph H associated with the Heaviside function, i.e.,

$$H(r) = 0 \quad \text{if } r < 0, \qquad H(r) = 1 \quad \text{if } r > 0, \qquad H(0) = [0,1]. \tag{2.5}$$

Replacing $\chi_{\mathcal{L}}$ by an appropriate function $\chi \in H(\tilde\theta)$, the weak formulation of problem (2.1) is then as follows.

$$\text{Find a pair} \quad (\tilde\theta, \chi) \in \big(L_2(0,T;H^1(\Omega)) \times L_2(Q)\big)$$

with $\tilde\theta \geq 0$ in Q which fulfills the following conditions.

$$\left|\begin{aligned}
&\chi \in H(\tilde\theta) \text{ a.e. in } Q, \quad \chi(x,0) = \chi_0 = \chi_{\Omega(0)}, \quad \tilde\theta = \gamma_D - \eta \text{ on } \Sigma_D, \\
&\int_Q [(f + \partial_t\lambda - q\,\eta)\,\phi - k\,\nabla\eta\,\nabla\phi]\,\chi\,dxdt + \int_{\Sigma_N} (\gamma_N - \kappa\,\eta)\,\phi\,dsdt \\
&= \int_Q (k\,\nabla\tilde\theta\,\nabla\phi + q\,\tilde\theta\,\phi)\,dxdt + \int_{\Sigma_N} \kappa\,\tilde\theta\,\phi\,dsdt - \int_Q \lambda\,\chi\,\partial_t\phi\,dxdt \\
&- \int_\Omega \chi_0\,(\lambda\phi)(0)\,dx \quad \forall\phi \in H^1(Q) \quad \text{with } \phi = 0 \text{ on } \Sigma_D \text{ and } \Omega \times \{T\}.
\end{aligned}\right| \tag{2.6}$$

Here, the initial region $\Omega(0)$ is given.

We mention that, together with $\nabla\eta = 0$, one can usually assume

$$f + \partial_t\lambda - q\,\eta = 0 \quad \text{in} \quad Q\backslash\mathcal{L} \quad \text{(i.e., in the second phase)}$$

(see the application problems in Section 2.2). Consequently, the first integral in (2.6) can be written without χ.

Remark 2.3 We recall that the temperature θ is a well-defined function of the enthalpy E in the classical one-phase Stefan problem. In contrast to this, the property that $\tilde\theta$ is not a single-valued function of the maximal monotone graph H defined in (2.5) is an essential feature of the weak formulation (2.6). This fact, caused by the vanishing time-derivative ('zero-specific heat'), represents a central difficulty in the considered problem.

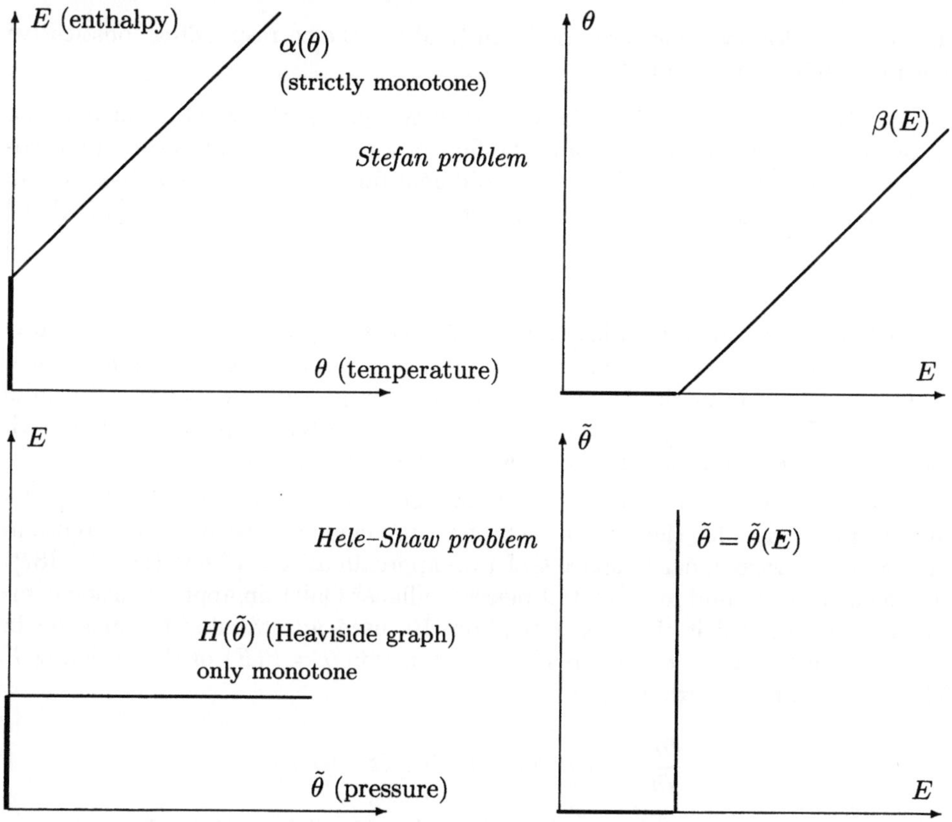

Figure 2.6: *Constitutive relations for the classical one-phase Stefan problem and the Hele–Shaw problem. Top: classical one-phase Stefan problem. Bottom: Hele–Shaw and related problems (see application problems in Section 2.2).*

To illustrate these properties let us assume just for the moment, that all material properties are normalized to one. Then, in the classical Stefan problem,

$$E \in \alpha(\theta)$$

with the maximal monotone graph

$$\alpha(r) = r + H(r).$$

Hence, there exists the state equation $\theta = \beta(E)$ with the inverse single-valued function

$$\beta(r) = \alpha^{-1}(r) = \min\{r, 0\} + \max\{r-1, 0\} \quad \forall r \in \mathbb{R}.$$

Here, β is a nondecreasing Lipschitz continuous function, whose graph has a flat part. Due to the fact that $H(\tilde{\theta})$ is only monotone, a corresponding function β

does not exist for the considered problem (2.6) (cf. the corresponding constitutive relations depicted in Figure 2.6).

Let us also recall the analogy between the two-phase Stefan problem and the porous medium equation describing the flow of a homogeneous gas through a homogeneous porous medium. Choosing a different function β, the two-phase Stefan problem and the porous medium equation formally coincide, see e.g., [VERD94] and [KORN97]. □

Remark 2.4 [i] As already indicated in the Application problem (2.1), the enthalpy formulation of one- and two-phase Stefan problems was studied from an analytical as well as from a numerical point of view in many papers; for theoretical results see e.g., [JERO83]. Here, we want to mention some essential papers dealing with the numerical treatment. Our overview is by far not complete.

Based on the standard enthalpy formulation, Jerome and Rose (cf. [JERO82]) derive error estimates for the two-phase Stefan problem using backward-difference in time and piecewise-linear in space Galerkin approximations. Elliott (cf. [ELLI87]) derived an error bound for a spatial piecewise-linear Galerkin approximation combined with an implicit time discretization. He used an enthalpy formulation in which the temperature θ is expressed as a function $\theta = \beta(E)$ of the enthalpy E (cf. the previous Remark 2.3), i.e.,

$$\frac{\partial E}{\partial t} - A\,\beta(E) = f \quad \text{in} \quad \Omega \times (0,T)$$

completed by corresponding initial and boundary conditions. Here, A stands for a linear elliptic differential operator (usually the Laplace operator $A = \triangle$).

Based on the just mentioned enthalpy formulation as a strongly nonlinear parabolic problem, a lot of papers are concerned with semi-discrete in time and fully discrete schemes arising in the theory of nonlinear semigroups of contractions (see e.g., [BEBR79], [MANV87], [NOVE88], [JAKA91]). This framework is so general that it allows us to include the Stefan problem and the porous medium equation. The theory of nonlinear semigroups leads not only to analytical results (existence, uniqueness, regularity results) but suggests also time-discrete algorithms to approximate the parabolic problem. Such linear approximation schemes are based on the so-called Chernoff formula and can be written as

$$\begin{aligned} \theta^n + \frac{\tau}{\mu}\, A\,\theta^n &= \beta(E^{n-1}) + \frac{\tau}{\mu}\, f, \\ E^n &= E^{n-1} + \mu\left(\theta^n - \beta(E^{n-1})\right). \end{aligned}$$

Here, τ is the time step, θ^n and E^n are the time-discrete solutions at time t^n. Further, $0 < \mu \leq L_\beta^{-1}$ denotes a relaxation parameter and L_β is the Lipschitz constant of β. Hence, such schemes are characterized by solving a linear elliptic

problem at each time step. Afterwards an algebraic correction is performed on account of the nonlinearity.
In [MANV87] several energy error estimates are derived for these time-discrete schemes. In [NOVE88] such a Chernoff formula is combined with a smoothing technique for the function β. More precisely, the non-decreasing function β is replaced by a strictly increasing function and then the above linearization algorithm is applied. A fully discrete scheme involving piecewise linear and constant finite elements is proposed and studied.
Furthermore, we refer to the paper [JAKA91] by Jäger and Kacur concerning the study of a fully discrete approximation of porous medium type systems. In contrast to the papers just mentioned, Jäger and Kacur use variable relaxation parameters μ which are determined by additional iterations. This is connected with the problem that sharp layers occuring in the exact solution may be smeared out in course of approximation.
Finally let us mention that various aspects of the numerical approximation of parabolic free boundary problems of Stefan and porous medium type are discussed in the research-expository work [VERD94].

[ii] The enthalpy method was extended to one-phase Stefan type problems with vanishing heat capacity e.g., by Crowley in [CROW79] (uniqueness result), Niezgodka in [NIEZ82] (existence) and Alt, Luckhaus in [ALLU83] (embedded in a general framework). □

2.3.3 The evolutionary variational inequality approach

Defining the differential operator A by

$$(Av)(x,t) = -\operatorname{div}\big(k(x,t)\ \nabla v(x,t)\big) + q(x,t)\ v(x,t),$$

the essence of the weak formulation (2.6) can be seen in its distributional form

$$f + \partial_t \lambda - A\ \eta = A\ \tilde{\theta} + \partial_t(\lambda\ \chi), \tag{2.7}$$

where χ is a function such that

$$\chi \in H(\tilde{\theta}) \quad \text{a.e. in } Q \quad \text{or equivalently} \quad 0 \leq \chi_{\{\tilde{\theta}>0\}} \leq \chi \leq 1 \quad \text{a.e. in } Q$$

(cf. (2.5)). Here, $\chi_{\{\tilde{\theta}>0\}}$ is the characteristic function related to the set $\{\tilde{\theta} > 0\}$. The crucial point in problem (2.7) is the time-derivative in connection with the set $\{\tilde{\theta} = 0\}$. Therefore, a natural way to overcome this difficulty and, furthermore, to construct another fixed domain formulation for (2.1), consists in the integration with respect to time.

Before we derive the corresponding evolutionary variational inequality, let us summarize some general remarks related to Baiocchi type transformations to free boundary problems.

Remark 2.5 We refer to [RODR87], Section 2.9 and [NAUM84], Section 4.1.2 concerning a general view of the application of Baiocchi type transformations to free boundary problems.
Variational inequality approaches to classical one- or two-phase Stefan problems are considered e.g., in [NAUM84], [RODR87], [RODR89] and [RODR94] from the analytical point of view (see also Remark 2.8[i]). □

Let us now perform the Baiocchi integral transformation for our considered case. For the moment we assume that the region $\Omega(t)$ and the free boundary $\Gamma_f(t)$ can be represented by a function $\omega(x)$ for each time $t \in (0, T]$, i.e.,

$$\Omega(t) = \{x \in \Omega : t > \omega(x)\}, \quad \Gamma_f(t) = \{x \in \Omega : t = \omega(x)\},$$
$$\omega(x) := 0 \quad \text{for } x \in \Omega(0).$$

Now, we introduce the Baiocchi transformation by

$$u(x,t) = \int_0^t \tilde{\theta}(x,t')\, dt' = \int_{\omega(x)}^t \tilde{\theta}(x,t')\, dt', \quad (x,t) \in \overline{Q}.$$

In the derivation of the relations satisfied by the new unknown u we have to consider the assumed time-dependency of the coefficients k and q.
At first, we obtain

$$\nabla u(x,t) = \int_{\omega(x)}^t \nabla\tilde{\theta}(x,t')\, dt' = \int_0^t \nabla\tilde{\theta}(x,t')\, dt'$$

using $\tilde{\theta} = 0$ on the free boundary $\Gamma_f(t)$. Now, taking into account the time-dependent coefficients k and q, integration by parts implies

$$(k\ \nabla u)(x,t) = \int_{\omega(x)}^t \left[(\partial_t k\ \nabla u)(x,t') + (k\ \nabla\tilde{\theta})(x,t')\right] dt' \quad \text{and}$$
$$(q\ u)(x,t) = \int_{\omega(x)}^t \left[(\partial_t q\ u)(x,t') + (q\ \tilde{\theta})(x,t')\right] dt'.$$

The next step consists in the computation of Au. Defining the differential operator B by

$$(Bv)(x,t') = -\operatorname{div}\left(\partial_t k(x,t')\ \nabla v(x,t')\right) + \partial_t q(x,t')\ v(x,t')$$

and using (2.1), in particular the second (jump) condition on $\Gamma_f(t)$, we derive the following relations for $(Au)(x,t)$.

$$Au = \begin{cases} \displaystyle\int_0^t (f + \partial_t\lambda - A\eta)(x,t')\, dt' + \int_0^t (Bu)(x,t')\, dt' - \lambda(x,t) + \lambda(x,0), & \text{(i)} \\ \displaystyle\int_{\omega(x)}^t (f + \partial_t\lambda - A\eta)(x,t')\, dt' + \int_{\omega(x)}^t (Bu)(x,t')\, dt' - \lambda(x,t), & \text{(ii)} \\ 0. & \text{(iii)} \end{cases}$$

Here, equation (i) corresponds to $x \in \Omega(0)$, $x \in \Omega(t)$, case (ii) is characterized by $x \notin \Omega(0)$, $x \in \Omega(t)$ ('phase transition') and situation (iii) is associated with $x \notin \Omega(0)$, $x \notin \Omega(t)$.

In order to summarize the cases (i), (ii) and (iii) in one relation, we use $u = 0$ in the 'second phase' $Q \backslash \mathcal{L}$. Furthermore, as indicated above, we can assume

$$f + \partial_t \lambda - A\eta = 0 \quad \text{in} \quad Q \backslash \mathcal{L}$$

(see also the application problems in Section 2.2). Consequently, transforming also the boundary conditions on Γ_D and Γ_N, we arrive at

$$\left.\begin{aligned} \lambda(x,t)\ \chi_{\Omega(t)} &= \int_0^t (f + \partial_t\lambda - A\eta)(x,t')\ dt' + \int_0^t (Bu)(x,t')\ dt' \\ &\quad + \lambda(x,0)\ \chi_{\Omega(0)} - (Au)(x,t) \quad \text{in } Q, \\ (k\ \nabla u \cdot \vec{n} + \kappa\ u)(x,t) &= \int_0^t (\partial_t k\ \nabla u \cdot \vec{n})(x,t')\ dt' \\ &\quad + \int_0^t (\gamma_N - \kappa\ \eta)(x,t')\ dt' \quad \text{on } \Sigma_N, \\ u(x,t) = \int_0^t (\gamma_D - \eta)(x,t')\ dt' &\quad \text{on } \Sigma_D, \quad u(0) = 0 \quad \text{in } \Omega, \quad \Omega(0) \text{ given.} \end{aligned}\right| \tag{2.8}$$

Now, we conclude that, if $\tilde{\theta} \geq 0$ (cf. Lemma 2.1) is a solution of the free boundary problem (2.1), then $u = u(t)$ will solve the following complementarity problem.

$$\left.\begin{aligned} &\lambda \geq \int_0^t (f + \partial_t\lambda - A\eta)\ dt' + \int_0^t (Bu)\ dt' + \lambda(0)\ \chi_{\Omega(0)} - Au, \quad u \geq 0, \\ &\left[\lambda + Au - \int_0^t (f + \partial_t\lambda - A\eta)\ dt' - \int_0^t (Bu)\ dt' - \lambda(0)\ \chi_{\Omega(0)}\right] u = 0 \end{aligned}\right| \tag{2.9}$$

a.e. in Ω and for each $t \in (0,T]$ together with the boundary and initial conditions from above (cf. (2.8)).

Introducing the bilinear forms a and b and further notation by

$$\left.\begin{aligned} a(t;v,w) &= \int_\Omega (k(t)\ \nabla v\ \nabla w + q(t)\ v\ w)\ dx + \int_{\Gamma_N} \kappa\ v\ w\ ds, \\ b(t';v,w) &= \int_\Omega (\partial_t k(t')\ \nabla v\ \nabla w + \partial_t q(t')\ v\ w)\ dx, \\ F(x,t) &= \lambda(x,0)\ \chi_{\Omega(0)} - \lambda(x,t) + \int_0^t (f + \partial_t\lambda - q\ \eta)(x,t')\ dt', \\ g_N(x,t) &= \int_0^t (\gamma_N - \kappa\ \eta)(x,t')\ dt', \quad g_D(x,t) = \int_0^t (\gamma_D - \eta)(x,t')\ dt' \end{aligned}\right| \tag{2.10}$$

as well as applying Green's formula to (2.9), we derive the following variational inequality of obstacle type.

$$\left.\begin{aligned}&\text{Find} \quad u(t) \in K(t) = \{w \in H^1(\Omega) : \ w \geq \varphi = 0 \text{ in } \Omega, \ w = g_D(t) \text{ on } \Gamma_D\}\\&a(t; u(t), v - u(t)) \geq (F(t), v - u(t)) + (g_N(t), v - u(t))_{\Gamma_N}\\&+ \int_0^t b(t'; u(t'), v - u(t))\, dt' \quad \forall v \in K(t), \quad t \in (0,T], \quad u(0) = 0.\end{aligned}\right| \tag{2.11}$$

This variational inequality of first kind is characterized by the memory term b and by the time-dependent convex subsets $K = K_{\varphi,g_D}(t) = K_{0,g_D}(t)$ of the Sobolev space $H^1(\Omega)$, which include the constraint (obstacle) $\varphi = 0$ and the imposed, time-dependent Dirichlet conditions $g_D = g_D(t)$.

Remark 2.6 We mention that, owing to $F(0) \leq 0$, $g_N(0) = 0$ and $g_D(0) = 0$, the initial condition $u(0) = 0$ in (2.11) will automatically be satisfied, when we solve the obstacle problem in (2.11) for $t = 0$. □

In the next remark we refer to some papers dealing with obstacle problems of the form (2.11) under the restriction of *time-independent* coefficients k and q. In this context see also Section 6.1.

Remark 2.7 It is easy to see that the bilinear form b (memory term) vanishes in (2.11) for $k = k(x)$ and $q = q(x)$. In this situation, a family of elliptic inequalities is obtained having time only as parameter. In [CIJA81], [ELJA81] and [ELJA83] such a family of elliptic inequalities is considered in connection with Hele–Shaw type flows in a closed mould (Neumann/Newton boundary conditions, cf. Application problems 2.2, 2.3). Gustafsson (see the paper [GUST85]) investigated the connections between some formulations (classical and weak) for this problem in an open mould. In [PRRO93], elliptic inequalities without a memory term but coupled with non-standard boundary conditions are discussed from the analytical point of view. The considered boundary conditions describe the total flux through $\partial\Omega_G$ together with the assumption that the pressure is constant, but unknown along $\partial\Omega_G$.

The elliptic inequalities with $b \equiv 0$ corresponding to the electro-chemical machining process ($k = \text{const}$) have been analysed in [LORO86], [RODR87], Section 9.5. Besides the proof of regularity results of the solution, the smoothness of the free boundary is studied in a special case (star-shaped configuration). Furthermore, convergence of the solution of the associated one-phase Stefan problem is discussed in the situation when the specific heat tends to zero.

In [BBT95], elliptic inequalities without a memory term are derived by means of the transformation

$$u(x,t) = \int_0^t k_2(t')\, p(x,t')\, dt',$$

under the assumptions $q = 0$, $\Gamma = \Gamma_D$ and, in particular, the separability of the coefficient k into $k(x,t) = k_1(x)\, k_2(t)$. □

At the end of this introductory chapter we briefly discuss some connections between various fixed domain formulations (variational inequalities of first and second kind, weak formulation).

Remark 2.8 [i] Based on the formulation (2.8) resulting from the integration of (2.7) in time, one can derive various kinds of variational inequalities by means of replacing $\chi_{\Omega(t)}$.

To obtain an inequality formulation of the second kind, which is also suitable for two-phase problems, one introduces the convex functional J by

$$J(v) = \int_\Omega \lambda(x,t)\, v^+(x)\, dx \quad \text{for} \quad v \in H^1(\Omega),$$

where $v^+ = \max\{v, 0\}$. Using $\partial_t u = \tilde{\theta} \le 0$ in $\Omega \backslash \Omega(t)$ and $\partial_t u = \tilde{\theta} \ge 0$ in $\Omega(t)$, one verifies

$$J(v) - J(\partial_t u) \ge \int_\Omega \lambda\, \chi\, (v - \partial_t u)\, dx$$

for fixed time t (χ from (2.6)). In other words, $\lambda\chi$ belongs to the subdifferential[1] $\partial J(\partial_t u)$ of the functional J. Note in this context, that $\partial j = H$ holds for the functional $j(r) = r^+$ with $r \in \mathbb{R}$ (the maximal monotone graph H is defined in (2.5)).

Connecting the previous relation and (2.8) as well as applying Green's formula, the following variational inequality of the second kind can be derived.

$$\left.\begin{aligned}
&\text{Find} \quad u(t), \partial_t u(t) \in H^1(\Omega), \quad u(t) = g_D(t) \quad \text{on } \Gamma_D \\
&a(t; u(t), v - \partial_t u(t)) + J(v) - J(\partial_t u(t)) \ge (F(t) + \lambda(t), v - \partial_t u(t)) \\
&+ (g_N(t), v - \partial_t u(t))_{\Gamma_N} + \int_0^t b(t'; u(t'), v - \partial_t u(t))\, dt' \\
&\forall v \in H^1(\Omega),\ v = \partial_t u = \partial_t g_D = (\gamma_D - \eta)(t) \quad \text{on } \Gamma_D, \\
&t \in (0,T]; \quad u(0) = 0.
\end{aligned}\right| \qquad (2.12)$$

But, as already mentioned above, this inequality formulation seems to be more useful for two-phase problems. A detailed discussion of the corresponding parabolic inequalities of the second kind related to two-phase Stefan problems (i.e., without memory term, but with a term containing the time-derivative of u) can be found in [RODR94].

[1] Let us recall from convex analysis that the subdifferential $\partial J(u)$ at $u \in V$ for a (convex, l.s.c. (lower semi-continuous)) functional $J : V \longrightarrow \mathbb{R}$ is defined by the set $\partial J(u) = \{u^* \in V^* : J(v) - J(u) \ge \langle u^*, v - u \rangle\ \forall v \in V\}$; cf. e.g., [EKTE76], Chapter I.5 and [ZEIDIII90], Chapter 47.

In contrast to this, we have the one-phase problem (2.1) under consideration. Hence, the set $\{\partial_t u < 0\}$ is empty and, owing to Lemma 2.1, an advancing moving boundary is guaranteed. Therefore, the functional J can be written also with respect to u, i.e.,

$$J(v) - J(u(t)) \geq \int_\Omega \chi\, \lambda\, (v - u(t))\, dx.$$

Moreover, due to $u \geq 0$, we have

$$J(v) - J(u(t)) = \int_\Omega \lambda\, (v - u(t))\, dx \geq \int_\Omega \chi\, \lambda\, (v - u(t))\, dx \quad \forall v \in H^1(\Omega),\ v \geq 0.$$

Connecting this with the inequality of the second kind (2.12), we arrive at the above obstacle problem (2.11).

[ii] Furthermore, it is worth remarking that (2.11) is equivalent to another variational inequality of the second kind, i.e.,

$$\begin{aligned} &\text{Find} \quad u(t) \in H^1(\Omega), \quad u(t) = g_D(t) \quad \text{on } \Gamma_D, \quad u(0) = 0, \\ &a(t; u(t), v - u(t)) + I_{K(t)}(v) - I_{K(t)}(u(t)) \geq (F(t), v - u(t)) \\ &+ \int_0^t b(t'; u(t'), v - u(t))\, dt' \quad \forall v \in H^1(\Omega), \quad v = g_D(t) \text{ on } \Gamma_D, \quad t \in (0,T]. \end{aligned}$$

In this context we mention that the non-empty, closed convex sets

$$K = K(t) \subset H^1(\Omega)$$

imply convex, l.s.c. and proper indicator functionals I_K defined by $I_K(v) = 0$ for $v \in K$ and $I_K(v) = +\infty$ for $v \notin K$ (cf. e.g. [GLOW84]). This inequality of the second kind can also be formulated as a multivalued operator equation (recall the definition of the subdifferential and see the book [ZEIDIII90], Chapter 54). □

Remark 2.9 It can be shown that, under the condition of an appropriate regularity of the variational inequality solution u of (2.11), the pair $(\tilde{\theta}, \chi)$ defined by

$$\tilde{\theta} = \partial_t u, \quad \chi = \lambda^{-1}\Big(F + \lambda + \int_0^t (Bu)(t')\, dt' - Au\Big),$$

solves the weak (enthalpy type) formulation (2.6). Roughly speaking, for this result, one needs a variational inequality solution u of (2.11) nondecreasing with respect to time and fulfilling

$$u \in \big(H^1(0,T; H^1(\Omega)) \cap L_2(0,T; H^2(\Omega))\big).$$

Furthermore, the function λ describing the jump at the interface should belong to $L_\infty(Q)$ and satisfy $\lambda \geq \lambda_0 > 0$. The proof of this result and also of the conversion that the solution pair of (2.6) solves the obstacle problem (2.11) (after the application of the Baiocchi transformation) can be found in the author's paper [STEIa97] for the inequality formulation related to injection/compression moulding (with Dirichlet boundary conditions). As a by-product of the proof we obtain that the solution u of (2.11) satisfies also the variational inequality (2.12) of the second kind. □

Remark 2.10 Elliptic variational inequalities can also be obtained by the application of a semi-discretization in time procedure (Rothe method, cf. [ROTH30]) to the enthalpy formulation of the two-phase Stcfan problem and related problems (e.g., (semi-discrete) porous medium equation).
More precisely, an elliptic differential inclusion or, in other words, a variational inequality of the second kind occurs after an implicit time discretization of the enthalpy formulation of the two-phase Stefan problem, see [JERO83], [ELLI81], [KORN97]. Jerome ([JERO83]) has shown convergence of the solution of time-discretized problems. Error estimates and a relaxation algorithm (SOR method) for spatial problems arising from time-discretized problems were given by Elliott in [ELLI81]. The monograph [KORN97] is devoted to the development of adaptive monotone multigrid methods for such nonlinear variational problems (minimization problems). □

International Series of Numerical Mathematics
Vol. 136, © 2002 Birkhäuser Verlag Basel/Switzerland

Chapter 3

Properties of the Variational Inequality Solution

In the previous chapter we saw that a variational inequality approach to the general moving free boundary problem (2.1) leads to obstacle problems of the form (2.11) which should be investigated in the sequel in a slightly more general setting. More precisely, we will consider a space-dependent obstacle $\varphi = \varphi(x)$ and space-time-dependent Dirichlet boundary conditions $g_D = g_D(x,t)$ characterizing the convex sets

$$K_{\varphi,g_D} = K_{\varphi,g_D}(t).$$

In the following analytical investigations we shall pay special attention to these *time-dependent* convex sets $K_{\varphi,g_D} = K_{\varphi,g_D}(t)$.

The content of this chapter is as follows. Firstly, we formulate the evolutionary variational inequality problem. Then, we recall some definitions and notations with respect to the (spatial) Sobolev spaces and several vector-valued function spaces. In Section 3.2 we prove the existence of a unique solution u of the variational inequality as a continuous mapping from the time interval $[0,T]$ to the Sobolev space $H^1(\Omega)$, i.e., u belongs to the space $C([0,T];H^1(\Omega))$. For this result we use a fixed-point argument in connection with a convergence result for convex sets in Mosco's sense.

In order to study the regularity of the solution with respect to time and its time evolution in Section 3.3, the inequality problem will be approximated by a Lewy–Stampacchia bounded penalization method (cf., in particular, Remark 3.7). As the main result we will show that the solution of the variational inequality is a Lipschitz continuous mapping from the time interval $[0,T]$ to the Sobolev space $H^1(\Omega)$. Furthermore, based on a maximum principle, there will be given upper bounds for the variational inequality solution.

In Section 3.4 we are concerned with a discussion of spatial regularity based again on investigation of the associated penalty problem. We exploit a semi-discretization in time procedure together with results from the elliptic regularity theory (L_p-theory) (see e.g., [GRIS85], [RODR87]). Depending on the smoothness of the data, we will prove that the variational inequality solution is an L_2-function (resp. L_∞-function) from $(0,T)$ to the Sobolev space $W_p^2(\Omega)$, $p < \infty$.

Finally, in Section 3.5 we summarize the regularity results obtained in the foregoing sections. Moreover, we discuss some consequences and establish further regularity statements which are of special interest for the numerical analysis (error estimates, see Section 5.2).

The author developed the line of reasoning used for existence, uniqueness and regularity results of this book mainly in the papers [STEIa98] and [STEIb98]. In this work we improve some regularity results and give a more comprehensive discussion of various aspects of the analytical investigation for the considered variational inequality problems.

3.1 Problem setting and general notations

Let us formulate the evolutionary obstacle problem to be considered.

$$\left.\begin{aligned} &\text{Find}\quad u(t) \in K_{\varphi,g_D}(t) = \{w \in H^1(\Omega) : w \geq \varphi \text{ in } \Omega, \quad w = g_D(t) \text{ on } \Gamma_D\}, \\ &a(t; u(t), v - u(t)) \geq (f(t), v - u(t)) + (g_N(t), v - u(t))_{\Gamma_N} \\ &+ \int_0^t b(t'; u(t'), v - u(t))\, dt' \quad \forall v \in K_{\varphi,g_D}(t), \quad t \in [0,T], \end{aligned}\right| \tag{3.1}$$

with the convex sets $K_{\varphi,g_D}(t) \neq \emptyset$. The bilinear forms a and b are defined by

$$\left.\begin{aligned} a(t; v(t), w) &= \int_\Omega [k(x,t)\ \nabla v(x,t)\ \nabla w(x) + q(x,t)\ v(x,t)\ w(x)]\ dx \\ &\quad + \int_{\Gamma_N} \kappa(x)\ v(x,t)\ w(x)\ ds, \\ b(t'; v(t'), w) &= \int_\Omega [\partial_t k(x,t')\ \nabla v(x,t')\ \nabla w(x) \\ &\quad + \partial_t q(x,t')\ v(x,t')\ w(x)]\ dx. \end{aligned}\right| \tag{3.2}$$

We note, that compared to (2.11), (2.10) from Section 2.3.3, we now denote the right-hand side by lower-case letter f.

Let Ω be a bounded domain of $\mathbb{R}^n$, $n \geq 1$ with a Lipschitz boundary

$$\Gamma = \partial\Omega = \Gamma_D \cup \Gamma_N \cup \Gamma_0.$$

Here, Γ_D and Γ_N are disjoint and open in $\partial\Omega$ (one of them can be empty) and Γ_0 with $\text{meas}_{n-1}(\Gamma_0) = 0$ is the set of those points, where one type of the boundary condition changes into another. Furthermore,

$$(v, w) = \int_\Omega v\, w\, dx$$

denotes the usual L_2-scalar product and

$$(v, w)_{\Gamma_N} = \int_{\Gamma_N} v\, w\, ds$$

is related to a Neumann/Newton (Robin) part Γ_N of the boundary $\partial\Omega$.

To ensure that the convex sets $K_{\varphi,g_D}(t)$ in (3.1) are non-empty, we assume

$$\varphi \in H^1(\Omega) \quad \text{with} \quad \varphi \leq g_D(t) \quad \text{on}\, \Gamma_D \quad \forall t \in [0, T]. \tag{3.3}$$

Furthermore, let us make some assumptions on the coefficients. In particular, we will work with a V-elliptic bilinear form a, where V denotes the subspace of the Sobolev space $H^1(\Omega)$, the trace of which is zero on the Dirichlet boundary Γ_D. More precisely, we have $V = \{w \in H^1(\Omega) : w = 0 \text{ on } \Gamma_D\}$, or $V = H_0^1(\Omega)$ if $\Gamma_D = \partial\Omega$, or else $V = H^1(\Omega)$ if $\Gamma_N = \partial\Omega$.

$$\left.\begin{array}{l}
k,\ q \in W^1_\infty(0, T; L_\infty(\Omega)), \quad \kappa \in L_\infty(\Gamma_N), \quad f \in C([0, T]; L_2(\Omega)),\\
g_N \in C([0, T]; L_2(\Gamma_N)), \quad g_D \in C([0, T]; H^1(\Omega)),\\
k(x, t) \geq k_0 > 0 \quad \text{for } x \in \overline{\Omega},\ t \in [0, T],\\
a(t; v, v) \geq m\ \|v\|_V^2\ \forall v \in V,\ t \in [0, T] \quad \text{with } m > 0 \ \text{ independent of } t.
\end{array}\right| \tag{3.4}$$

We emphasize, that according to the application problems introduced in Section 2.2 we do *not* assume any condition on the sign of $\partial_t k$ and $\partial_t q$.

To preserve the just mentioned ellipticity of the bilinear form a in (3.4) we assume

$$\begin{array}{l}
q(x, t) \geq 0 \quad \text{for } x \in \Omega,\ t \in [0, T], \quad \kappa(x) \geq 0 \quad \text{for } x \in \Gamma_N,\\
\text{meas}_{n-1}(\Gamma_D) + \text{meas}_{n-1}(\Gamma_\kappa) + \text{meas}_n(\Omega_q) > 0 \quad \text{with}\\
\Omega_q = \{x \in \Omega : q(x, t) \geq q_0 > 0\ \forall t \in [0, T]\}, \quad \Gamma_\kappa = \{x \in \overline{\Gamma_N} : \kappa(x) \geq \kappa_0 > 0\}.
\end{array}$$

Owing to Poincaré type inequalities (cf. e.g., [RODR87], Chapter 3, Theorem 5.8 and [GGZ74], Chapter II, Lemma 1.36), in the case of $\int_{\Gamma_D} ds > 0$, one considers the equivalent norm defined by $\|v\|_V = \|\nabla v\|_{L_2(\Omega)}$ in V. Otherwise, V is endowed with the norm induced from that of $H^1(\Omega)$. We shall use the same notation m for the ellipticity constant of the bilinear form a with respect to $H^1(\Omega)$.

By V^* we denote the dual space of V and $\langle .,. \rangle$ is the duality pairing between V^* and V.

There are several textbooks on Sobolev spaces and other function spaces including trace and embedding theorems, see e.g., [ADAM75], [LSU68], [GRIS85] and [ZEIDII90]. In what follows we give only a brief overview of the framework of the spaces which we will use. More precise references for the properties and results exploited in our work will be given at the relevant places in the text.

In this work we use the (spatial) Sobolev spaces

$$W_p^k(\Omega) \quad (H^k(\Omega) = W_2^k(\Omega) \quad \text{for} \quad p = 2 \quad \text{is abbreviated}),$$

which are, roughly speaking, the spaces of real functions whose generalized derivatives up to the k-th order $(k \in \{0, 1, 2, \ldots\})$ belong to $L_p(\Omega)$ with $p \in [1, \infty]$.

The boundary condition $w = g_D$ on Γ_D in the inequality formulation (3.1) is to be understood in the sense of traces (operator of restriction to the boundary of Ω). We shall exploit results of trace theorems several times, such that at some places Sobolev spaces of fractional order (boundary space $W_p^r(\partial\Omega)$ for $0 < r < 1$ and for noninteger $r > 1$) will be used. Concerning the concept of traces (characterization of generalized boundary values) we refer to [RODR87], Section 3.5, [GRIS85], Section 1.5 and [ZEIDII90], Section 21.3 and Appendix. For example, the boundary functions $g \in L_p(\partial\Omega)$ which correspond to functions $w \in W_p^1(\Omega)$ form the subset $W_p^{1-1/p}(\partial\Omega)$ of $L_p(\partial\Omega)$.

We shall deal with evolutionary problems such that, besides the (spatial) Sobolev spaces, we use the notations

$$C([0, T]; X)$$

for the Banach space consisting of all continuous functions $u : [0, T] \longrightarrow X$ (X as Banach space) and

$$L_p(0, T; X), \quad 1 \le p \le \infty$$

for the Lebesgue space of vector-valued functions (see e.g., [ZEIDII90], Chapter 23 and [GGZ74], Chapter IV). Furthermore, we will work with

$$W_p^1(0, T; X) = \{v | \ \exists w \in L_p(0, T; X) : \ v(t) = v(0) + \int_0^t w(t') \, dt' \quad \forall t \in [0, T]\}$$

for $p = 2$ (then $H^1(0, T; X) = W_2^1(0, T; X)$ is abbreviated) and $p = \infty$, where w is identified with the generalized time derivative $w = \partial_t v$ for $v \in W_p^1(0, T; X)$, see [RODR87], Section 9.5 and [RODR94], Section 2.1.

In this context we recall the inclusion (in general strict)

$$W_\infty^1(0, T; X) \subset C^{0,1}([0, T]; X).$$

However, if X is a Hilbert space, these spaces can be identified, see [ZEIDII90], Section 23.5 and [BREZ73], Appendix.

At this stage let us mention that throughout this work the symbols C, c, C_i, c_i, $\bar{C}_i$, $\tilde{C}_i$, $i = 1, 2, \ldots$ will be used as generic positive constants, independent of the related functions and discretization parameters. They are not the same at each occurence.

Furthermore, let us emphasize that we use the notations w^+ and w^- for the positive and negative part of a function w, i.e.,

$$w^+ = w \vee 0 = \max\{w, 0\} \geq 0 \quad \text{and} \quad w^- = w \wedge 0 = \min\{w, 0\} \leq 0,$$

such that $w = w^+ + w^-$ holds. We note the important relations

$$a(t; v^+, v^-) = 0, \quad a(t; v, v^-) = a(t; v^-, v^-), \quad a(t; v, v^+) = a(t; v^+, v^+).$$

(cf. [BELI82], Section 5.2, [KIST80], Chapter II.A and [RODR87], Section 3.3).

At this point let us anticipate that in the sequel we will use Gronwall's inequality several times both in its continuous version (see e.g., [GGZ74], Chapter V, Lemma 1.3 and [CANN84], Lemma 8.4.1), but also in the following discrete version. Several discrete inequalities of Gronwall's type can be found in [JERO83], Section 2.2.

Lemma 3.1 *Suppose, that the discrete functions*

$$d^j = d(t^j) \geq 0 \quad \textit{and} \quad r^j = r(t^j) \geq 0$$

satisfy

$$d^0 \leq r^0 \quad \textit{and} \quad d^j \leq c \sum_{i=0}^{j-1} \tau\, d^i + r^j \quad \textit{for all} \quad j = 1, \ldots, N,$$

$0 \leq c =$ *const. Then, the estimate*

$$d^j \leq r^j + c \exp(c\, t^{j-1}) \sum_{i=0}^{j-1} \tau\, r^i$$

holds for all $j = 1, \ldots, N$, *where* $t^j = j\tau$ *for* $j = 0, \ldots, N$, $N\tau = T$.
If additionally $r^j \leq r^{j+1}$ *holds for all* $j = 0, \ldots, N-1$, *one will get*

$$d^j \leq \exp(c\, t^j)\, r^j \quad \textit{for all} \quad j = 0, \ldots, N.$$

3.2 Existence and uniqueness result

By means of a (Banach) fixed-point argument (cf. [ZEIDI93], Chapter 1) and the theory of elliptic variational inequalities, we prove the existence of a unique solution for problem (3.1), (3.2). Due to the time-dependent convex sets $K_{\varphi,g_D} = K_{\varphi,g_D}(t)$ we need, additionally, a convergence result for convex sets (in Mosco's sense, see [MOSC69] and [RODR87], Section 4.4). This result by Mosco allows us to study the stability of the solution not only with respect to the data but also to perturbations in the convex sets.

For the purpose of investigation it will be sometimes convenient to consider a problem equivalent to (3.1), which is obtained by a simple translation trick with respect to the non-homogeneous Dirichlet boundary conditions, i.e.,

$$u(t) = \tilde{u}(t) + g_D(t) \quad \forall t \in [0,T].$$

Then, the inequality problem for $\tilde{u}$ is given by

$$\left.\begin{aligned} &\text{Find} \quad \tilde{u}(t) \in K_{\tilde{\varphi},0}(t) = \{w \in V: \ w \geq \tilde{\varphi}(t) := \varphi - g_D(t) \quad \text{in}\, \Omega\}\,, \\ &a(t; \tilde{u}(t) + g_D(t), v - \tilde{u}(t)) \geq (f(t), v - \tilde{u}(t)) \ + (g_N(t), v - \tilde{u}(t))_{\Gamma_N} \\ &+ \int_0^t b(t'; \tilde{u}(t') + g_D(t'), v - \tilde{u}(t))\, dt' \quad \forall v \in K_{\tilde{\varphi},0}(t), \quad t \in [0,T]. \end{aligned}\right| \tag{3.5}$$

In order to exploit a fixed-point argument for proving the existence of a unique solution, we consider at first problem (3.1), (3.2) without the memory term. In other words, $b \equiv 0$ is assumed.

Lemma 3.2 *Let the assumptions (3.4) and (3.3) together with* $\partial_t k = \partial_t q \equiv 0$ *be fulfilled. Then, for every* $f \in C([0,T]; V^*)$ *and* $g_N \in C([0,T]; L_2(\Gamma_N))$, *there exists a unique solution*

$$u \in C([0,T]; H^1(\Omega))$$

of (3.1), (3.2) (resp. $\tilde{u} \in C([0,T]; V)$ *of problem (3.5), (3.2)). Furthermore, the (nonlinear) solution mapping* $S: \ (f, g_N) \longrightarrow u = S(f, g_N)$ *is Lipschitz continuous (for fixed* φ *and* g_D*), such that an a priori estimate of the form*

$$\left.\begin{aligned} \|u_1 - u_2\|_{C([0,T];H^1(\Omega))} \ &\leq M \ \big(\|f_1 - f_2\|_{C([0,T];V^*)} \\ &+ C_1 \,\|(g_N)_1 - (g_N)_2\|_{C([0,T];L_2(\Gamma_N))}\big) \end{aligned}\right| \tag{3.6}$$

holds for solutions $u_i = S(f_i, (g_N)_i), \ i = 1,2$ *with* $M = m^{-1}$.

Proof. The existence and uniqueness of the solution $\tilde{u}$ of problem (3.5) (resp. u of (3.1)) for each fixed $t \in [0,T]$ is an immediate consequence of the elliptic variational inequality theory (Lions–Stampacchia–theorem, cf. [RODR87], [CHIP84], [KIST80] and [GLOW84]).

To show the continuity of the function $\tilde{u}: [0,T] \ni t \longrightarrow V$, we examine the assumptions which are necessary for an abstract stability result (see Theorem 4.4.1 in [RODR87]; convergence of convex sets in Mosco's sense, see also [MOSC69]).

Let a sequence $\{t_n\} \subset [0,T]$ with $t_n \longrightarrow t \in [0,T]$ be given. Owing to the continuity of g_D with respect to time, we have $\tilde{\varphi}(t_n) \longrightarrow \tilde{\varphi}(t)$ in $H^1(\Omega)$. Now, we conclude that

$$K_{\tilde{\varphi},0}(t_n) \longrightarrow K_{\tilde{\varphi},0}(t)$$

holds in Mosco's sense. More precisely, on the one hand, for all $v \in K_{\tilde{\varphi},0}(t)$ there exists a sequence $\{v_n\}$ with $v_n \in K_{\tilde{\varphi},0}(t_n)$, such that $v_n \longrightarrow v$ in V (take e.g., $v_n = v - (v - \tilde{\varphi}(t_n))^- = \max\{v, \tilde{\varphi}(t_n)\}$). On the other hand, for any sequence $\{v_n\}$ with $v_n \in K_{\tilde{\varphi},0}(t_n)$ and $v(t_n) \rightharpoonup v$ in V (weakly), one has $v \in K_{\tilde{\varphi},0}(t)$.

Furthermore, we verify the convergence $a(t_n; v_n, w) \longrightarrow a(t; v(t), w)$ for any sequence $\{v_n\}$ with $v_n \in K_{\tilde{\varphi},0}(t_n)$, such that $v_n \longrightarrow v \in K_{\tilde{\varphi},0}(t)$. Taking the previous properties together with $g_D(t_n) \longrightarrow g_D(t)$ in $H^1(\Omega)$, $f(t_n) \longrightarrow f(t)$ in $L_2(\Omega)$ and $g_N(t_n) \longrightarrow g_N(t)$ in $L_2(\Gamma_N)$, all the assumptions of the above mentioned Mosco theorem are fulfilled. Therefore, we get

$$\tilde{u}(t_n) \longrightarrow \tilde{u}(t) \quad \text{in} \quad V \subset H^1(\Omega) \quad \text{for} \quad \tilde{u}(t_n) \quad \text{and} \quad \tilde{u}(t)$$

as solutions of (3.5) for t_n and t, respectively.

Finally, the estimate (3.6) is obtained by taking $v = u_{3-i}(t)$ in the inequality for $u_i(t)$, $i = 1,2$. ∎

Based on the previous lemma and a (Banach) fixed-point argument the existence of a unique solution $u \in C([0,T]; H^1(\Omega))$ for problem (3.1), (3.2) can be proved.

Theorem 3.3 *Under the above assumptions (3.4), (3.3) there exists a unique solution* u *for problem (3.1), (3.2) with*

$$u \in C([0,T]; H^1(\Omega)).$$

Moreover, the a priori estimate (3.6) holds with a constant $M = M(m,T,L)$, *where* $L = L(\|k\|_{W^1_\infty(0,T;L_\infty(\Omega))}, \|q\|_{W^1_\infty(0,T;L_\infty(\Omega))})$.

Proof. For the translated problem (3.5), (3.2) let us consider the mapping

$$\bar{w} = U_{\delta_j,\delta}\, w, \quad j = 0,1,2,\ldots,$$

which is defined in the following way.

$$\left.\begin{aligned}
&\text{Find} \quad \bar{w}(t) \in K_{\tilde{\varphi},0}(t): \\
&a(t; \bar{w}(t), v - \bar{w}(t)) - \delta_j \int_0^t b(t'; (\bar{w} + g_D)(t'), v - \bar{w}(t'))\, dt' \\
&\geq (f(t), v - \bar{w}(t)) + (g_N(t), v - \bar{w}(t))_{\Gamma_N} - a(t, g_D(t), v - \bar{w}(t)) \\
&+ (\delta - \delta_j) \int_0^t b(t; (w + g_D)(t'), v - \bar{w}(t))\, dt' \quad \forall v \in K_{\tilde{\varphi},0}(t), \quad t \in [0,T].
\end{aligned}\right| \quad (3.7)$$

First of all, we investigate the case $\delta_0 = 0,\ j = 0,\ 0 < \delta \leq 1$ for this mapping (3.7).

Taking an element

$$w \in \mathcal{K} = \{v \in C([0,T];V) :\ v(t) \in K_{\tilde{\varphi},0}(t) \quad \forall t \in [0,T]\},$$

we obtain $f_w \in C([0,T];V^*)$, where f_w is given by

$$\begin{aligned}\langle f_w(t), v\rangle &= (f(t),v) + (g_N(t),v)_{\Gamma_N} - a(t; g_D(t), v)\\ &\quad + (\delta - \delta_0) \int_0^t b(t'; (w + g_D)(t'), v)\ dt'.\end{aligned}$$

Consequently, owing to Lemma 3.2, the relation $\mathcal{K} \ni \bar{w} = U_{0,\delta}\ w$ is guaranteed. To exploit the Banach fixed-point theorem (cf. [ZEIDI93], Chapter 1), we examine the contractivity of the mapping $U_{0,\delta}$. Using the estimate (3.6) for functions $\bar{w}_i$ corresponding to f_{w_i}, $i = 1,2$, we derive

$$\begin{aligned}&\|\bar{w}_1 - \bar{w}_2\|_{C([0,T];H^1(\Omega))}\\ &\leq M\ (\delta - \delta_0) \max_{t\in[0,T]} \sup_{v\in V} \|v\|_V^{-1} \left| \int_0^t b(t'; w_1(t') - w_2(t'), v)\ dt' \right|\\ &\leq M\ (\delta - \delta_0)\ L \int_0^T \|w_1(t') - w_2(t')\|_{H^1(\Omega)}\ dt'\\ &\leq M\ (\delta - \delta_0)\ T\ L\ \|w_1 - w_2\|_{C([0,T];H^1(\Omega))}\,.\end{aligned}$$

Thus, for all $\delta \in (0, (LTM)^{-1})$ we get a fixed-point mapping $\bar{w} = U_{\delta_0=0,\delta}\ w$. Therefore, there exists a unique solution u of the evolutionary inequality problem

$$\begin{aligned}&a(t; u(t), v - u(t)) \geq (f(t), v - u(t)) + (g_N(t), v - u(t))_{\Gamma_N}\\ &+ \delta \int_0^t b(t'; u(t'), v - u(t))\ dt' \quad \forall v \in K_{\varphi,g_D}(t),\ t \in [0,T],\ \delta \in (0,(LTM)^{-1}).\end{aligned}$$

The a priori estimate (3.6) for this inequality problem can be derived in the following way. Let us take $v = u_{3-i}(t)$ in the inequality related to u_i, $i = 1,2$. The addition of both inequalities leads to

$$\begin{aligned}m\,\|u_1(t) - u_2(t)\|_{H^1(\Omega)} &\leq \|f_1 - f_2\|_{C([0,T];V^*)}\\ &\quad + \|(g_N)_1 - (g_N)_2\|_{C([0,T];L_2(\Gamma_N))}\\ &\quad + L \int_0^t \|u_1(t') - u_2(t')\|_{H^1(\Omega)}\ dt' \quad \forall t \in [0,T],\end{aligned}$$

such that Gronwall's inequality implies (3.6) with $M = M(T, L, m)$.

If $(MLT)^{-1} > 1$ holds, the proof is complete. Otherwise, we proceed by successive repetition of the considerations for

$$\bar{w} = U_{\delta_{j+1},\delta}\, w \quad \text{with} \quad \delta_j < \delta_{j+1} \leq \delta_j + (MLT)^{-1}$$

until, after a finite number of steps, the situation $\delta = 1$ is fulfilled. ∎

Remark 3.4 In the absence of Dirichlet boundary conditions, or provided that $\varphi_i = (g_D)_i,\ i = 1, 2$ on Γ_D holds, the a priori estimate (3.6) can be extended to a variation of compatible obstacles such that

$$\begin{aligned} \|u_1 - u_2\|_{C([0,T];H^1(\Omega))} \ &\leq M\ \big(\|f_1 - f_2\|_{C([0,T];V^*)} + C_2\, \|\varphi_1 - \varphi_2\|_{H^1(\Omega)} \\ &\quad + C_1\, \|(g_N)_1 - (g_N)_2\|_{C([0,T];L_2(\Gamma_N))} \big). \end{aligned}$$

Furthermore, it should be mentioned that we have certainly not looked for minimal assumptions on f, g_N and φ to ensure the validity of the results of Theorem 3.3. For instance, owing to the Sobolev embeddings $H^1(\Omega) \subset C(\overline{\Omega})$ for $n = 1$ and $H^1(\Omega) \subset L_q(\Omega)$ with the critical exponent $q < \infty$ for $n = 2$ and $q \leq 2n/(n-2)$ for $n \geq 3$, it is sufficient to assume $f \in C([0,T]; L_p(\Omega))$, where $p = 1$ for $n = 1$, $p > 1$ for $n = 2$ and $p \geq 2n/(n+2)$ for $n \geq 3$. □

Remark 3.5 [i] Our existence result of Theorem 3.3 is essentially based on the V-ellipticity of the bilinear form a (cf. assumption (3.4)). But it seems to be reasonable that, under certain assumptions, also the semi-coercive evolutionary inequality problem possesses a solution. The corresponding semi-coercive formulation is characterized by $q = 0$ and exclusive boundary conditions of Neumann type (i.e., $\Gamma = \Gamma_N$ and $\kappa = 0$) in problem (3.1), (3.2). In this context we refer to [RODR87], Section 8.6, where a semi-coercive Signorini problem is treated.

[ii] The existence of a unique solution can also be proved for a more abstract evolutionary inequality formulation considered in a general Hilbert space V. In [STEI91] we have proved the existence of a unique solution $u \in C([0,T]; V)$ of the following problem.

$$\begin{aligned} &\text{Find} \quad u \in \mathcal{K}_C = \{ w \in C([0,T]; V) : \ w(t) \in K \quad \forall t \in [0,T] \}, \\ &\langle A(t)\, u(t), v - u(t) \rangle \geq \langle f(t), v - u(t) \rangle + \Big\langle \int_0^t (Bu)(t')\, dt', v - u(t) \Big\rangle, \\ &\forall v \in K,\ t \in [0,T], \end{aligned}$$

where K is a convex, closed and nonempty subset of a Hilbert space V.

The (nonlinear) operator $A(t) : V \longrightarrow V^*$ should be semicontinuous (i.e., $u_n \to u$ in V (strongly) implies $A(t)u_n \rightharpoonup A(t)u$ in V^* (weakly)) and strongly monotone for all $t \in [0,T]$. Furthermore, the mapping $[0,T] \ni t \longrightarrow (A(t)\, w) \in C([0,T];V^*)$ should be continuous for all $w \in V$.

The (nonlinear) operator $B : L_2(0,T;V) \longrightarrow L_2(0,T;V^*)$ is assumed to satisfy a Lipschitz condition and $f \in C([0,T];V^*)$.

Moreover, the just mentioned abstract variational inequalities, integrated over the time interval $[0,T]$ are discussed in [STEI91].

However, we emphasize that, due to the time-dependent convex sets $K_{\varphi,g_D}(t)$, the variational inequality (3.1), (3.2) is *not* a special case of the abstract formulations mentioned above. □

3.3 Monotonicity properties and regularity with respect to time

In this section we shall study the regularity of the solution of problem (3.1), (3.2) with respect to time and the time evolution of this solution. Furthermore, an upper bound for the variational inequality solution will be discussed using a maximum principle. An essential tool in these investigations is a penalization (regularization) method of Lewy–Stampacchia type.

At first, we will consider the evolutionary inequality formulation (3.1), (3.2) with a convex set $K = K_{\varphi,g_D}$ independent of time (corresponding to the injection, compression type inequalities, see Applications 2.2 and 2.3. Afterwards, the more complicated situation of time-dependent convex sets $K = K_{\varphi,g_D}(t)$ (cf. the ECM Application 2.4) will be investigated in Section 3.3.2.

3.3.1 Time-independent convex sets

In this section we consider the inequality problem (3.1), (3.2) characterized by a constraint set K_{φ,g_D} independent of time t. Therefore, we assume the imposed Dirichlet boundary conditions $g_D = g_D(x)$ on Γ_D to be independent of time.

Let us now prove that the solution u of (3.1), (3.2) is actually a Lipschitz continuous function from the time interval $[0,T]$ to the Sobolev space $H^1(\Omega)$.

Theorem 3.6 *Additionally to (3.3), (3.4) let the assumptions*

$$f \in W^1_\infty(0,T;L_2(\Omega)) \quad \text{and} \quad g_N \in W^1_\infty(0,T;L_2(\Gamma_N))$$

be fulfilled. Then, the unique solution u of (3.1), (3.2) is such that

$$u \in W^1_\infty(0,T;H^1(\Omega)) = C^{0,1}([0,T];H^1(\Omega)).$$

Proof. To compare two solutions $u(t^i)$ at times $t^i \in [0,T]$, $i=1,2$, $t^1 \neq t^2$, we take $v = u(t^{3-i})$ in the inequality corresponding to t^i, $i = 1,2$. By subtraction we obtain

$$
\begin{aligned}
&m \left\|u(t^1) - u(t^2)\right\|^2_{H^1(\Omega)} \leq (f(t^1) - f(t^2), u(t^1) - u(t^2))\\
&+ (g_N(t^1) - g_N(t^2), u(t^1) - u(t^2))_{\Gamma_N} + \int_{t^2}^{t^1} b(t'; u(t'), u(t^1) - u(t^2))\, dt'\\
&+ a(t^2; u(t^2), u(t^1) - u(t^2)) - a(t^1; u(t^2), u(t^1) - u(t^2)).
\end{aligned}
$$

The Lipschitz continuity is now concluded by means of the estimate

$$
\begin{aligned}
&m \ \left\|u(t^1) - u(t^2)\right\|_{H^1(\Omega)} \leq |t^1 - t^2| \left\{ \left\| \frac{(f(t^1) - (t^2)}{t^1 - t^2} \right\|_{L_2(\Omega)} \right.\\
&+ C \left\| \frac{g_N(t^1) - g_N(t^2)}{t^1 - t^2} \right\|_{L_2(\Gamma_N)} + L \ \|u\|_{C([0,T];H^1(\Omega))}\\
&+ \left. \left(\left\| \frac{k(t^1) - k(t^2)}{t^1 - t^2} \right\|_{L_\infty(\Omega)} + \left\| \frac{q(t^1) - q(t^2)}{t^1 - t^2} \right\|_{L_\infty(\Omega)} \right) \ \left\|u(t^2)\right\|_{H^1(\Omega)} \right\}.
\end{aligned}
$$

Consequently, the desired result is proved. ∎

In order to investigate monotonicity properties and later the spatial regularity (see Section 3.4) we shall consider a Lewy–Stampacchia type penalization problem for the evolutionary inequality (3.1), (3.2). This regularization problem is defined by

$$
\left|
\begin{aligned}
&\text{Find} \quad z_\varepsilon(t) \in H^1(\Omega), \quad z_\varepsilon(t) = g_D \text{ on } \Gamma_D :\\
&a(t; z_\varepsilon(t), v) + (\mu \ \beta_\varepsilon(z_\varepsilon(t) - \varphi), v) = (f(t) + \mu, v)\\
&+ (g_N(t), v)_{\Gamma_N} + \int_0^t b(t'; z_\varepsilon(t'), v)\, dt' \quad \forall v \in V, \quad t \in [0,T]\\
&\text{with} \quad \mu = \mu(x) \geq 0, \quad \mu \in L_2(\Omega)\\
&\text{and} \quad \beta_\varepsilon(w) = \beta\Big(\frac{w}{\varepsilon}\Big) = \frac{(w^+/\varepsilon)^2}{(w/\varepsilon + 0.5)^2} = \frac{(w^+)^2}{(w + \varepsilon/2)^2} \quad \text{for } \varepsilon > 0.
\end{aligned}
\right| \tag{3.8}
$$

This problem corresponds to the boundary value problem

$$
\left|
\begin{aligned}
&A(t)\ z_\varepsilon(t) + \mu\ \beta_\varepsilon(z_\varepsilon(t) - \varphi) = f(t) + \mu + \int_0^t (B z_\varepsilon)(t')\, dt' \quad \text{in } \Omega,\\
&l(t)\ z_\varepsilon(t) = g_N(t) + \int_0^t (\partial_t k\ \nabla z_\varepsilon \cdot \vec{n})(t')\, dt' \quad \text{on } \Gamma_N,\\
&z_\varepsilon(t) = g_D \quad \text{on } \Gamma_D
\end{aligned}
\right| \tag{3.9}
$$

with the second order differential operators A, B and the boundary operator l defined by

$$\left.\begin{aligned} A(t)\,w &= -\operatorname{div}\left(k(t)\,\nabla w\right) + q(t)\,w, \\ B(t')\,w &= -\operatorname{div}\left(\partial_t k(t')\,\nabla w\right) + \partial_t q(t')\,w, \\ l(t)\,w &= k(t)\,\nabla w \cdot \vec{n} + \kappa\,w. \end{aligned}\right| \qquad (3.10)$$

Remark 3.7 It is easy to see (cf. Figure 3.1) that

$$\beta = \beta(r) = \frac{(r^+)^2}{(r+0.5)^2}, \quad r \in \mathbb{R}$$

is a nondecreasing Lipschitz function (actually a C^1-function) from $\mathbb{R} \longrightarrow [0,1]$ with

$$\beta(r) = 0 \quad \text{for } r \leq 0, \quad \beta(+\infty) = 1 \quad \text{and} \quad \exists\, C_\beta > 0 : (1-\beta(r))\,r \leq C_\beta \quad \text{for } r \geq 0.$$

These properties imply in a sense the convergence $\beta_\varepsilon(r) = \beta(r/\varepsilon)$ as $\varepsilon \longrightarrow 0$ to the multivalued Heaviside graph defined in (2.5). It is not difficult to construct other functions β, which satisfy the just mentioned properties of this Lewy–Stampacchia type penalization (see [RODR87], Section 5.3 for a general discussion and further references as well as [BELI84], Section 2.5.4).
The penalization defined in (3.8) can be interpreted as a regularization of

$$\left(f(x,t) + \int_0^t (Bu)(x,t')\,dt' - (Au)(x,t)\right) \in \partial j\left(u(x,t) - \varphi(x)\right),$$

where A and B are the second order operators (cf. (3.10)) associated with the bilinear forms a and b, respectively. Here, $j = j(r)$ denotes the functional

$$j(r) = +\infty \quad \text{for} \quad r < 0 \quad \text{and} \quad j(r) = 0 \quad \text{for} \quad r \geq 0.$$

Its subdifferential ∂j is given by $\partial j(r) = \emptyset$ for $r < 0$, $\partial j(r) = (-\infty, 0]$ for $r = 0$ and $\partial j(r) = 0$ for $r > 0$.

The auxiliary function μ in (3.8) should be taken as a consequence of the related Lewy–Stampacchia inequalities (see Section 3.5 and [RODR87], Section 5.3, [BELI84], Section 2.5.4). Then, the term $(\beta_\varepsilon - 1)\,\mu$ is a regularization of the subdifferential of the indicator functional of the convex set K (cf. Remark 2.8[ii]).

Compared to the frequently exploited penalty term $\varepsilon^{-1}\,\min\{z_\varepsilon - \varphi, 0\}$ becoming progressively larger as the solution moves away from the convex set K_{φ,g_D}, the just discussed regularization displays some advantages with respect to the considered obstacle problem, in particular in the treatment of Neumann/Newton boundary conditions. □

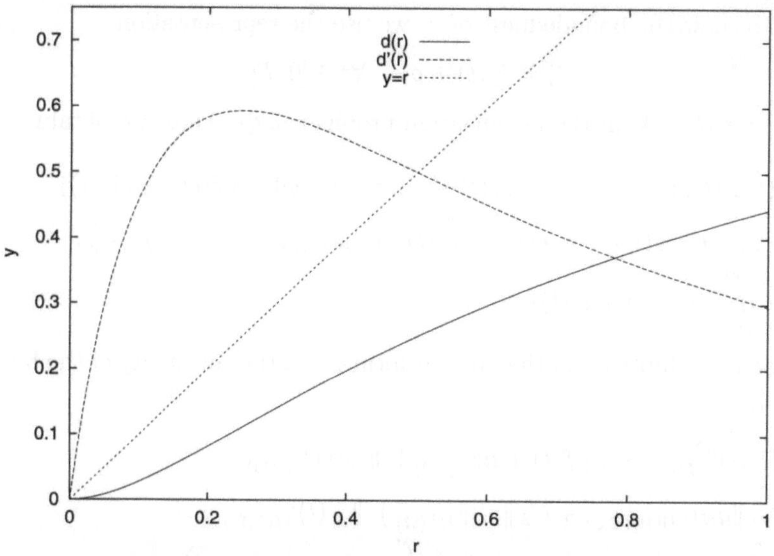

Figure 3.1: *Illustration of the penalization function. Solid line: penalization function* $\beta(r) = d(r) = (r^+)^2/(r+0.5)^2$. *Dashed line:* $\beta'(r) = d'(r) = r^+/((r+0.5)^3)$.
Dotted line: graph $y = r$ *(for comparison).*

Owing to the properties of the function β just mentioned in Remark 3.7, a unique solution of problem (3.8) is obtained in the same way as in Lemma 3.2, Theorem 3.3, now based on a fixed point argument and the elliptic theory for monotone operators.

Lemma 3.8 [i] *Under the above assumptions (3.3), (3.4) there exists a unique solution*

$$z_\varepsilon \in C([0,T]; H^1(\Omega))$$

of problem (3.8). Moreover, one has

$$\|z_\varepsilon\|_{C([0,T];H^1(\Omega))} \le C \quad \textit{independent of} \quad \varepsilon > 0. \tag{3.11}$$

[ii] *If additionally the assumptions of Theorem 3.6 are satisfied, the solution* z_ε *of (3.8) is such that*

$$z_\varepsilon \in W^1_\infty(0,T; H^1(\Omega)) \quad \textit{uniformly in} \quad \varepsilon > 0.$$

[iii] *Let in addition to [ii] the conditions* $\mu \in L_p(\Omega)$ *with* $p \ge \max\{2, n/2\}$ *(n as space dimension) and* $\partial_t f(t) \ge 0$ *a.e. in* Ω, $\partial_t g_N(t) \ge 0$ *a.e. on* Γ_N *and a.e. in* $(0,T)$ *be fulfilled. Then, the solution* z_ε *of (3.8) satisfies* $\partial_t z_\varepsilon \ge 0$ *a.e. in* Ω, $t \in (0,T)$.

Proof. *[i]*. To show the boundedness of z_ε we use the representation

$$z_\varepsilon(t) = \tilde{z}_\varepsilon(t) + g_D \quad \forall t \in [0,T]$$

and take $v = \tilde{z}_\varepsilon(t) \in V$ in the penalization problem (3.8). Thus, we obtain

$$\begin{aligned} & a(t; \tilde{z}_\varepsilon(t), \tilde{z}_\varepsilon(t)) + (\mu\ \beta_\varepsilon(z_\varepsilon(t) - \varphi), z_\varepsilon(t) - \varphi) = (f(t) + \mu, \tilde{z}_\varepsilon(t)) \\ & + (g_N(t), \tilde{z}_\varepsilon(t))_{\Gamma_N} - a(t; g_D, \tilde{z}_\varepsilon(t)) + (\mu\ \beta_\varepsilon(z_\varepsilon(t) - \varphi), g_D - \varphi) \\ & + \int_0^t b(t'; z_\varepsilon(t'), \tilde{z}_\varepsilon(t))\ dt'. \end{aligned}$$

Now, using the V-ellipticity of the bilinear form a and the continuity of the bilinear form b, we get

$$\begin{aligned} & m\ \|\tilde{z}_\varepsilon(t)\|^2_{H^1(\Omega)} \le \left(\|f(t) + \mu\|_{L_2(\Omega)} \right)\ \|\tilde{z}_\varepsilon(t)\|_{L_2(\Omega)} \\ & + \left(C_1\, \|g_N(t)\|_{L_2(\Gamma_N)} + C_2\, \|g_D\|_{H^1(\Omega)} \right)\ \|\tilde{z}_\varepsilon(t)\|_{H^1(\Omega)} \\ & + \|\mu\|_{L_2(\Omega)}\ \|g_D - \varphi\|_{L_2(\Omega)} + L \left(\int_0^t \|z_\varepsilon(t')\|_{H^1(\Omega)}\ dt' \right)\ \|\tilde{z}_\varepsilon(t)\|_{H^1(\Omega)} \end{aligned}$$

for all $t \in [0,T]$. Hence, the statement follows by the application of Young's and Gronwall's inequality.

[ii]. The result is obtained by the same technique as in the proof of Theorem 3.6 and taking into consideration (3.11) and the monotonicity of β_ε.

[iii]. A nondecreasing solution $z_\varepsilon = z_\varepsilon(t)$ of (3.8) is deduced by differentiating problem (3.8) with respect to time t, i.e.,

$$\left. \begin{aligned} & \text{Find} \quad w_\varepsilon(t) \in V: \quad a(t; w_\varepsilon(t), v) + (\mu\ \beta'_\varepsilon(z_\varepsilon(t) - \varphi)\ w_\varepsilon(t), v) \\ & = \ (\partial_t f(t), v) + (\partial_t g_N(t), v)_{\Gamma_N} \qquad \forall v \in V, \text{ a.e. in } (0,T). \end{aligned} \right| \tag{3.12}$$

Due to $\beta'_\varepsilon(z_\varepsilon(t) - \varphi) \ge 0$, $\mu \ge 0$ and the assumption $\mu \in L_p(\Omega)$ with $p = \max\{2, n/2\}$ (note the Sobolev embedding $H^1(\Omega) \subset L_q(\Omega)$, cf. Remark 3.4), the function $w_\varepsilon = \partial_t z_\varepsilon$ is the unique solution of problem (3.12).
To show $\partial_t z_\varepsilon(t) \ge 0$ we apply a weak maximum principle. Taking

$$V \ni v = (\partial_t z_\varepsilon(t))^- \le 0$$

in (3.12) we obtain

$$\begin{aligned} & a(t; \partial_t z_\varepsilon(t),\ (\partial_t z_\varepsilon(t))^-) + (\mu\ \beta'_\varepsilon(z_\varepsilon(t) - \varphi)\ \partial_t z_\varepsilon(t)\ ,\ (\partial_t z_\varepsilon(t))^-) \\ & = \ (\partial_t f(t),\ (\partial_t z_\varepsilon(t))^-) + (\partial_t g_N(t),\ (\partial_t z_\varepsilon(t))^-)_{\Gamma_N} \le 0. \end{aligned}$$

Since $\mu \ge 0$ and $\beta'_\varepsilon \ge 0$, we verify $\|(\partial_t z_\varepsilon(t))^-\|_{H^1(\Omega)} \le 0$. Hence, we have $(\partial_t z_\varepsilon(t))^- = 0$ in $H^1(\Omega)$. Consequently, $\partial_t z_\varepsilon(t) \ge 0$ holds. ■

Let us now demonstrate that the penalty solutions z_ε of (3.8) converge to the solution u of the obstacle problem (3.1), (3.2) in the passage to the limit as $\varepsilon \longrightarrow 0$. As consequence we obtain a variational inequality solution u of (3.1), (3.2) which is also nondecreasing with respect to time.

In the proof of the following theorem we apply Green's formula at the initial time $t = 0$. Therefore, we assume $\varphi \in H^2(\Omega)$ and $k(0) \in C^1(\overline{\Omega})$. Here, we will not discuss slightly weaker assumptions (see e.g., [RODR87], Section 3.6 and 5.3).

Theorem 3.9 *Let the assumptions of Lemma 3.8[iii] together with the conditions* $\varphi \in H^2(\Omega)$, $k(0) \in C^1(\overline{\Omega})$, $\mu \geq (A(0)\,\varphi - f(0))^+$ *a.e. in* Ω *and* $l(0)\,\varphi \leq g_N(0)$ *on* Γ_N *be fulfilled.*

[i] *The unique solution* $z_\varepsilon \in W^1_\infty(0,T;H^1(\Omega))$ *of (3.8) preserves the obstacle, i.e.,*

$$z_\varepsilon(t) \in K_{\varphi,g_D} \quad \textit{for each} \quad \varepsilon > 0 \quad \textit{and for all} \quad t \in [0,T].$$

[ii] *Furthermore,* z_ε *converges as* $\varepsilon \longrightarrow 0$ *to the solution* u *of the obstacle problem (3.1), (3.2) in* $C([0,T];H^1(\Omega))$ *(strongly). The error estimate*

$$\|u - z_\varepsilon\|_{C([0,T];H^1(\Omega))} \leq C\,\sqrt{\varepsilon}\,\|\mu\|^{1/2}_{L_1(\Omega)}\,.$$

holds.

[iii] *The unique solution* $u \in W^1_\infty(0,T;H^1(\Omega))$ *of the variational inequality (3.1), (3.2) is such that* $\partial_t u(t) \geq 0$ *a.e. in* Ω, $t \in (0,T)$.

Proof. [i]. For $t = 0$ problem (3.8) coincides with the elliptic equation

$$a(0;z_\varepsilon(0),v) + (\mu\,\beta_\varepsilon(z_\varepsilon(0)-\varphi),v) = (f(0)+\mu,v) + (g_N(0),v)_{\Gamma_N} \quad \forall v \in V.$$

Now, we take $V \ni v = (z_\varepsilon(0)-\varphi)^- \leq 0$ in this equation, subtract $a(0;\varphi,v)$ and apply Green's formula. Furthermore, we use the inequalities

$$\mu \geq (A(0)\,\varphi - f(0))^+ \geq A(0)\,\varphi - f(0), \quad g_N(0) - l(0)\,\varphi \geq 0$$

according to the assumptions and take into account $\beta_\varepsilon(z_\varepsilon(0)-\varphi)\,(z_\varepsilon(0)-\varphi)^- = 0$. As result we deduce $\|(z_\varepsilon(0)-\varphi)^-\|_{H^1(\Omega)} \leq 0$. Consequently, we get $z_\varepsilon(0) \geq \varphi$, which implies $z_\varepsilon(0) \in K_{\varphi,g_D}$.
By means of Lemma 3.8[iii] we have $z_\varepsilon(t) \geq z_\varepsilon(0)$. Hence $z_\varepsilon(t) \in K_{\varphi,g_D}$ holds for all $t \in [0,T]$.

[ii]. Due to statement [i], the choice $v = z_\varepsilon(t) \in K_{\varphi,g_D}$ in the inequality problem (3.1) is permissible. Combining this with $v = (u - z_\varepsilon)(t) \in V$ in the regularization problem (3.8) we obtain

$$\begin{aligned} m\,\|(z_\varepsilon - u)(t)\|^2_{H^1(\Omega)} &\leq (\mu\,[1-\beta_\varepsilon(z_\varepsilon(t)-\varphi)],\ (z_\varepsilon - u)(t)) \\ &\quad + \int_0^t b(t';(z_\varepsilon - u)(t'),(z_\varepsilon - u)(t))\,dt' \end{aligned}$$

by subtraction. Recalling the estimate $(1-\beta(r))\, r \leq 1$ for $r \geq 0$ from Remark 3.7, the first term on the right-hand side can be estimated by

$$\left(\mu\, [1 - \beta_\varepsilon(z_\varepsilon(t) - \varphi)]\ ,\ z_\varepsilon(t) - \varphi + \varphi - u(t)\right) \leq \varepsilon\ (\mu, 1).$$

Now, applying Young's inequality we arrive at

$$m\ \|(z_\varepsilon - u)(t)\|^2_{H^1(\Omega)} \leq \varepsilon\ \|\mu\|_{L_1(\Omega)} + \frac{L\,\delta}{2}\ \|(z_\varepsilon - u)(t)\|^2_{H^1(\Omega)}$$
$$+ \frac{L\,T}{2\delta} \int_0^t \|(z_\varepsilon - u)(t')\|^2_{H^1(\Omega)}\ dt' \quad \text{for all } t \in [0,T],\ \delta > 0.$$

Now, choosing a suitable $\delta > 0$ and applying Gronwall's inequality, the statement is obtained.

[iii]. Due to Lemma 3.8[ii] there exist a subsequence $\tilde{z}_{\varepsilon'} = z_{\varepsilon'} - g_D$ and a function $z \in H^1(0,T;V)$ such that $\tilde{z}_{\varepsilon'} \rightharpoonup z$ and $\partial_t \tilde{z}_{\varepsilon'} \rightharpoonup \partial_t z$ (weakly) in $L_2(0,T;V)$. On the other hand, owing to part [ii], we have $z_{\varepsilon'} \longrightarrow u$ strongly in the space $C([0,T];H^1(\Omega))$, such that $z = u - g_D$ holds. Finally, Lemma 3.8[iii] guarantees $\partial_t u \geq 0$. ■

Remark 3.10 Recalling the injection and compression moulding process as application problems (see Applications 2.2 and 2.3) of the general obstacle problem (2.11), (2.10), we have

$$\varphi = 0, \quad g_N(0) = 0, \quad \partial_t g_N \geq 0, \quad \partial_t f \geq 0.$$

Thus, $\mu = (-f(0))^+$ can be chosen and the assumptions of Theorem 3.9 are satisfied. Consequently, the statement $\partial_t u \geq 0$ is guaranteed. This situation reflects the underlying physical processes in the proper way, i.e., liquid (melt) is injected and the two mould halves are brought together, respectively. □

3.3.2 Time-dependent convex sets

In Application 2.4 we have seen that the mathematical modelling of the electrochemical machining process leads to obstacle problems with time-dependent convex sets caused by Dirichlet boundary conditions depending on time. Therefore, compared to the previous Section 3.3.1, we will now study the inequality problem (3.1), (3.2) with time-dependent convex sets $K = K_{\varphi,g_D}(t)$. Due to the structure of these constraint sets we do not immediately obtain that $u \in W^1_\infty(0,T;H^1(\Omega))$ holds as in Theorem 3.6. For this result a more comprehensive investigation will be necessary.

In this section we use again the penalization problem (3.8), but now with Dirichlet conditions g_D depending both on the space variable x and time variable t.

Lemma 3.11 [i] *Let the assumptions (3.3) and (3.4) be fulfilled. Then there exists a unique solution*

$$z_\varepsilon \in C([0,T]; H^1(\Omega))$$

of problem (3.8). Moreover, the solution is bounded such that

$$\|z_\varepsilon\|_{C([0,T];H^1(\Omega))} \le C \quad \textit{uniformly in} \quad \varepsilon > 0.$$

[ii] *Together with the conditions (3.3) and (3.4) let the following assumptions be satisfied, i.e.,* $\mu \in L_p(\Omega)$, $p \ge \max\{2, n/2\}$ *and*

$$f \in H^1(0,T; L_2(\Omega)), \quad g_N \in H^1(0,T; L_2(\Gamma_N)), \quad g_D \in H^1(0,T; H^1(\Omega))$$

with $\partial_t f(t) \ge 0$ *a.e. in* Ω, $\partial_t g_N(t) \ge 0$ *a.e. on* Γ_N, $\partial_t g_D(t) \ge 0$ *a.e. on* Γ_D *and a.e. in* $(0,T)$ *as well as* $\partial_t g_D \in L_\infty(Q)$.
Then,

$$\partial_t z_\varepsilon(t) \ge 0 \quad \textit{a.e. in} \quad \Omega,\ t \in (0,T)$$

and the time-derivative of the penalty solution z_ε *is bounded, i.e.,*

$$\|\partial_t z_\varepsilon\|_{L_2(0,T;H^1(\Omega))} \le C \quad \textit{independent of} \quad \varepsilon > 0.$$

Proof. [i]. This part is proved with minor modifications compared to the statement of Lemma 3.8[i].

[ii]. Similar to the proof of Lemma 3.8[iii] we differentiate the penalization problem (3.8), but now, due to the time-dependent Dirichlet conditions, we obtain

$$\left.\begin{aligned} &\text{Find} \quad w_\varepsilon(t) \in H^1(\Omega), \quad w_\varepsilon(t) = \partial_t g_D(t) \quad \text{on } \Gamma_D, \\ &a(t; w_\varepsilon(t), v) + \ (\mu\ \beta_\varepsilon'(z_\varepsilon(t) - \varphi)\ w_\varepsilon(t)\ ,\ v) \\ &= \ (\partial_t f(t), v) + (\partial_t g_N(t), v)_{\Gamma_N} \quad \forall v \in V, \text{ a.e. in } (0,T). \end{aligned}\right| \tag{3.13}$$

Due to $\beta_\varepsilon' \ge 0$ and $\mu \ge 0$, problem (3.13) has the unique solution $w_\varepsilon = \partial_t z_\varepsilon$ belonging to $L_2(0,T; H^1(\Omega))$ (for fixed $\varepsilon > 0$).
Applying a weak maximum principle we recall the assumption $\partial_t g_D \ge 0$. Hence, we have $\partial_t z_\varepsilon \ge 0$ on the Dirichlet part Γ_D of the boundary $\partial\Omega$. Therefore, it is allowed to take $0 \ge v = (\partial_t z_\varepsilon(t))^- \in V$ in problem (3.13). With this choice we get

$$\begin{aligned} &a(t; \partial_t z_\varepsilon(t), (\partial_t z_\varepsilon(t))^-) + \big(\mu\ \beta_\varepsilon'(z_\varepsilon(t) - \varphi)\ \partial_t z_\varepsilon(t),\ (\partial_t z_\varepsilon(t))^-\big) \\ &= \ (\partial_t f(t),\ (\partial_t z_\varepsilon(t))^-) + \big(\partial_t g_N(t),\ (\partial_t z_\varepsilon(t))^-\big)_{\Gamma_N} \le 0. \end{aligned}$$

Consequently, we obtain $\partial_t z_\varepsilon \ge 0$.

In order to show the boundedness of $\partial_t z_\varepsilon$ in $L_2(0,T;H^1(\Omega))$ independent of $\varepsilon > 0$, we take $v = \partial_t(z_\varepsilon(t) - g_D(t)) \in V$ in problem (3.13). We get

$$a(t; \partial_t z_\varepsilon(t), \partial_t(z_\varepsilon(t) - g_D(t))) + \big(\mu\, \beta_\varepsilon'(z_\varepsilon(t) - \varphi)\, \partial_t z_\varepsilon(t),\ \partial_t(z_\varepsilon(t) - g_D(t))\big)$$
$$= \big(\partial_t f(t),\ \partial_t(z_\varepsilon(t) - g_D(t))\big) + \big(\partial_t g_N(t),\ \partial_t(z_\varepsilon(t) - g_D(t))\big)_{\Gamma_N}.$$

The application of the ellipticity of the bilinear form a and Young's inequality as well as the integration in time lead to

$$C_1\ \|\partial_t(z_\varepsilon - g_D)\|^2_{L_2(0,T;H^1(\Omega))} \le C_2\ \|\partial_t f\|^2_{L_2(0,T;L_2(\Omega))}$$
$$+ C_3\ \|\partial_t g_N\|^2_{L_2(0,T;L_2(\Gamma_N))} + C_4\ \|\partial_t g_D\|^2_{L_2(0,T;H^1(\Omega))}$$
$$+ \int_0^T \big(\mu\, \beta_\varepsilon'(z_\varepsilon(t) - \varphi)\, \partial_t z_\varepsilon(t),\ \partial_t g_D(t)\big)\, dt.$$

Owing to $0 \le \beta_\varepsilon'(z_\varepsilon(t) - \varphi)\, \partial_t z_\varepsilon(t) = \partial_t(\beta_\varepsilon(z_\varepsilon(t) - \varphi))$ and $\mu \ge 0$, the application of integration by parts to the last term gives

$$\int_0^T \big(\mu\, \beta_\varepsilon'(z_\varepsilon(t) - \varphi)\, \partial_t z_\varepsilon(t),\ \partial_t g_D(t)\big) \le C\, \|\partial_t g_D\|_{L_\infty(Q)}\ \|\mu\|_{L_1(\Omega)}.$$

Hence, we have shown the boundedness of $\partial_t z_\varepsilon$ in the space $L_2(0,T;H^1(\Omega))$ independent of $\varepsilon > 0$. ■

Let us mention that in the electro-chemical machining process as an application problem (cf. Application 2.4) the Dirichlet conditions g_D depend only on the time t. Then, the assumption on g_D made in Lemma 3.11[ii] will be fulfilled if

$$g_D \in C^{0,1}[0,T] \quad \text{with} \quad \partial_t g_D \ge 0$$

holds.

Remark 3.12 Assuming the conditions (3.3), (3.4) and

$$f \in C^{0,1/2}([0,T]; L_2(\Omega)),\ g_N \in C^{0,1/2}([0,T]; L_2(\Gamma_N)),\ g_D \in W^1_\infty(0,T; H^1(\Omega)),$$

we obtain for the penalty solution

$$z_\varepsilon \in C^{0,1/2}([0,T]; H^1(\Omega)) \quad \text{with} \quad \|z_\varepsilon\|_{C^{0,1/2}([0,T];H^1(\Omega))} \le C$$

uniformly in $\varepsilon > 0$. Supposing additionally

$$f \in W^1_\infty(0,T; L_2(\Omega)), \quad g_N \in W^1_\infty(0,T; L_2(\Gamma_N)), \quad \mu \in L_p(\Omega),\ p \ge \max\{2, n/2\}$$

one has the Lipschitz continuity of z_ε, i.e., $z_\varepsilon \in W^1_\infty(0,T; H^1(\Omega))$, but not uniformly in $\varepsilon > 0$.

These two results can be verified by the subtraction of two penalty equations corresponding to time instants $t^i \in [0,T]$, $i = 1,2$. We take the test function $v = (\tilde{z}_\varepsilon(t^1) - \tilde{z}_\varepsilon(t^2)) \in V$ with $z_\varepsilon(t^i) = \tilde{z}_\varepsilon(t^i) + g_D(t^i)$, $i = 1,2$. This leads to

$$
\begin{aligned}
& a(t^1; \tilde{z}_\varepsilon(t^1) - \tilde{z}_\varepsilon(t^2), \tilde{z}_\varepsilon(t^1) - \tilde{z}_\varepsilon(t^2)) \\
& \quad + (\mu\ [\beta_\varepsilon(z_\varepsilon(t^1) - \varphi) - \beta_\varepsilon(z_\varepsilon(t^2) - \varphi)],\ z_\varepsilon(t^1) - z_\varepsilon(t^2)) \\
& = \ (f(t^1) - f(t^2),\ \tilde{z}_\varepsilon(t^1) - \tilde{z}_\varepsilon(t^2)) + (g_N(t^1) - g_N(t^2),\ \tilde{z}_\varepsilon(t^1) - \tilde{z}_\varepsilon(t^2))_{\Gamma_N} \\
& \quad + a(t^2; z_\varepsilon(t^2),\ \tilde{z}_\varepsilon(t^1) - \tilde{z}_\varepsilon(t^2)) - a(t^1; z_\varepsilon(t^2),\ \tilde{z}_\varepsilon(t^1) - \tilde{z}_\varepsilon(t^2)) \\
& \quad + a(t^1; g_D(t^2) - g_D(t^1),\ \tilde{z}_\varepsilon(t^1) - \tilde{z}_\varepsilon(t^2)) \\
& \quad + \int_0^t b(t'; z_\varepsilon(t'), \tilde{z}_\varepsilon(t^1) - \tilde{z}_\varepsilon(t^2))\ dt' \\
& \quad + (\mu\ [\beta_\varepsilon(z_\varepsilon(t^1) - \varphi) - \beta_\varepsilon(z_\varepsilon(t^2) - \varphi)],\ g_D(t^1) - g_D(t^2)).
\end{aligned}
$$

The crucial point is now the last term. We conclude $z_\varepsilon \in C^{0,1/2}([0,T]; H^1(\Omega))$ uniformly in $\varepsilon > 0$ by means of

$$
\left\| \beta_\varepsilon(z_\varepsilon(t^1) - \varphi) - \beta_\varepsilon(z_\varepsilon(t^2) - \varphi) \right\|_{L_\infty(\Omega)} \leq C
$$

and the application of Young's inequality.
On the other hand, using the Sobolev embedding $H^1(\Omega) \subset L_q(\Omega)$ (cf. Remark 3.4) and the properties of the regularization term β_ε (cf. Remark 3.7) we derive

$$
\begin{aligned}
& (\mu\ [\beta_\varepsilon(z_\varepsilon(t^1) - \varphi) - \beta_\varepsilon(z_\varepsilon(t^2) - \varphi)],\ g_D(t^1) - g_D(t^2)) \leq C\ \varepsilon^{-1}\ \|\mu\|_{L_p(\Omega)} \\
& * \ \left\|g_D(t^1) - g_D(t^2)\right\|_{H^1(\Omega)} \left(\left\|\tilde{z}_\varepsilon(t^1) - \tilde{z}_\varepsilon(t^2)\right\|_{H^1(\Omega)} + \left\|g_D(t^1) - g_D(t^2)\right\|_{H^1(\Omega)} \right).
\end{aligned}
$$

Applying again Young's inequality we obtain $z_\varepsilon \in W^1_\infty(0,T; H^1(\Omega))$ for each (fixed) $\varepsilon > 0$. □

In order to derive monotonicity and regularity properties of the solution of inequality (3.1), (3.2), we perform now the passage to the limit as $\varepsilon \longrightarrow 0$ in the regularization problem (3.8). This will be done similar to Section 3.3.1 (cf. Theorem 3.9).

Theorem 3.13 *Let the assumptions of Lemma 3.11[ii] together with the conditions* $\varphi \in H^2(\Omega)$, $k(0) \in C^1(\overline{\Omega})$, $\mu \geq (A(0)\ \varphi - f(0))^+$ *a.e. in* Ω *and* $l(0)\ \varphi \leq g_N(0)$ *on* Γ_N *be fulfilled.*

[i] *The unique solution* $z_\varepsilon = z_\varepsilon(t)$ *of (3.8) belongs to* $K_{\varphi,g_D}(t)$ *for each* $\varepsilon > 0$ *and for all* $t \in [0,T]$.

[ii] *Moreover, the estimate*

$$\|u - z_\varepsilon\|_{C([0,T];H^1(\Omega))} \leq C\ \sqrt{\varepsilon}\ \|\mu\|^{1/2}_{L_1(\Omega)}$$

holds, where u is the unique solution of the evolutionary obstacle problem (3.1), (3.2).

[iii] *The unique solution u of (3.1), (3.2) is such that*

$$u \in H^1(0,T;H^1(\Omega)) \subset C^{0,1/2}([0,T];H^1(\Omega))$$

and $\partial_t u(t) \geq 0$ a.e. in Ω, $t \in (0,T)$ holds.

Proof. Taking into consideration the time-dependent convex sets K_{φ,g_D}, parts [i] and [ii] are proved in a similar way as for the statements of Theorem 3.9[i], [ii].

[iii]. Owing to the statements of Lemma 3.11 we can extract a subsequence $z_{\varepsilon'}$ of z_ε with the property that, for $\tilde{z}_{\varepsilon'} = z_{\varepsilon'} - g_D$, one has $\tilde{z}_{\varepsilon'} \rightharpoonup z$ and $\partial_t \tilde{z}_{\varepsilon'} \rightharpoonup \partial_t z$ in $L_2(0,T;H^1(\Omega))$ (weakly) with a function $z \in H^1(0,T;V)$.

But in the previous part [ii], we have shown the convergence $z_\varepsilon \longrightarrow u$ in the space $C([0,T];H^1(\Omega))$ (strongly). Therefore, z and $u - g_D$ coincide as mappings from $[0,T] \longrightarrow V$ and hence $u \in H^1(0,T;H^1(\Omega))$ holds.

Finally, $\partial_t u \geq 0$ follows from Lemma 3.11[ii]. ∎

Recalling the application problems introduced in Section 2.2 (in particular the ECM Application 2.4), we note that in the case of an obstacle φ identical with zero, one can take $\mu(x) = (-f(x,0))^+$.

Based on comparison arguments, let us briefly discuss an upper bound for the time-derivative of the solution u of our evolutionary obstacle problem (3.1), (3.2).

Corollary 3.14 *Additionally to the assumptions of Theorem 3.13, let $q(x,t) \geq q_0(t) > 0$ for $x \in \Omega$, $t \in (0,T)$ and $\kappa(x) \geq \kappa_0 > 0$ for $x \in \Gamma_N$ be fulfilled. Then, the solutions z_ε and u of the penalty problem (3.8) and the evolutionary inequality (3.1), (3.2), respectively, satisfy*

$$0 \leq \partial_t z_\varepsilon(t) \leq R(t) := \max\Big\{ \sup_{\Gamma_D} \partial_t g_D(t),\ \frac{1}{q_0(t)} \sup_{\Omega} \partial_t f(t),\ \frac{1}{\kappa_0} \sup_{\Gamma_N} \partial_t g_N(t) \Big\}$$

and $0 \leq \partial_t u(t) \leq R(t)$ *a.e. in* Ω, $t \in (0,T)$.

Proof. In order to prove the upper bound for the penalty solution $\partial_t z_\varepsilon$ of (3.13), we take $v = (\partial_t z_\varepsilon(t) - R(t))^+ \in V$ in problem (3.13). This choice leads to

$$\begin{aligned}
& a(t; \partial_t z_\varepsilon(t) - R(t),\ (\partial_t z_\varepsilon(t) - R(t))^+) \\
& \quad + \left(\mu\, \beta'_\varepsilon(z_\varepsilon(t) - \varphi)\, \partial_t z_\varepsilon(t),\ (\partial_t z_\varepsilon(t) - R(t))^+\right) \\
& = \int_\Omega \left(\partial_t f(x,t) - q(x,t)\, R(t)\right) \left(\partial_t z_\varepsilon(x,t) - R(t)\right)^+ dx \\
& \quad + \int_{\Gamma_N} \left(\partial_t g_N(x,t) - \kappa(x)\, R(t)\right) \left(\partial_t z_\varepsilon(x,t) - R(t)\right)^+ ds \\
& \leq 0.
\end{aligned}$$

Consequently, we obtain $\|(\partial_t z_\varepsilon(t) - R(t))^+\|_{H^1(\Omega)} \leq 0$, which implies the inequality $\partial_t z_\varepsilon(t) \leq R(t)$.
The statement for the solution u of the obstacle problem (3.1), (3.2) is concluded by performing the passage to the limit $\varepsilon \longrightarrow 0$ (cf. proof of part[iii] of Theorem 3.13). ■

Remark 3.15 Let us mention that, if $\partial_t f = 0$ holds, the estimates of the previous Corollary 3.14 will already be valid for $q(x,t) \geq 0$. Analogously, in the situation $\partial_t g_N = 0$ one needs only $\kappa(x) \geq 0$.

Furthermore, let us emphasize that the solution $u(0)$ of the inequality problem (3.1), (3.2) for the initial time $t = 0$ fulfills the following weak maximum principle (cf. [RODR87], Section 4.5 and 4.6).

$$\begin{aligned}
& r_0 := \min\left\{0,\ \inf_{\Gamma_D} g_D(0),\ \frac{1}{q_0(0)} \inf_\Omega f(0),\ \frac{1}{\kappa_0} \inf_{\Gamma_N} g_N(0)\right\} \leq\ u(x,0) \\
& \leq\ R_0 := \max\left\{0,\ \sup_{\Gamma_D} g_D(0),\ \frac{1}{q_0(0)} \sup_\Omega f(0),\ \frac{1}{\kappa_0} \sup_{\Gamma_N} g_N(0),\ \sup_\Omega \varphi\right\} \quad \text{a.e. in } \Omega.
\end{aligned}$$

For this comparison result one needs $q(x,0) \geq q_0(0) > 0$ and $\kappa(x) \geq \kappa_0 > 0$. If $f(0) \leq 0$ (resp. $f(0) \geq 0$), then $q(x,0) \geq 0$ will be sufficient and if $g_N(0) \leq 0$ (resp. $g_N(0) \geq 0$), then $\kappa(x) \geq 0$ will be sufficient for the upper bound R_0 (resp. for the lower bound r_0).

Therefore, the solution $u = u(t)$ of the variational inequality (3.1), (3.2) satifies

$$r_0 \leq u(t) \leq R_0 + \int_0^t R(t')\, dt'.$$

For example, the solution u of (3.1), (3.2) related to the electro-chemical machining problem of Application 2.4 is such that

$$0 \leq u(t) \leq g_D(t) = \int_0^t \gamma_A(t')\, dt' \quad \text{a.e. in} \quad \Omega, \quad t \in [0,T].$$

We note that the corresponding bounds are assumed to be finite in the present remark. □

In the remaining part of this section we will discuss under what conditions the solution u of the evolutionary variational inequality (3.1), (3.2) with time-dependent convex sets K_{0,g_D} is also a Lipschitz continuous function from the time interval $[0,T]$ to the Sobolev space $H^1(\Omega)$. According to the derivation of the inequality problem in Chapter 2.3 (cf. (2.11)) we concentrate on the situation $\varphi = 0$ and, in particular, to the ECM Application 2.4 (see, however, Remark 3.18[ii]).

To demonstrate the Lipschitz continuity we discuss at first a general criteria (special kind of transformation of Dirichlet boundary conditions) and then we give some conditions on the data such that this criteria is fulfilled.

Lemma 3.16 *Let the assumptions (3.3), (3.4) together with*

$$f \in W^1_\infty(0,T;L_2(\Omega)), \quad g_N \in W^1_\infty(0,T;L_2(\Gamma_N)), \quad g_D \in W^1_\infty(0,T;H^1(\Omega))$$

be fulfilled. Furthermore, let a function $\delta \in W^1_\infty(0,T;H^1(\Omega))$ *with*

$$\delta(x,t) = -g_D(x,t) \quad \text{for } x \in \Gamma_D, \quad \delta(x,t) \le 0 \quad \text{for } x \in \Omega,$$

for all $t \in [0,T]$ *be given, such that for the function* η *defined by*

$$\eta = \eta(x,t) := u(x,t) + \delta(x,t),$$

it holds that $\eta(t) \ge 0$ *a.e. in* Ω *and* $\eta(t) \in V$ *for all* $t \in [0,T]$.
Then, the solution u *of the problem (3.1), (3.2) (with the obstacle* $\varphi = 0$*) belongs to* $W^1_\infty(0,T;H^1(\Omega))$.

Proof. To verify the assertion we substitute $u = \eta - \delta$ into (3.1), which leads to

$$a(t;\eta(t),w(t)-\eta(t)) \ge (f(t),w(t)-\eta(t)) + (g_N(t),w(t)-\eta(t))_{\Gamma_N}$$
$$+\ a(t;\delta(t),w(t)-\eta(t)) + \int_0^t b(t';u(t'),w(t)-\eta(t))\,dt'$$

for all functions $w(t) = v + \delta(t)$ with $v \in K_{0,g_D}(t)$. Consequently, we have $w(t) \in V$ and, in particular, $w(t) \ge \delta(t)$ for $t \in [0,T]$.
But now, owing to $\delta(t) \le 0$ and $V \ni \eta(t) \ge 0$, it is allowed to take $w(t^i) = \eta(t^{3-i})$ for $t^i \in [0,T]$, $i = 1,2$. Adding the inequalities for t^1 and t^2 we arrive at

$$a(t^1;\eta(t^1)-\eta(t^2),\eta(t^1)-\eta(t^2)) \le (f(t^1)-f(t^2),\eta(t^1)-\eta(t^2))$$
$$+\ (g_N(t^1)-g_N(t^2),\eta(t^1)-\eta(t^2))_{\Gamma_N} + \int_{t^2}^{t^1} b(t';u(t'),\eta(t^1)-\eta(t^2))\,dt'$$
$$+\ a(t^2;\eta(t^2),\eta(t^1)-\eta(t^2)) - a(t^1;\eta(t^2),\eta(t^1)-\eta(t^2))$$
$$+\ a(t^1;\delta(t^1)-\delta(t^2),\eta(t^1)-\eta(t^2)) + a(t^1;\delta(t^2),\eta(t^1)-\eta(t^2))$$
$$-\ a(t^2;\delta(t^2),\eta(t^1)-\eta(t^2)).$$

Now, similar to the proof of Theorem 3.6, we show $\eta \in W^1_\infty(0,T;H^1(\Omega))$. Consequently, we obtain $u \in W^1_\infty(0,T;H^1(\Omega))$. ■

Based on an auxiliary problem and the Lewy–Stampacchia penalization technique, let us now construct corresponding functions δ and η for the ECM Application 2.4. We recall that for this problem the related evolutionary variational inequality (3.1), (3.2) is characterized by $\varphi = 0,\ q = 0,\ \partial\Omega = \Gamma_D$. Furthermore, we have Dirichlet boundary conditions

$$g_D(t) = \int_0^t \gamma_A(t')\,dt', \quad \gamma_A = \gamma_A(t) > 0 \quad \text{(independent of space)},$$

where $\gamma_A(t)$ is the potential at the anode. The right-hand side f in (3.1), (3.2) is given by

$$f(x) = -\lambda_E\,(\chi_{\Omega(0)} - 1) = \lambda_E\,\chi_{A(0)} \le 0 \quad \text{(independent of time)},$$

where $\lambda_E < 0$ denotes the electrochemical equivalent and $A(0)$ is the anode; cf. the ECM Application 2.4 and, in particular, Figure 2.5. An essential point for the construction of the corresponding function δ for the ECM problem is the assumption of a smooth, closed subset G such that

$$A(0) \subset G \subset \Omega \quad \text{and} \quad \{x \in \Omega : f(x) < 0\} \subset G.$$

Theorem 3.17 *Let the assumptions of Theorem 3.13 together with $\gamma_A \in C[0,T]$ be fulfilled. Then, the solution u of the inequality problem (3.1), (3.2) related to the ECM Application 2.4 is such that*

$$u \in W^1_\infty(0,T;H^1(\Omega)).$$

Proof. Based on the penalization problem (3.8) and its time derivative (3.13) we shall construct a function $V \ni \eta_\varepsilon(t) = z_\varepsilon(t) + \delta(t)$ with $\eta_\varepsilon(t) \ge 0$ for all $t \in [0,T]$ (due to $\partial\Omega = \Gamma_D$ we have $V = H^1_0(\Omega)$). Then the desired function $0 \le \eta = u + \delta$ will be obtained in passing to the limit as $\varepsilon \longrightarrow 0$.
We define the function δ by

$$\delta(t) = \int_0^t \rho(t')\,dt',$$

where $\rho = \rho(x,t)$ is the solution of the following auxiliary problem.

$$\left.\begin{aligned}
&\text{Find} \quad \rho(t) \in H^1(\Omega) \quad \text{with}\\
&\rho(t) = -\gamma_A(t) < 0 \text{ on } \Gamma_D, \quad \rho(t) = 0 \quad \text{on } \partial G,\\
&\int_{\Omega\backslash G} k(x,t)\,\nabla\rho(x,t)\,\nabla v(x)\,dx = 0 \quad \forall v \in H^1(\Omega\backslash G),\ v = 0 \text{ on } \Gamma_D,\ \partial G.
\end{aligned}\right| \tag{3.14}$$

Here, G is a smooth closed subset of Ω such that $A(0) \subset G \subset \Omega$, cf. Figure 2.5. The function $\rho(t)$ is extended by 0 in the subset $G \subset \Omega$ such that the maximum principle implies

$$0 \geq \rho(x,t) \geq -\gamma_A(t).$$

Furthermore, we have $\rho \in C([0,T]; H^1(\Omega))$ and consequently

$$\delta \in W^1_\infty(0,T; H^1(\Omega)), \quad \delta(x,0) = 0, \quad \delta \leq 0.$$

Now, by means of Theorem 3.13[i] we have $V \ni \eta_\varepsilon(0) = z_\varepsilon(0) + \delta(0) = z_\varepsilon(0) \geq 0$. Owing to the representation

$$\eta_\varepsilon(t) = \eta_\varepsilon(0) + \int_0^t \partial_t \eta_\varepsilon(t')\, dt',$$

it remains to show $\partial_t \eta_\varepsilon(t) \geq 0$ in Ω.

By virtue of $\partial_t \eta_\varepsilon = \partial_t z_\varepsilon \geq 0$ in G and $\partial_t \eta_\varepsilon = 0$ on Γ_D, we have $(\partial_t \eta_\varepsilon(t))^- = 0$ in G and on Γ_D. Consequently, we derive

$$\begin{aligned} m \left\| (\partial_t \eta_\varepsilon(t))^- \right\|^2_{H^1(\Omega)} &\leq a(t; (\partial_t \eta_\varepsilon(t))^-\ ,\ (\partial_t \eta_\varepsilon(t))^-) \\ &= a(t; \partial_t z_\varepsilon(t)\ ,\ (\partial_t \eta_\varepsilon(t))^-) + a(t; \partial_t \delta(t)\ ,\ (\partial_t \eta_\varepsilon(t))^-). \end{aligned}$$

By means of problem (3.13) we deduce that, due to $\partial_t f = 0$,

$$a(t; \partial_t z_\varepsilon(t)\ ,\ (\partial_t \eta_\varepsilon(t))^-) \leq \int_\Omega \beta'_\varepsilon(z_\varepsilon(t))\ \partial_t z_\varepsilon(t)\ \mu\ (\partial_t \eta_\varepsilon(t))^-\ dx = 0.$$

Note, that owing to $\mu = (-f(0))^+$ and the assumption on the subset G, we have $\mu\ (\partial_t \eta_\varepsilon(t))^- = 0$.

Furthermore, due to $(\partial_t \eta_\varepsilon(t))^- = 0$ in G and on Γ_D, we can take $v = (\partial_t \eta_\varepsilon(t))^-$ in the auxiliary problem (3.14) introduced above. Recalling $\partial_t \delta = \rho$, the problem (3.14) implies

$$a(t; \partial_t \delta(t)\ ,\ (\partial_t \eta_\varepsilon(t))^-) = \int_{\Omega \setminus G} k(x,t)\ \nabla(\partial_t \delta(x,t))\ \nabla((\partial_t \eta_\varepsilon(x,t))^-)\ dx\ = 0.$$

Consequently, $\partial_t \eta_\varepsilon(x,t) \geq 0$ is shown, which completes the proof. ■

Remark 3.18 [i] Let us emphasize that the underlying idea for the definition of the function ρ (auxiliary problem (3.14)) consists in a generalization of an argument which has already been used in [LORO86], [RODR87], Section 9.5 for the ECM problem with constant conductivity of the electrolyte.

[ii] Modifying the proof of the previous Theorem 3.17 we can obtain the regularity $u \in W^1_\infty(0,T; H^1(\Omega))$ for the solution u of evolutionary inequalities (3.1), (3.2), which are, compared to the considered ECM inequality, more general (in particular, with respect to the Dirichlet boundary conditions and the right-hand side f). At this point, we omit further details. □

3.4 Regularity with respect to space variables

In this section we concentrate our study on the spatial smoothness of the solution of the variational inequality problem (3.1), (3.2).

To obtain such regularity results with respect to space we shall employ a semi-discretization in time procedure (Rothe's method, see [ROTH30], [KACU85]) in connection with the elliptic regularity theory and a priori estimates for second order elliptic equations (see [GRIS85] for a self-contained study and [RODR87], Section 3.7 for a summary).

The purpose of the application of Rothe's method is twofold. On the one hand, we use it for treatment of the memory term and, on the other hand, by means of this semi-discretization method we show that the solution u of the inequality problem is a (Bochner) measurable function from the time interval $(0,T)$ to the Sobolev space $W_p^2(\Omega)$.

The mathematical tools just mentioned are applied to the penalization problems (3.8) and then a passage to the limit as $\varepsilon \longrightarrow 0$ yields assertions for the variational inequalities. Owing to the well established elliptic regularity for problems with Dirichlet and Neumann/Newton boundary conditions we shall consider the cases $\partial\Omega = \Gamma_D$ and $\partial\Omega = \Gamma_N$ in problems (3.1), (3.2) in a separate way. In particular, we focus our attention on a bounded domain $\Omega \subset \mathbb{R}^n$ with a $C^{1,1}$-boundary. For the elliptic regularity in the case of *convex* domains (i.e., $H^2(\Omega)$) we refer to [GRIS85], Chapter 3 and [RODR87], Section 3.7 (in particular, Remark 3.7.7).

We use the regularity statement that the solution w of the Dirichlet (respectively Neumann/Newton) boundary value problem

$$\left.\begin{aligned} &-\operatorname{div}(a_1(x)\ \nabla w) + a_2(x)\ w = f(x) \quad \text{in } \Omega \quad \text{with} \\ &\text{either} \quad w = g_1(x) \quad \text{on } \Gamma = \partial\Omega \quad \text{or} \\ &l_2 w + a_0(x)\ w = g_2(x) \quad \text{on } \Gamma = \partial\Omega \end{aligned}\right| \tag{3.15}$$

is such that

$$w \in W_p^2(\Omega), \quad 2 \le p < \infty,$$

provided that $a_1 \in C^1(\overline{\Omega})$, $a_2 \in C(\overline{\Omega})$, $f \in L_p(\Omega)$, $a_0 \in C^{0,1}(\overline{\Omega})$ and $g_1 \in W_p^{2-1/p}(\partial\Omega)$, $g_2 \in W_p^{1-1/p}(\partial\Omega)$ as well as the associated bilinear form is H^1-elliptic (see e.g., [GRIS85], Section 2 and [RODR87], Section 3.7). Here, $l_2 w$ denotes $l_2 w = a_1 \nabla w \cdot \vec{n}$.

Besides this regularity result, in what follows we shall explicitly exploit a general a priori estimate (see [GRIS85], Section 2.3.3) of the form that a constant $C > 0$ exists such that

$$\|w\|_{W_p^2(\Omega)} \le C \left(\|\operatorname{div}(a_1 \nabla w)\|_{L_p(\Omega)} + \|l_0\ w\|_{W_p^{2-d-1/p}(\partial\Omega)} + \|w\|_{W_p^1(\Omega)} \right) \tag{3.16}$$

for all $w \in W_p^2(\Omega)$, where l_0 is either the identity operator (then $d = 0$) or $l_0 = l_2$ (then $d = 1$).

3.4.1 Dirichlet boundary conditions

At first we investigate the inequalities (3.1), (3.2) with Dirichlet boundary conditions which correspond, for instance, to the electro-chemical machining problem (see Application 2.4) and in injection/compression moulding (Applications 2.2, 2.3) to an open mould.

Let us consider the following auxiliary problem.

$$\left.\begin{aligned} &\text{Find} \quad z(t) \in H^1(\Omega), \quad z(t) = g_D(t) \quad \text{on } \Gamma_D, \\ &a(t; z(t), v) = (f(t), v) + \int_0^t b(t'; z(t'), v)\, dt' \quad \forall v \in H_0^1(\Omega),\ t \in [0, T] \end{aligned}\right| \tag{3.17}$$

with the bilinear forms a, b defined in (3.2) (but with $\Gamma_D = \partial\Omega$, $\Gamma_N = \emptyset$). Furthermore, let us assume

$$\left.\begin{aligned} &k \in W_\infty^1(0, T; C^1(\overline{\Omega})), \quad q \in W_\infty^1(0, T; C(\overline{\Omega})), \\ &\text{boundary } \partial\Omega \text{ of } \Omega \text{ is of the class } C^{1,1}. \end{aligned}\right| \tag{3.18}$$

Note that the auxiliary problem (3.17) corresponds to the boundary value problem

$$\left.\begin{aligned} &A(t)\ z(t) = f(t) + \int_0^t B(t')\ z(t')\ dt' \quad \text{in } \Omega, \\ &z(t) = g_D(t) \quad \text{on } \partial\Omega \end{aligned}\right| \tag{3.19}$$

with the differential operators A and B defined in (3.10).

Lemma 3.19 *Suppose that the conditions (3.4) and (3.18) hold. Then, for every right-hand side* $f \in L_2(0, T; L_p(\Omega))$ *and every* $g_D \in L_2(0, T; W_p^{2-1/p}(\partial\Omega))$, *there exists a function*

$$z \in \big(C([0, T]; H^1(\Omega)) \cap L_2(0, T; W_p^2(\Omega))\big),$$

which is the unique solution of problem (3.17) and satisfies

$$\begin{aligned} \|z\|_{L_2(0,T;W_p^2(\Omega))} \le C_1\ \big[&\|f\|_{L_2(0,T;L_p(\Omega))} + \|f\|_{C([0,T];L_2(\Omega))} \\ &+ \|g_D\|_{L_2(0,T;W_p^{2-1/p}(\partial\Omega))} + \|g_D\|_{C([0,T];H^1(\Omega))} \big], \quad 2 \le p < \infty. \end{aligned}$$

Assume additionally that $f \in L_\infty(0, T; L_p(\Omega))$ *and* $g_D \in L_\infty(0, T; W_p^{2-1/p}(\partial\Omega))$ *hold. Then, the solution* z *of (3.17) belongs to* $L_\infty(0, T; W_p^2(\Omega))$ *and there exists a constant* $C_2 > 0$ *such that*

$$\begin{aligned} \|z\|_{L_\infty(0,T;W_p^2(\Omega))} &\leq C_2 \; \big[\|f\|_{L_\infty(0,T;L_p(\Omega))} + \|f\|_{C([0,T];L_2(\Omega))} \\ &+ \|g_D\|_{L_\infty(0,T;W_p^{2-1/p}(\partial\Omega))} + \|g_D\|_{C([0,T];H^1(\Omega))} \big], \quad 2 \leq p < \infty. \end{aligned}$$

Proof. Recalling the existence result of Theorem 3.3, which is also valid for problem (3.17), it remains to show the spatial regularity. Consequently, at first we prove $z \in L_2(0,T;W_p^2(\Omega))$ and after that we conclude $z \in L_\infty(0,T;W_p^2(\Omega))$. We use a time discretization procedure and discretize $[0,T]$ into $\{t^j = j\tau,\ j = 0,\ldots,N\}$, $N\tau = T$, where N will later tend towards $+\infty$.
Let us define a sequence $\{w_j\}_{j=0}^{N-1}$ by solving the elliptic problems

$$\left. \begin{aligned} a(t^j; w_j, v) &= (f_j, v) + \sum_{i=0}^{j-1} \int_{i\tau}^{(i+1)\tau} b(t'; w_i, v)\, dt' \quad \forall v \in H_0^1(\Omega), \\ w_j = (g_D)_j &:= \frac{1}{\tau} \int_{j\tau}^{(j+1)\tau} g_D(t')\, dt' \quad \text{on } \partial\Omega, \\ f_j &:= \frac{1}{\tau} \int_{j\tau}^{(j+1)\tau} f(t')\, dt'. \end{aligned} \right| \tag{3.20}$$

On the one hand, taking $v = (w_j - (g_D)_j) \in H_0^1(\Omega)$ in (3.20) and applying Gronwall's inequality in its discrete version (cf. Lemma 3.1), we get

$$\left. \begin{aligned} \|w_j\|_{H^1(\Omega)} &\leq c_1 \left[\|f_j\|_{L_2(\Omega)} + \|(g_D)_j\|_{H^1(\Omega)} \right] \\ &\quad + c_2 \left[\|f\|_{L_2(0,T;L_2(\Omega))} + \|g_D\|_{L_2(0,T;H^1(\Omega))} \right]. \end{aligned} \right| \tag{3.21}$$

Here, c_1, c_2 are positive constants independent of j and τ. On the other hand, due to the elliptic regularity theory (see problem (3.15) and, in particular, cf. [GRIS85], Theorem 2.4.2.5 and [RODR87], Theorem 3.7.4), we conclude successively the spatial regularity $w_j \in W_p^2(\Omega)$ for each $j = 0,\ldots,N-1$.
Furthermore, an a priori estimate of the form

$$\|w_j\|_{W_p^2(\Omega)} \leq C \left(\|\hat{f}_j\|_{L_p(\Omega)} + \|(g_D)_j\|_{W_p^{2-1/p}(\partial\Omega)} + \|w_j\|_{H^1(\Omega)} \right) \tag{3.22}$$

holds with

$$\hat{f}_j = f_j + \sum_{i=0}^{j-1} \int_{i\tau}^{(i+1)\tau} B(t')\, w_i\, dt'$$

and the differential operator B defined in (3.10). In particular, we mention that the constant C in (3.22) can be chosen independent of j and τ (see the general a priori estimate (3.16) and cf. Remark 3.20 below). Owing to $w_i \in W_p^2(\Omega)$ we have

$$\|\hat{f}_j\|_{L_p(\Omega)} \leq \|f_j\|_{L_p(\Omega)} + \bar{C}\, \tau \sum_{i=0}^{j-1} \|w_i\|_{W_p^2(\Omega)}$$

by Minkowski's inequality. Therefore, by means of (3.21), the estimate (3.22) is rewritten as

$$\begin{aligned}\|w_j\|_{W_p^2(\Omega)} &\le C_1\left(\|f_j\|_{L_p(\Omega)} + \|(g_D)_j\|_{H^1(\Omega)} + \|(g_D)_j\|_{W_p^{2-1/p}(\partial\Omega)}\right)\\ &+ C_2\left(\|f\|_{L_2(0,T;L_2(\Omega))} + \|g_D\|_{L_2(0,T;H^1(\Omega))}\right)\\ &+ C_3\,\tau\sum_{i=0}^{j-1}\|w_i\|_{W_p^2(\Omega)} \qquad j=0,\dots,N-1.\end{aligned}$$

With the help of Gronwall's inequality in its discrete version (see Lemma 3.1) we derive

$$\begin{aligned}\|w_j\|_{W_p^2(\Omega)} &\le \tilde{C}_1\left(\|f_j\|_{L_p(\Omega)} + \|(g_D)_j\|_{H^1(\Omega)} + \|(g_D)_j\|_{W_p^{2-1/p}(\partial\Omega)}\right)\\ &+ \tilde{C}_2\left(\|f\|_{L_2(0,T;L_2(\Omega))} + \|g_D\|_{L_2(0,T;H^1(\Omega))}\right)\\ &+ \tilde{C}_3\,\tau\sum_{i=0}^{j-1}\left(\|f_i\|_{L_p(\Omega)} + \|(g_D)_i\|_{H^1(\Omega)} + \|(g_D)_i\|_{W_p^{2-1/p}(\partial\Omega)}\right),\end{aligned} \tag{3.23}$$

where all the constants are independent of j and τ. Now, we define step functions by

$$\begin{aligned} f_\tau(t) &= f_j & &\text{for}\quad t\in[j\tau,\ (j+1)\tau), & f_\tau(T) &= f_{N-1},\\ (g_D)_\tau(t) &= (g_D)_j & &\text{for}\quad t\in[j\tau,\ (j+1)\tau), & (g_D)_\tau(T) &= (g_D)_{N-1},\\ w_\tau(t) &= w_j & &\text{for}\quad t\in[j\tau,\ (j+1)\tau), & w_\tau(T) &= w_{N-1}\end{aligned}$$

with $j=0,\dots,N-1$. Using the Hölder inequality several times we get

$$\begin{aligned}\|f_\tau\|_{L_2(0,T;L_p(\Omega))} &= \left(\tau\sum_{j=0}^{N-1}\|f_j\|^2_{L_p(\Omega)}\right)^{1/2} \le \|f\|_{L_2(0,T;L_p(\Omega))},\\ \tau\sum_{i=0}^{j-1}\|f_i\|_{L_p(\Omega)} &\le \sqrt{t^j}\ \|f\|_{L_2((0,t^j);L_p(\Omega))},\end{aligned}$$

as well as analogous estimates for g_D. Combining these estimates with (3.23) and performimg some calculations, we find that w_τ can be bounded by

$$\begin{aligned}\|w_\tau\|_{L_2(0,T;W_p^2(\Omega))} &\le \bar{C}_1\ \|f\|_{L_2(0,T;L_p(\Omega))} + \bar{C}_2\ \|g_D\|_{L_2(0,T;H^1(\Omega))}\\ &+ \bar{C}_3\ \|g_D\|_{L_2(0,T;W_p^{2-1/p}(\partial\Omega))}\,.\end{aligned}$$

Therefore, w_τ remains in a bounded subset of $L_2(0,T;W_p^2(\Omega))$. Hence, extracting a subsequence again denoted by w_τ, we get

$$w_\tau \rightharpoonup w \quad \text{in} \quad L_2(0,T;W_p^2(\Omega)) \quad \text{(weakly)}.$$

Moreover, by means of direct calculations we can show the strong convergence $f_\tau \longrightarrow f$ in $L_2(0,T;L_p(\Omega))$ and $(g_D)_\tau \longrightarrow g_D$ in $L_2(0,T;W_p^{2-1/p}(\partial\Omega))$ (resp. in $L_2(0,T;H^1(\Omega))$).
Integrating (3.20) in time and multiplying it by $v \in L_2(0,T;H_0^1(\Omega))$, we derive

$$\begin{aligned}
&\int_0^T a(t;w_\tau(t),v(t))\,dt - \int_0^T \Big(a(t;w_\tau(t),v(t)) - a(\tilde{t};w_\tau(t),v(t))\Big)\,dt \\
&= \int_0^T \Big(\int_0^t b(t';w_\tau(t'),v(t))\,dt' - \int_{\tilde{t}}^t b(t';w_\tau(t'),v(t))\,dt'\Big)\,dt \\
&+ \int_0^T (f_\tau(t),v(t))\,dt,
\end{aligned}$$

where $\tilde{t} = [\frac{t}{\tau}]\,\tau$ for $t \in [0,T)$ ($[\frac{t}{\tau}]$ as the integer part of $\frac{t}{\tau}$) and $\tilde{t} = T-\tau$ for $t=T$.

Now, performing the passage to the limit as $\tau \longrightarrow 0$ we obtain

$$\int_0^T a(t;w(t),v(t))\,dt = \int_0^T (f(t),v(t))\,dt + \int_0^T \Big(\int_0^t b(t';w(t'),v(t))\,dt'\Big)\,dt$$

for all $v \in L_2(0,T;H_0^1(\Omega))$ and $w(t) = g_D(t)$ a.e. in $(0,T)$ (note the linear continuous trace operator from $W_p^2(\Omega)$ to $W_p^{2-1/p}(\partial\Omega)$; see [RODR87], Theorem 3.5.7 and [GRIS85], Theorem 1.5.1.2). Consequently, it follows that

$$a(t;w(t),v) = (f(t),v) + \int_0^t b(t',w(t'),v)\,dt' \quad \forall v \in H_0^1(\Omega), \text{ a.e. in } (0,T),$$

which, together with (3.17), implies $w(t) = z(t)$ a.e. in $(0,T)$. Consequently, we have proved $z \in L_2(0,T;W_p^2(\Omega))$.

Let us now prove the estimates for z. At first, we take $v = (z(t) - g_D(t)) \in H_0^1(\Omega)$ in (3.17) and apply Gronwall's inequality. This leads to

$$\|z\|_{C([0,T];H^1(\Omega))} \le c_1\, \|f\|_{C([0,T];L_2(\Omega))} + c_2\, \|g_D\|_{C([0,T];H^1(\Omega))}.$$

Owing to the just proved regularity $z \in L_2(0,T;W_p^2(\Omega))$ for the solution z of problem (3.17), we have the a priori estimate

$$\left.\begin{aligned}
\|z(t)\|_{W_p^2(\Omega)} &\le C\,\Big(\|z(t)\|_{H^1(\Omega)} + \|f(t)\|_{L_p(\Omega)} + \int_0^t \|(Bz)(t')\|_{L_p(\Omega)}\,dt' \\
&\quad + \|g_D(t)\|_{W_p^{2-1/p}(\partial\Omega)}\Big) \quad \text{a.e. in } (0,T),
\end{aligned}\right| \tag{3.24}$$

where the constant C does not depend on t (see (3.16) and cf. Remark 3.20). Combining the last two inequalities, we obtain

$$\begin{aligned} \|z(t)\|^2_{W^2_p(\Omega)} &\le C_1 \left(\|f\|^2_{C([0,T];L_2(\Omega))} + \|g_D\|^2_{C([0,T];H^1(\Omega))} \right) \\ &+ C_2 \left(\|f(t)\|^2_{L_p(\Omega)} + \|g_D(t)\|^2_{W_p^{2-1/p}(\partial\Omega)} \right) \\ &+ C_3 \int_0^t \|z(t')\|^2_{W^2_p(\Omega)} \, dt' \quad \text{a.e. in } (0,T). \end{aligned} \tag{3.25}$$

Integrating (3.25) in time and using again Gronwall's inequality, we get

$$\begin{aligned} &\int_0^t \|z(t')\|^2_{W^2_p(\Omega)} \, dt' \le \bar{C}_1 \left(\|f\|^2_{C([0,T];L_2(\Omega))} + \|g_D\|^2_{C([0,T];H^1(\Omega))} \right) \\ &+ \bar{C}_2 \int_0^t \left(\|f(t')\|^2_{L_p(\Omega)} + \|g_D(t)\|^2_{W_p^{2-1/p}(\partial\Omega)} \right) dt' \quad \forall t \in [0,T]. \end{aligned} \tag{3.26}$$

Now, we conclude the estimate for z in the norm of $L_2(0,T;W^2_p(\Omega))$ by means of (3.26) for $t = T$.
Moreover, substituting (3.26) into (3.25), we obtain $z \in L_\infty(0,T;W^2_p(\Omega))$ and the desired estimate for $\|z\|_{L_\infty(0,T;W^2_p(\Omega))}$. ■

Remark 3.20 Let us briefly point out that the constants in the a priori estimates (3.22), (3.24) can be chosen independent of t and τ. Considering the elliptic problem

$$-\operatorname{div}(k(x,t)\,\nabla v) + q(x,t)\,v = f(x,t) \quad \text{in } \Omega, \qquad v = g_D(x,t) \quad \text{on } \partial\Omega$$

for fixed $t \in [0,T]$, we have $v \in W^2_p(\Omega)$ resulting from the elliptic regularity (cf. [GRIS85], Theorem 2.4.2.5 and [RODR87], Theorem 3.7.4). Thus, the differential equation can be multiplied by

$$|\Delta v|^{p-1} \in L_{\frac{p}{p-1}}(\Omega).$$

Integrating over Ω as well as using Young's and Hölder's inequality, we obtain

$$\|\Delta v\|_{L_p(\Omega)} \le C\big(\|v\|_{W^1_p(\Omega)} + \|f(t)\|_{L_p(\Omega)} \big), \quad C = C(p, k_0, \|k(t)\|_{C^1(\overline{\Omega})}, \|q(t)\|_{C(\overline{\Omega})}).$$

Therefore, owing to the assumptions (3.18), the constant C can be taken independent of the time t. Now, applying the general a priori estimate (3.16) to Δv with $v \in W^2_p(\Omega)$ and $v = g_D(t)$ on $\partial\Omega$, we have

$$\begin{aligned} &\|v\|_{W^2_p(\Omega)} \le c_1 \|v\|_{W^1_p(\Omega)} + c_2 \|f(t)\|_{L_p(\Omega)} + c_3 \|g_D(t)\|_{W_p^{2-1/p}(\partial\Omega)} \quad \text{and} \\ &\|v\|_{W^2_p(\Omega)} \le C_1 \|v\|_{L_p(\Omega)} + C_2 \|f(t)\|_{L_p(\Omega)} + C_3 \|g_D(t)\|_{W_p^{2-1/p}(\partial\Omega)}. \end{aligned} \tag{3.27}$$

The second estimate in (3.27) is obtained from the first one using a Gagliardo–Nirenberg inequality (cf. [ZEIDII90], Section 21.17, 21.19; especially (63) on pages 284, 285).

We combine the second relation of (3.27) and the Sobolev embeddings

$$W_r^j(\Omega) \subset L_q(\Omega), \quad j = 1, 2$$

for corresponding values of r and q (depending on $2 \leq p < \infty$ and the spatial dimension n, a multiple application can be necessary). Finally, we arrive at

$$\|v\|_{W_p^2(\Omega)} \leq \tilde{C}_1 \ \|v\|_{H^1(\Omega)} + \tilde{C}_2 \ \|f(t)\|_{L_p(\Omega)} + \tilde{C}_3 \ \|g_D(t)\|_{W_p^{2-1/p}(\partial\Omega)}, \ 2 \leq p < \infty,$$

where the constants do not depend on the time t. □

We are now able to extend the results of Lemma 3.19 to the solution z_ε of the regularization problem (3.8).

Corollary 3.21 *Let the assumptions (3.4), (3.3), (3.18) and $\mu \in L_p(\Omega)$ be fulfilled.*

[i] *If $f \in L_2(0,T; L_p(\Omega))$ and $g_D \in L_2(0,T; W_p^{2-1/p}(\partial\Omega))$ hold, then the unique solution z_ε of the penalty problem (3.8) with $\partial\Omega = \Gamma_D$ will be such that*

$$z_\varepsilon \in \big(C([0,T]; H^1(\Omega)) \ \cap \ L_2(0,T; W_p^2(\Omega))\big),$$
$$\|z_\varepsilon\|_{L_2(0,T;W_p^2(\Omega))} \leq C_1 \quad \textit{uniformly in } \varepsilon > 0, \quad 2 \leq p < \infty.$$

[ii] *If $f \in L_\infty(0,T; L_p(\Omega))$ and $g_D \in L_\infty(0,T; W_p^{2-1/p}(\partial\Omega))$ are satisfied, then the unique solution z_ε of (3.8) will be such that*

$$z_\varepsilon \in \big(C([0,T]; H^1(\Omega)) \ \cap \ L_\infty(0,T; W_p^2(\Omega))\big),$$
$$\|z_\varepsilon\|_{L_\infty(0,T;W_p^2(\Omega))} \leq C_2 \quad \textit{uniformly in } \varepsilon > 0, \quad 2 \leq p < \infty.$$

Proof. The statement $z_\varepsilon \in L_2(0,T; W_p^2(\Omega))$ follows from Lemma 3.19, which can be applied with a modified right-hand side

$$\breve{f}(t) = f(t) + [1 - \beta_\varepsilon(z_\varepsilon(t) - \varphi)] \ \mu$$

to problem (3.8). Under the assumptions $f \in L_2(0,T; L_p(\Omega))$ and $\mu \in L_p(\Omega)$ the just defined function $\breve{f}$ also belongs to $L_2(0,T; L_p(\Omega))$.

The boundedness of z_ε in $L_2(0,T; W_p^2(\Omega))$ is obtained by similar arguments as in the proof of the foregoing Lemma 3.19 (see (3.24), (3.25), (3.26)).

Finally, the assertions of part [ii] are concluded again by applying Lemma 3.19. ■

Now we come to the global regularity of the solution u of the obstacle problem (3.1), (3.2) with Dirichlet boundary conditions. These main statements are obtained connecting the results of Theorems 3.9, 3.13 (strong convergence of z_ε to u) and the foregoing Corollary 3.21 (boundedness of z_ε).

Theorem 3.22 *Let (3.18) and $\varphi \in W_p^2(\Omega)$, $f(0) \in L_p(\Omega)$ be fulfilled.*
[i] *Under the assumptions of Theorem 3.13 and*

$$f \in L_2(0,T;L_p(\Omega)), \quad g_D \in L_2(0,T;W_p^{2-1/p}(\partial\Omega))$$

the solution u of the evolutionary variational inequality (3.1), (3.2) with $\Gamma_D = \partial\Omega$ is such that

$$u \in \left(H^1(0,T;H^1(\Omega)) \cap L_2(0,T;W_p^2(\Omega))\right), \quad 2 \le p < \infty.$$

Under the additional assumptions

$$f \in L_\infty(0,T;L_p(\Omega)), \quad g_D \in L_\infty(0,T;W_p^{2-1/p}(\partial\Omega)),$$

it follows that

$$u \in \left(H^1(0,T;H^1(\Omega)) \cap L_\infty(0,T;W_p^2(\Omega))\right), \quad 2 \le p < \infty$$

holds for the solution u of (3.1), (3.2).

[ii] *Let the assumptions of Theorem 3.9 (resp. Theorem 3.17) be satisfied. Then, the solution u of the inequality problem (3.1), (3.2) with $\Gamma_D = \partial\Omega$ will be such that*

$$u \in \left(W_\infty^1(0,T;H^1(\Omega)) \cap L_2(0,T;W_p^2(\Omega))\right), \quad 2 \le p < \infty,$$

if

$$f \in L_2(0,T;L_p(\Omega)), \quad g_D \in L_2(0,T;W_p^{2-1/p}(\partial\Omega))$$

is fulfilled.
Under the assumptions

$$f \in L_\infty(0,T;L_p(\Omega)), \quad g_D \in L_\infty(0,T;W_p^{2-1/p}(\partial\Omega)),$$

it holds that

$$u \in \left(W_\infty^1(0,T;H^1(\Omega)) \cap L_\infty(0,T;W_p^2(\Omega))\right), \quad 2 \le p < \infty.$$

Proof. Owing to Corollary 3.21[i] the family $\{z_\varepsilon\}$ of penalty solutions is bounded in $L_2(0,T;W_p^2(\Omega))$ (independent of ε). Hence, there exists a subsequence $z_{\varepsilon'}$, which converges weakly in the space $L_2(0,T;W_p^2(\Omega))$ to an element w with $w \in L_2(0,T;W_p^2(\Omega))$.

Thus, we have $z_{\varepsilon'} \rightharpoonup w$ in $L_2(0,T;H^1(\Omega))$ (weakly). Moreover, owing to Theorem 3.13, one has $z_\varepsilon \longrightarrow u$ in $C([0,T];H^1(\Omega))$ (strongly). Therefore, we conclude $u = w$.

The statement $u \in L_\infty(0,T;W_p^2(\Omega))$ is proved by means of the weak* convergence. We observe that $L_1(0,T;(W_p^2(\Omega))^*)$ is a separable Banach space (cf. [ZEIDII90], Chapter 23, in particular, Problem 23.12).

Hence, due to $\|z_\varepsilon\|_{L_\infty(0,T;W_p^2(\Omega))} \leq C$ proved in Corollary 3.21[ii], there exists a subsequence $z_{\varepsilon'}$ of z_ε with $z_{\varepsilon'} \overset{*}{\rightharpoonup} w$ in $L_\infty(0,T;W_p^2(\Omega))$, where w belongs to $L_\infty(0,T;W_p^2(\Omega))$.

Again, the strong convergence $z_\varepsilon \longrightarrow u$ in $C([0,T];H^1(\Omega))$ (proved in Theorem 3.13) implies that $w = u$ solves the variational inequality (3.1), (3.2). ■

3.4.2 Boundary conditions of Neumann/Newton type

Let us now study the spatial smoothness of the solution of the variational inequality (3.1), (3.2) with natural boundary conditions, i.e., $\partial\Omega = \Gamma_N$.

We shall make use of ideas similar to those of Section 3.4.1 for the problem with Dirichlet conditions. But, compared to (3.17), we need now an auxiliary problem of the form

$$\left.\begin{aligned} &\text{Find}\quad z(t) \in H^1(\Omega), \quad t \in [0,T],\\ &a(t;z(t),v) = (f(t),v) + (g_N(t),v)_{\partial\Omega} + \int_0^t b(t';z(t'),v)\,dt' \quad \forall v \in H^1(\Omega) \end{aligned}\right| \tag{3.28}$$

with the bilinear forms a and b defined in (3.2) ($\partial\Omega = \Gamma_N$). This corresponds to the boundary value problem

$$\left.\begin{aligned} &A(t)\,z(t) = f(t) + \int_0^t B(t')\,z(t')\,dt' \quad \text{in } \Omega,\\ &(k\,\nabla z \cdot \vec{n})(t) + \kappa\,z(t) = \int_0^t (\partial_t k\,\nabla z \cdot \vec{n})(t')\,dt' + g_N(t) \quad \text{on } \partial\Omega \end{aligned}\right| \tag{3.29}$$

with the differential operators A and B defined in (3.10).

In the following lemma we summarize the (spatial) regularity statements for the solution of problem (3.28) in an analogous way as in Lemma 3.19 for the problem with Dirichlet boundary conditions.

Lemma 3.23 *Let the assumptions (3.4), (3.18) and* $\kappa \in C^{0,1}(\overline{\Omega})$ *be fulfilled and let* $\partial\Omega = \Gamma_N$.

[i] *Then, for every* $f \in L_2(0,T;L_p(\Omega))$ *and every* $g_N \in L_2(0,T;W_p^{1-1/p}(\partial\Omega))$, *the unique solution* z *of problem (3.28) is such that*

$$\begin{aligned}
&z \in \big(C([0,T];H^1(\Omega)) \ \cap\ L_2(0,T;W_p^2(\Omega))\big),\\
&\|z\|_{L_2(0,T;W_p^2(\Omega))} \le C\ \big(\|f\|_{L_2(0,T;L_p(\Omega))} + \|f\|_{C([0,T];L_2(\Omega))}\\
&+\ \|g_N\|_{L_2(0,T;W_p^{1-1/p}(\partial\Omega))} + \|g_N\|_{C([0,T];L_2(\partial\Omega))}\big),\quad 2 \le p < \infty.
\end{aligned}$$

[ii] *Suppose, additionally to [i], that*

$$f \in L_\infty(0,T;L_p(\Omega)) \quad \textit{and} \quad g_N \in L_\infty(0,T;W_p^{1-1/p}(\partial\Omega))$$

hold. Then, the unique solution z *of problem (3.28) is such that*

$$\begin{aligned}
&z \in \big(C([0,T];H^1(\Omega)) \ \cap\ L_\infty(0,T;W_p^2(\Omega))\big),\\
&\|z\|_{L_\infty(0,T;W_p^2(\Omega))} \le C\ \big(\|f\|_{L_\infty(0,T;L_p(\Omega))} + \|f\|_{C([0,T];L_2(\Omega))}\\
&+\ \|g_N\|_{L_\infty(0,T;W_p^{1-1/p}(\partial\Omega))} + \|g_N\|_{C([0,T];L_2(\partial\Omega))}\big),\quad 2 \le p < \infty.
\end{aligned}$$

Proof. We restrict ourselves to the essential differences to the proof of Lemma 3.19. In place of (3.20) the sequence $\{w_j\}_{j=0}^{N-1}$ is now defined by the solution of the elliptic problems

$$\left.\begin{aligned}
&a(t^j;w_j,v) = (f_j,v) + ((g_N)_j,v)_{\partial\Omega} + \sum_{i=0}^{j-1}\int_{i\tau}^{(i+1)\tau} b(t';w_i,v)\,dt' \quad \forall v \in H^1(\Omega),\\
&f_j = \frac{1}{\tau}\int_{j\tau}^{(j+1)\tau} f(t')\,dt', \quad (g_N)_j = \frac{1}{\tau}\int_{j\tau}^{(j+1)\tau} g_N(t')\,dt'.
\end{aligned}\right| \tag{3.30}$$

Taking $v = w_j$ in (3.30), applying the discrete Gronwall inequality (see Lemma 3.1) and performing some calculations, we obtain

$$\left.\begin{aligned}
\|w_j\|_{H^1(\Omega)} \ &\le c\,\Big(\|f_j\|_{L_2(\Omega)} + \|f\|_{L_2(0,T;L_2(\Omega))}\\
&+\ \|(g_N)_j\|_{L_2(\partial\Omega)} + \|g_N\|_{L_2(0,T;L_2(\partial\Omega))}\Big)
\end{aligned}\right| \tag{3.31}$$

with the constant $c > 0$ independent of j and τ. Let us introduce the notations

$$\begin{aligned}
\hat{f}_j &= \Big(f_j + \sum_{i=0}^{j-1}\int_{i\tau}^{(i+1)\tau} B(t')\,w_i\,dt'\Big),\\
\widehat{(g_N)}_j &= \Big((g_N)_j + \sum_{i=0}^{j-1}\int_{i\tau}^{(i+1)\tau} \partial_t k(t')\,\nabla w_i\cdot\vec{n}\,dt'\Big).
\end{aligned}$$

Owing to the elliptic regularity theory (cf. problem (3.15) and, in particular, [GRIS85], Theorems 2.4.2.6, 2.4.2.7 and [RODR87], Theorem 3.7.8), we conclude successively $w_j \in W_p^2(\Omega)$ for $j = 0, \ldots, N-1$. Here, we use $\hat{f}_j \in L_p(\Omega)$ and $\widehat{(g_N)}_j \in W_p^{1-1/p}(\partial\Omega)$. In this context we also refer to Remark 3.24 and to the trace operator

$$W_p^2(\Omega) \ni u \longrightarrow \left\{\gamma u, \gamma\frac{\partial u}{\partial n}\right\} \in \left(W_p^{2-1/p}(\partial\Omega) \times W_p^{1-1/p}(\partial\Omega)\right)$$

(cf. [GRIS85], Theorem 1.5.1.2 and [RODR87], Section 3.5).

Based on the general a priori estimate (3.16) (see also Remark 3.24) there holds an a priori estimate

$$\|w_j\|_{W_p^2(\Omega)} \leq C\left(\|\hat{f}_j\|_{L_p(\Omega)} + \|\widehat{(g_N)}_j\|_{W_p^{1-1/p}(\partial\Omega)} + \|w_j\|_{H^1(\Omega)}\right) \tag{3.32}$$

for the sequence $\{w_j\}_{j=0}^{N-1}$ defined in problem (3.30), where C does not depend on j and τ. Owing to the smoothness $w_i \in W_p^2(\Omega)$, it follows that

$$\|\hat{f}_j\|_{L_p(\Omega)} \leq \|f_j\|_{L_p(\Omega)} + \bar{C}_1\,\tau \sum_{i=0}^{j-1} \|w_i\|_{W_p^2(\Omega)} \quad \text{and}$$

$$\|\widehat{(g_N)}_j\|_{W_p^{1-1/p}(\partial\Omega)} \leq \|(g_N)_j\|_{W_p^{1-1/p}(\partial\Omega)} + \bar{C}_2\,\tau \sum_{i=0}^{j-1} \|w_i\|_{W_p^2(\Omega)}\,.$$

Here, the above mentioned trace theorem and a result concerning the multiplication of Sobolev functions (cf. Remark 3.24, in particular, (3.37)) are exploited. Now, we substitute the last two estimates and (3.31) into (3.32). Similar to (3.23) we deduce

$$\left.\begin{aligned} \|w_j\|_{W_p^2(\Omega)} &\leq \tilde{C}_1\left(\|f\|_{L_2(0,T;L_2(\Omega))} + \|g_N\|_{L_2(0,T;L_2(\partial\Omega))}\right) \\ &\quad + \tilde{C}_2\left(\|f_j\|_{L_p(\Omega)} + \|(g_N)_j\|_{W_p^{1-1/p}(\partial\Omega)}\right) \\ &\quad + \tilde{C}_3\,\tau\sum_{i=0}^{j-1}\left(\|f_i\|_{L_p(\Omega)} + \|(g_N)_i\|_{W_p^{1-1/p}(\partial\Omega)}\right) \end{aligned}\right| \tag{3.33}$$

by means of Gronwall's inequality in its discrete form (Lemma 3.1).

Now, we define step functions f_τ, $(g_N)_\tau$ and w_τ in the same way as in the proof of Lemma 3.19. After some calculations (similar to the proof of Lemma 3.19) we derive that w_τ remains in a bounded subset of $L_2(0,T;W_p^2(\Omega))$. Hence, extracting a subsequence again denoted by w_τ, we have $w_\tau \rightharpoonup w$ weakly in $L_2(0,T;W_p^2(\Omega))$.

Integrating (3.30) in time and multiplying it by $v \in L_2(0,T;H^1(\Omega))$, we deduce

$$\begin{aligned}
&\int_0^T a(t;w_\tau(t),v(t))\,dt - \int_0^T \left[a(t;w_\tau(t),v(t)) - a(\tilde{t};w_\tau(t),v(t))\right] dt \\
&= \int_0^T (f_\tau(t),v(t))\,dt + \int_0^T (g_N)_\tau(t),v(t))_{\partial\Omega}\,dt \\
&+ \int_0^T \Big(\int_0^t b(t';w_\tau(t'),v(t))\,dt' - \int_{\tilde{t}}^t b(t';w_\tau(t'),v(t))\,dt' \Big)\,dt,
\end{aligned}$$

where $\tilde{t} = [\frac{t}{\tau}]\tau$ for $t \in [0,T)$ ($[\frac{t}{\tau}]$ as the integer part of $\frac{t}{\tau}$) and $\tilde{t} = T - \tau$ for $t = T$.

Now, performing the passage to the limit as $\tau \longrightarrow 0$, we obtain $w(t) = z(t)$ a.e. in $(0,T)$ by the same line of reasoning as in the proof of Lemma 3.19. Therefore, we have $z \in L_2(0,T;W_p^2(\Omega))$.

In order to show the claimed a priori estimates for the solution z of (3.28), we take at first $v = z(t) \in H^1(\Omega)$ in (3.28). The application of Gronwall's inequality leads to

$$\|z\|_{C([0,T];H^1(\Omega))} \le c\,\left(\|f\|_{C([0,T];L_2(\Omega))} + \|g_N\|_{C([0,T];L_2(\partial\Omega))}\right).$$

Taking advantage of the proved regularity $z \in L_2(0,T;W_p^2(\Omega))$ as solution of (3.28), we deduce the a priori estimate

$$\begin{aligned}
&\|z(t)\|_{W_p^2(\Omega)} \le C\,\Big(\|f(t)\|_{L_p(\Omega)} + \|g_N(t)\|_{W_p^{1-1/p}(\partial\Omega)} + \|z(t)\|_{H^1(\Omega)} \\
&+ \int_0^t \left(\|(Bz)(t')\|_{L_p(\Omega)} + \|(\partial_t k\ \nabla z \cdot \vec{n})(t')\|_{W_p^{1-1/p}(\partial\Omega)} \right) dt' \Big) \quad \text{a.e. in } (0,T),
\end{aligned} \tag{3.34}$$

where C can be chosen independent of t (cf. Remark 3.24 and (3.16)). Again, combining the last two inequalities as well as using the trace theorem and the relation (3.37) (given below), we infer

$$\begin{aligned}
\|z(t)\|^2_{W_p^2(\Omega)} &\le C_1\,\left(\|f\|_{C([0,T];L_2(\Omega))} + \|g_N\|_{C([0,T];L_2(\partial\Omega))}\right)^2 \\
&+ C_2\,\left(\|f(t)\|^2_{L_p(\Omega)} + \|g_N(t)\|^2_{W_p^{1-1/p}(\partial\Omega)}\right) \\
&+ C_3 \int_0^t \|z(t')\|^2_{W_p^2(\Omega)}\,dt' \quad \text{a.e. in } (0,T).
\end{aligned} \tag{3.35}$$

Finally, integrating (3.35) in time and applying Gronwall's inequality, we obtain

$$\begin{aligned}
&\int_0^t \|z(t')\|^2_{W_p^2(\Omega)}\,dt' \le \bar{C}_1\left(\|f\|_{C([0,T];L_2(\Omega))} + \|g_N\|_{C([0,T];L_2(\partial\Omega))}\right)^2 \\
&+ \bar{C}_2 \int_0^t \left(\|f(t')\|^2_{L_p(\Omega)} + \|g_N(t')\|^2_{W_p^{1-1/p}(\partial\Omega)}\right) dt' \quad \forall t \in [0,T].
\end{aligned} \tag{3.36}$$

Taking $t = T$ in (3.36), the bound for $\|z\|_{L_2(0,T;W_p^2(\Omega))}$ follows.

By substitution of (3.36) into (3.35) we get the assertions of part [ii] of this lemma for z as solution of (3.28). ■

Remark 3.24 In the previous proof we use the following result concerning the multiplication of Sobolev functions at several places.
Assuming a function $\kappa \in C^{0,1}(\overline{\Omega})$ (or $\kappa \in C^1(\overline{\Omega})$), it holds that $\kappa\, w \in W_p^r(\partial\Omega)$ for each function $w \in W_p^r(\partial\Omega)$, $0 \le r < 1$. Furthermore, there exists a constant c, independent of κ, such that

$$\|\kappa\, w\|_{W_p^r(\partial\Omega)} \le c\ \|\kappa\|_{C^{0,1}(\overline{\Omega})}\ \|w\|_{W_p^r(\partial\Omega)}\,, \quad 0 \le r < 1, \quad 2 \le p < \infty. \tag{3.37}$$

This estimate can be checked by a direct calculation of the norm in $W_p^r(\partial\Omega)$ using a corresponding finite covering of the boundary $\partial\Omega$. The norm in $W_p^r(\partial\Omega)$ can be defined by

$$\|v\|_{W_p^r(\partial\Omega)} = \Big(\int_{\partial\Omega} |v|^p\, ds + \int_{\partial\Omega}\int_{\partial\Omega} \frac{|v(x)-v(y)|^p}{|x-y|^{n-1+rp}}\, ds(x)\, ds(y) \Big)^{1/p}, \quad 0 < r < 1$$

cf. [GRIS85], Section 1.3 and [ZEIDII90], Appendix.

Furthermore, let us emphasize that the constants in the a priori estimates (3.32) and (3.34) can be chosen independent of t and τ. In order to examine this property we consider, similar to Remark 3.20, an elliptic problem for fixed time $t \in [0,T]$, but now with a Neumann/Newton boundary condition, i.e.,

$$-\operatorname{div}(k(x,t)\ \nabla v) + q(x,t)\ v = f(x,t) \quad \text{in } \Omega,$$

$$k(x,t)\ \nabla v\cdot\vec{n} + \kappa(x)\ v = g_N(x,t) \quad \text{on } \partial\Omega.$$

Owing to the elliptic regularity (cf. [RODR87], Theorem 3.7.8 and [GRIS85], Theorems 2.4.2.6, 2.4.2.7), we have $v \in W_p^2(\Omega)$. Consequently, we obtain

$$\|\Delta v\|_{L_p(\Omega)} \le C\ \big(\|v\|_{W_p^1(\Omega)} + \|f(t)\|_{L_p(\Omega)} \big)$$

with a constant $C > 0$ independent of the time t (cf. Remark 3.20).
Rewriting the boundary condition as $\nabla v\cdot\vec{n} = (g_N - \kappa\, v)/k$ (note $k(x,t) \ge k_0 > 0$), the general a priori estimate (3.16) is applied to Δv with $v \in W_p^2(\Omega)$ and $\nabla v\cdot\vec{n}$ on $\partial\Omega$. Therefore, we obtain

$$\|v\|_{W_p^2(\Omega)} \le C\ \Big(\|\Delta v\|_{L_p(\Omega)} + \|v\|_{W_p^1(\Omega)} + \big\|(k(t))^{-1}\ (g_N(t) - \kappa\ v)\big\|_{W_p^{1-1/p}(\partial\Omega)} \Big).$$

Now, using (3.37), the trace theorem and the above estimate for Δv, we derive

$$\|v\|_{W_p^2(\Omega)} \le \bar{C}\ \big(\|v\|_{W_p^1(\Omega)} + \|f(t)\|_{L_p(\Omega)} + \|g_N(t)\|_{W_p^{1-1/p}(\partial\Omega)} \big).$$

By means of a Gagliardo–Nirenberg inequality (cf. [ZEIDII90], Section 21.17, 21.19) we get

$$\|v\|_{W_p^2(\Omega)} \le \tilde{C} \left(\|v\|_{L_p(\Omega)} + \|f(t)\|_{L_p(\Omega)} + \|g_N(t)\|_{W_p^{1-1/p}(\partial\Omega)} \right).$$

Hence, the desired estimate

$$\|v\|_{W_p^2(\Omega)} \le C \left(\|v\|_{H^1(\Omega)} + \|f(t)\|_{L_p(\Omega)} + \|g_N(t)\|_{W_p^{1-1/p}(\partial\Omega)} \right), \quad 2 \le p < \infty$$

with C independent of the time t is obtained similar to Remark 3.20 (see, in particular, (3.27)). □

Similar to Corollary 3.21 we are now able to establish spatial smoothness for the solution of the penalization problem (3.8) with Neumann/Newton boundary conditions.

Corollary 3.25 *Let the assumptions (3.4), (3.3), (3.18), and* $\kappa \in C^{0,1}(\overline{\Omega})$ *as well as* $\mu \in L_p(\Omega)$ *be fulfilled.*

[i] *If* $f \in L_2(0,T;L_p(\Omega))$ *and* $g_N \in L_2(0,T;W_p^{1-1/p}(\partial\Omega))$ *hold, then the unique solution* z_ε *of the penalty problem (3.8) with* $\partial\Omega = \Gamma_N$ *will be such that*

$$z_\varepsilon \in \left(C([0,T];H^1(\Omega)) \cap L_2(0,T;W_p^2(\Omega)) \right),$$

$$\|z_\varepsilon\|_{L_2(0,T;W_p^2(\Omega))} \le C_1 \quad \textit{uniformly in } \varepsilon > 0, \quad 2 \le p < \infty.$$

[ii] *If* $f \in L_\infty(0,T;L_p(\Omega))$ *and* $g_N \in L_\infty(0,T;W_p^{1-1/p}(\partial\Omega))$ *are satisfied, then the unique solution* z_ε *of (3.8) will be such that*

$$z_\varepsilon \in \left(C([0,T];H^1(\Omega)) \cap L_\infty(0,T;W_p^2(\Omega)) \right),$$

$$\|z_\varepsilon\|_{L_\infty(0,T;W_p^2(\Omega))} \le C_2 \quad \textit{uniformly in } \varepsilon > 0, \quad 2 \le p < \infty.$$

Considering a modified right-hand side term

$$\breve{f}(t) = f(t) + [1 - \beta_\varepsilon(z_\varepsilon(t) - \varphi)]\, \mu,$$

the results of this corollary can be proved by the same arguments used in the proof of Corollary 3.21, but now based on Lemma 3.23.

Now we state the global regularity results for the evolutionary variational inequality (3.1), (3.2) with natural boundary conditions of Neumann/Newton type. The same line of reasoning is exploited as for the inequality problem with Dirichlet conditions (cf. Theorem 3.22). To be more precise, we connect the results of Theorem 3.9 and Corollary 3.25.

Theorem 3.26 *Let the assumptions of Theorem 3.9,* $\varphi \in W_p^2(\Omega)$, $f(0) \in L_p(\Omega)$ *and (3.18) be satisfied.*
Then, assuming

$$f \in L_2(0,T;L_p(\Omega)), \quad g_N \in L_2(0,T;W_p^{1-1/p}(\partial\Omega)),$$

the solution u *of the evolutionary inequality problem (3.1), (3.2) will be such that*

$$u \in \left(W_\infty^1(0,T;H^1(\Omega)) \cap L_2(0,T;W_p^2(\Omega))\right), \quad 2 \le p < \infty.$$

If

$$f \in L_\infty(0,T;L_p(\Omega)), \quad g_N \in L_\infty(0,T;W_p^{1-1/p}(\partial\Omega))$$

is fulfilled, then it holds that

$$u \in \left(W_\infty^1(0,T;H^1(\Omega)) \cap L_\infty(0,T;W_p^2(\Omega))\right), \quad 2 \le p < \infty.$$

3.5 Some remarks on further regularity results

In the previous sections we have shown that, under corresponding assumptions, the solution u of the evolutionary variational inequality (3.1), (3.2) fulfills

$$u \in \left(W_\infty^1(0,T;H^1(\Omega)) \cap L_\infty(0,T;W_p^2(\Omega))\right), \quad 2 \le p < \infty.$$

To prove the spatial regularity, we have considered the problems with Dirichlet and Neumann/Newton boundary conditions separately.
In what follows we will discuss some consequences of this regularity which are, in particular, essential for the numerical analysis of the evolutionary variational inequality (see the error analysis in Section 5.2).

Continuity and Lipschitz continuity

Let us introduce the space $W_r^{2,1}(Q)$ as (Banach) space of the functions $z \in L_r(Q)$ such that the (generalized) derivatives $\partial_t z$, $\partial_{x_i} z$ and $\partial^2 z/\partial_{x_i}\partial_{x_j}$ belong to $L_r(Q)$ for $i,j = 1,\dots,n$, where Q is the space-time domain $Q = \Omega \times (0,T)$. We define

$$W_r^{2,1}(Q) = \left\{ z \in L_r(Q) : \ \frac{\partial z}{\partial x_i}, \ \frac{\partial^2 z}{\partial x_i \partial x_j}, \ \frac{\partial z}{\partial t} \in L_r(Q) \right\},$$

cf. e.g., [LSU68], Chapter I, [BENS82], Section II.3.1 and [MEIR92], Section I.2.2. From these references we recall the following embeddings:

$$\text{if} \quad z \in W_r^{2,1}(Q) \quad \text{for} \quad r > n/2+1, \quad \text{then} \quad z \in C(\overline{Q}),$$

$$\text{if} \quad z \in W_r^{2,1}(Q) \quad \text{for} \quad r > n+2, \quad \text{then} \quad z, \ \partial_{x_i} z \in C(\overline{Q}),$$

cf. e.g., [LSU68], Section II.3, [BENS82], Section II.3.1, page 78 and [MEIR92], Section I.2.3.

Now, the above mentioned regularity of the variational inequality solution u implies

$$u \in W_r^{2,1}(Q) \quad \text{with} \quad r = \min\{p, q\},$$

where q is determined by the Sobolev embedding $H^1(\Omega) \subset L_q(\Omega)$. Consequently, restricting ourselves to the spatial dimension $n \leq 3$ (hence, $q = 6$ is guaranteed), we have a continuous solution

$$u \in C(\overline{Q}) \quad \text{for} \quad p \geq 3.$$

Moreover, $\partial_{x_i} u \in C(\overline{Q})$ holds for $p \geq 6$.

Concerning the Lipschitz continuity of u with respect to the space-time domain Q we can establish the following result.
Owing to the Sobolev embedding

$$W_p^2(\Omega) \subset C^{1,\alpha}(\overline{\Omega}) \quad \text{for} \quad p > n, \quad \alpha = 1 - n/p,$$

the proved regularity $u \in L_\infty(0,T; W_p^2(\Omega))$ implies that the variational inequality solution u also belongs to $L_\infty(0,T; C^{1,\alpha}(\overline{\Omega}))$. Connecting this result with the estimate for $\partial_t u$ from Corollary 3.14, we obtain that all (first) derivatives of u are bounded. Hence, in this case, we have

$$u \in C^{0,1}(\overline{Q}) = W_\infty^1(Q).$$

Lewy–Stampacchia inequalities and complementarity problem

For each fixed time $t \in [0,T]$ the evolutionary variational inequality (3.1), (3.2) represents an elliptic variational inequality with a right-hand side $\hat{f}$ given by

$$\hat{f}(u,t) = f(t) + \int_0^t (Bu)(t')\, dt' \quad \text{in } \Omega$$

as well as a right-hand side $\widehat{g_N}$ of the Neumann/Newton boundary condition defined by

$$\widehat{g_N}(u,t) = g_N(t) + \int_0^t (\partial_t k\ \nabla u \cdot \vec{n})(t')\, dt' \quad \text{on } \Gamma_N$$

and Dirichlet boundary conditions $u(t) = g_D(t)$ on Γ_D. In this context see the problems (3.9), (3.19) and (3.29).
Owing to the regularity $u \in L_\infty(0,T; W_p^2(\Omega))$ which was shown for either Dirichlet or Neumann/Newton boundary conditions in Section 3.4, we get $\hat{f}(u,t) \in L_p(\Omega)$ for all $t \in [0,T]$, provided that $f \in C([0,T]; L_p(\Omega))$ holds additionally.

At first, let us consider the evolutionary variational inequality (3.1), (3.2) with Dirichlet boundary conditions on $\partial\Omega$. Assuming $g_D \in C([0,T]; W_p^{2-1/p}(\partial\Omega))$ we obtain not only $u \in L_\infty(0,T; W_p^2(\Omega))$, but also

$$u(t) \in W_p^2(\Omega) \quad \text{for all} \quad t \in [0,T]$$

for the solution u of the evolutionary variational inequality as an immediate consequence of the regularity theory for elliptic variational inequalities, see [GRIS85] and [RODR87], Section 5.3.

The assumption

$$l(t)\ \varphi \leq\ \widehat{g_N}(\varphi,t) \quad \text{on} \quad \partial\Omega$$

is required to obtain the analogous regularity for the evolutionary variational inequality (3.1), (3.2) with Neumann/Newton boundary conditions on $\partial\Omega$. This condition implies the equality

$$l(t)\ u(t) =\ \widehat{g_N}(u,t) \quad \text{on} \quad \partial\Omega$$

(cf. [RODR87], Section 5.3). From (3.10) we recall the boundary operator $l = l(x,t)$ given by

$$l(x,t)\ u(x,t) = (k\ \nabla u \cdot \vec{n})(x,t) + \kappa(x)\ u(x,t).$$

Due to the time-independent obstacle $\varphi = \varphi(x)$ and the assumptions $l(0)\ \varphi \leq g_N(0)$ (cf. Theorems 3.9 and 3.13), $g_N(t) \geq g_N(0)$ (cf. Lemmas 3.8 and 3.11), we are in the above mentioned situation $l(t)\ \varphi \leq\ \widehat{g_N}(\varphi,t)$. Thus,

$$(k\ \nabla u \cdot \vec{n})(x,t) + \kappa(x)\ u(x,t) = g_N(x,t) + \int_0^t (\partial_t k\ \nabla u \cdot \vec{n})(x,t')\ dt'$$

holds on the boundary $\partial\Omega$.

We mention that this equality has several important consequences. Firstly, there is no necessity to introduce a boundary penalization term on $\partial\Omega$ in the regularization problem (3.8). Secondly, we can apply the elliptic regularity theory also for the evolutionary inequality with Neumann/Newton boundary conditions. Finally, the numerical (error) analysis is simplified (cf. the proofs of Theorems 5.16, 5.17).

Summarizing, under the additional assumption $g_N \in C([0,T]; W_p^{1-1/p}(\partial\Omega))$, the proved regularity $u \in L_\infty(0,T; W_p^2(\Omega))$ guarantees $\widehat{g_N}(u,t) \in W_p^{1-1/p}(\partial\Omega)$ for all $t \in [0,T]$. Consequently, we obtain

$$u(t) \in W_p^2(\Omega) \quad \text{for all} \quad t \in [0,T]$$

for the solution u of the evolutionary variational inequality (3.1), (3.2) with Neumann/Newton boundary conditions on $\partial\Omega$.

Arguing again that for each fixed time $t \in [0,T]$ the evolutionary variational inequality (3.1), (3.2) represents an elliptic variational inequality, the associated

Lewy–Stampacchia inequalities (cf. [RODR87], Section 5.3 and [BELI84], Section 2.5.4) take the following form.

$$f(t) + \int_0^t (Bu)(t')dt' \leq (Au)(t) \leq \max\{f(t), A(0)\varphi\} + \int_0^t (Bu)(t')dt' \text{ a.e. in } \Omega,$$

$$(k\ \nabla u \cdot \vec{n})(t) + \kappa\ u(t) = g_N(t) + \int_0^t (\partial_t k\ \nabla u \cdot \vec{n})(t')\ dt' \quad \text{a.e. on} \quad \partial\Omega$$

$$\text{or} \quad u(t) = g_D(t) \quad \text{on} \quad \partial\Omega, \qquad \forall\, t \in [0,T].$$

In consequence of $u \in L_\infty(0,T;W_p^2(\Omega))$ and the just given bound for $(Au)(t)$, we can deduce that the norm

$$\|u(t)\|_{W_p^2(\Omega)}$$

is bounded for all $t \in [0,T]$ (*independent* of the time $t \in [0,T]$). For the verification of this statement we use similar arguments as in the proofs of Lemmas 3.19 and 3.23 (cf., in particular, Remarks 3.20 and 3.24).

Remark 3.27 We refer to [RODR87], Section 5.4 concerning the continuous dependence of the solutions of *elliptic* variational inequalities in the space $W_p^2(\Omega)$. Taking into account the statements developed there, it seems to be reasonable, that, under corresponding assumptions, the solution u of the evolutionary inequality (3.1), (3.2) fulfills not only

$$u \in L_\infty(0,T;W_p^2(\Omega)) \quad \text{and} \quad u(t) \in W_p^2(\Omega)\ \forall t \in [0,T],$$

but is also a *continuous* mapping from the time interval $[0,T]$ to the Sobolev space $W_p^2(\Omega)$, $2 \leq p < \infty$. However, at this stage we will not discuss further details. □

Moreover, due to the just discussed equality in the Neumann/Newton boundary condition, the complementarity problem related to the evolutionary inequality (3.1), (3.2) reads as follows.

$$(Au)(t) \geq f(t) + \int_0^t (Bu)(t')\ dt', \quad u(t) \geq \varphi,$$

$$\left(\ (Au)(t) - f(t) - \int_0^t (Bu)(t')\ dt'\ \right)\ (u(t) - \varphi) = 0 \quad \text{a.e. in } \Omega;$$

$$(k\ \nabla u \cdot \vec{n})(t) + \kappa\ u(t) = g_N(t) + \int_0^t (\partial_t k\ \nabla u \cdot \vec{n})(t')\ dt' \quad \text{a.e. on } \partial\Omega$$

$$\text{or} \quad u(t) = g_D(t) \quad \text{on} \quad \partial\Omega; \qquad \forall\, t \in [0,T].$$

We will return to this complementarity problem in the numerical error analysis which will be performed in Section 5.2 for a finite element and finite volume approximation of the evolutionary inequality (3.1), (3.2).

Finally, we mention that our regularity results are suited to show that the solution of the variational inequality satisfies the weak formulation introduced in Section 2.3.2 (see also Remark 2.9).

International Series of Numerical Mathematics
Vol. 136, © 2002 Birkhäuser Verlag Basel/Switzerland

Chapter 4

Finite Volume Approximations for Elliptic Variational Inequalities

This chapter is aimed at providing a systematic treatment of the finite volume method (often also called box method) as a discretization method for the numerical solution of interior and boundary obstacle problems of *elliptic* type with mixed boundary conditions (Dirichlet, Neumann, Newton) in two and three space dimensions. The investigations will be done in a variational (finite element) framework. Therefore, besides the direct study of the finite volume approximation (solvability, stability, maximum principle, error analysis, penalization methods for their solution) also a detailed comparison with a (piecewise linear) finite element approximation will be performed.

In this chapter we will make use of some results reported by the author in the papers [STEIa96] and [STEIb97]), but here we will give a representation in a more general framework, especially with regard to the evolutionary variational inequalities their numerical treatment will be considered in Chapter 5. More precisely, the purpose of this chapter is twofold. On the one hand, the numerical analysis of finite volume schemes applied to *elliptic* variational inequalities is of interest for its own sake. On the other hand, the results of these investigations are basic tools for the numerical analysis of the *evolutionary* inequality problems (see Chapter 5).

The present chapter is organized as follows. After introducing the elliptic obstacle problem and its finite element approximation in Sections 4.1.1, 4.1.2, we construct in Section 4.1.3 general box schemes in two and three dimensions for the approximation of second order elliptic boundary value problems. The well-known (classical) circumcentric (Voronoi diagram, cell) and barycentric boxes (Donald diagram) can be regarded as special cases of the constructed boxes. From the literature we briefly point out some connections to computational geometry and mesh generation. The finite volume approximation for interior obstacle problems with mixed boundary conditions is formulated in Section 4.1.4. Furthermore, some basic properties of the proposed finite volume (box) schemes will be proved.

In Section 4.2 the (discrete) variational inequalities obtained by the box method are compared to finite element discretizations and as the main result we will prove first and second order convergence in the energy norm between the box and finite element solutions depending on the choice of boxes and smoothness of the coefficients. The comparable accuracy of box and finite element solutions shows that box schemes are also well suited for the application of fast solvers (adaptive multilevel techniques; see, in particular, Remark 4.14).

In Section 4.3 we will derive error estimates in the energy norm for box solutions. More precisely, for elliptic obstacle problems we discuss the convergence rate for solutions of the box schemes in a variational framework. We refer to Remark 4.19 concerning error estimates in a generalized finite difference context.

In Section 4.4 we propose and analyse two penalization methods for solution of the discrete interior obstacle problems obtained by means of finite volumes. The properties of the penalization problems essentially depend on a discrete maximum principle, which is one of the reasons to study finite volume approximations. Based on the just mentioned comparison principle, the penalization error is estimated both in the energy and maximum norm. Consequently, by coupling the discretization and penalization parameters the overall error resulting from spatial discretization and penalization can be reduced (cf. Remark 4.29[i]). Furthermore, iterative schemes (Newton type methods) are presented for solving the penalization problems.

Section 4.5 is concerned with a brief discussion about the extension of the results for interior obstacle problems to boundary obstacle problems of Signorini type. Finally, results from numerical experiments for three obstacle type problems are presented in Section 4.6 illustrating the statements of the numerical analysis. In particular, we discuss the numerically determined convergence behaviour between the exact, Galerkin and box method solution as well as the coupling of penalization and discretization parameters.

4.1 Finite element and finite volume approximations for the obstacle problem

4.1.1 The elliptic obstacle problem

Both stationary and evolutionary variational inequalities of obstacle type can be encountered in many engineering application problems modelled by means of mathematical physics, but there also exist numerous applications in operations research. Examples for the first type are treated e.g., in [DULI76], [KIST80], [GLT81], [ELOC82], [BACA84], [CHIP84], [CRAN84], [GLOW84], [RODR87], [FRIE88], [KIOD88], [FRSP93] and [HANE96]. The application of obstacle problems in stochastic control and operations research can be found e.g., in [BENS82], [BELI82], [BELI84] and [CPS92].

Let us consider the following elliptic obstacle problem with a convex set $K = K_{\varphi,g_D}$ characterized by a lower (interior) obstacle φ and Dirichlet boundary conditions on a part Γ_D of the boundary $\partial\Omega$.

$$\begin{aligned} &\text{Find} \quad u \in K_{\varphi,g_D} = \{w \in H^1(\Omega) : w \geq \varphi \text{ in } \Omega, \quad w = g_D \text{ on } \Gamma_D\} \neq \emptyset, \\ &a(u, v-u) \geq (f, v-u) + (g_N, v-u)_{\Gamma_N} \quad \forall\, v \in K \end{aligned} \tag{4.1}$$

with the bilinear form a defined by

$$a(u,v) = \int_\Omega (k\ \nabla u\ \nabla v + q\ u\ v)\ dx + \int_{\Gamma_N} \kappa\ u\ v\ ds. \tag{4.2}$$

As in Chapter 3, let Ω be a bounded domain of $\mathbb{R}^n$, $n \geq 1$ with a Lipschitz boundary $\partial\Omega = \Gamma_D \cup \Gamma_N \cup \Gamma_0$, where Γ_D, Γ_N are disjoint and open in $\partial\Omega$ (one of them can be empty). Again, Γ_0 with $\text{meas}_{n-1}(\Gamma_0) = 0$ is the set of those points, where one type of boundary condition changes into another (cf. e.g., [KRNE90], p. 17). Similar to Chapter 3 (see (3.10)), but now with only space-dependent coefficients, the second order differential operator A and the boundary operator l associated with the bilinear form a are given by

$$\begin{aligned} &A\,u = -\operatorname{div}(k\ \nabla u) + q\ u \quad \text{in } \Omega, \\ &l\ u = k\ \nabla u \cdot \vec{n} + \kappa\ u \quad \text{on } \Gamma_N, \quad l\ u = u \quad \text{on } \Gamma_D. \end{aligned} \tag{4.3}$$

In order to have a unique solution $u \in K_{\varphi,g_D}$ for problem (4.1) we make the following assumptions on the data.

$$\begin{aligned} &k,\ q \in L_\infty(\Omega), \quad k(x) \geq k_0 > 0, \quad q(x) \geq 0, \quad x \in \Omega; \\ &\kappa \in L_\infty(\Gamma_N), \quad \kappa(x) \geq 0, \quad x \in \Gamma_N; \\ &\text{meas}_{n-1}(\Gamma_D) + \text{meas}_{n-1}(\Gamma_\kappa) + \text{meas}_n(\Omega_q) > 0 \\ &\text{with} \quad \Omega_q = \{x \in \Omega : q(x) \geq q_0 > 0\}, \quad \Gamma_\kappa = \{x \in \overline{\Gamma_N} : \kappa(x) \geq \kappa_0 > 0\}; \\ &f \in L_2(\Omega); \quad g_N \in L_2(\Gamma_N); \quad g_D \in H^1(\Omega) \cap C(\overline{\Gamma_D}). \end{aligned} \tag{4.4}$$

Using a Friedrichs–Poincaré type inequality, these assumptions ensure the V-ellipticity

$$a(v,v) \geq m\ \|v\|_V^2 \quad \forall\, v \in V, \quad m > 0$$

of the bilinear form a. Here, V again denotes the subspace of the Sobolev space $H^1(\Omega)$, the trace of which is zero on Γ_D.

To control the obstacle φ pointwise, we suppose

$$\varphi \in W_p^1(\Omega), \quad p > n; \quad \varphi \leq g_D \quad \text{on } \Gamma_D. \tag{4.5}$$

The set $K = K_{\varphi,g_D}$ is a closed convex nonempty subset of $H^1(\Omega)$ such that problem (4.1), (4.2) has a unique solution u applying the general theory of variational inequalities (Lions–Stampacchia theorem, see e.g., [KIST80], [RODR87], Section 4.2 or [GLOW84], Section I.3).

We refer to the monographs [KIST80], Chapter 4, [BELI82], Chapter 3, [BELI84], Chapter 2.5, [CHIP84], Chapter 3 and [RODR87], Chapter 5 concerning the study of the regularity of the solution of the above obstacle problem. Penalization techniques are often used in connection with the regularity theory for elliptic boundary value problems (for the latter see [GRIS85] and [RODR87], Section 3.7 as well as our overview given at the beginning of Section 3.4).

Let us recall the well-known fact that the obstacle does not allow us to surpass a certain threshold of regularity of the solution regardless of the smoothness of the data. In general $u \in W_p^2(\Omega)$ and $u \in C^{1,1}(\overline{\Omega})$, respectively are possible, but *not* $u \in C^2(\overline{\Omega})$; see Remark 4.18 and Figure 4.2 for an illustration of this property.

Roughly speaking, the regularity properties of the solution u of the obstacle problem (4.1), (4.2) can be summarized in the following statement. The solution u belongs to $W_p^2(\Omega)$, $2 \le p < \infty$, provided that $\varphi \in W_p^2(\Omega)$ and the solution w of the corresponding boundary value problem

$$A\,w = f \quad \text{in } \Omega, \qquad l\,w = g_N \quad \text{on } \Gamma_N, \qquad w = g_D \quad \text{on } \Gamma_D \tag{4.6}$$

fulfills $w \in W_p^2(\Omega)$. Note, that in the case of Neumann/Newton conditions on Γ_N, the assumption $l\,\varphi \le g_N$ on Γ_N is essential because this condition implies $l\,u = g_N$ on Γ_N in the Lewy–Stampacchia inequalities (cf. [RODR87], Section 5.3 and see also our assumptions made in Theorems 3.9 and 3.13). Then, the related complementarity problem reads as follows.

$$\left.\begin{aligned} & f \le A\,u, \quad (A\,u - f)\,(\varphi - u) = 0, \quad u \ge \varphi \quad \text{a.e. in } \Omega; \\ & l\,u = g_N \quad \text{a.e. on } \Gamma_N; \quad u = g_D \quad \text{on } \Gamma_D. \end{aligned}\right| \tag{4.7}$$

4.1.2 Finite element approximations for the obstacle problem

In order to construct finite element and finite volume approximations for the problem (4.1), (4.2) we use a usual triangulation $\mathcal{S}_h$ over the region $\overline{\Omega}$. In what follows we restrict ourselves to the case that Ω is a bounded *polygonal (polyhedral)* domain in $\mathbb{R}^2$ ($\mathbb{R}^3$). More precisely, we consider *admissible* triangulations $\mathcal{S}_h$ as finite sets of *closed* n-simplices S (triangles for $n = 2$ and tetrahedra for $n = 3$) such that

$$\overline{\Omega} = \bigcup_{S \in \mathcal{S}_h} S; \quad \operatorname{int} S_1 \cap \operatorname{int} S_2 = \emptyset \quad \text{and the sets} \quad S_1 \cap S_2 \quad \text{are either empty}$$

$$\text{or a common } (n-k) - \text{simplex}, \; k = 1, \ldots, n \quad \forall S_1, S_2 \in \mathcal{S}_h \text{ with } S_1 \ne S_2.$$

Here, a point is considered as 0-simplex.

For each n-simplex S the parameters h_S, ρ_S and for any triangulation $\mathcal{S}_h$, the discretization parameters h are given by

$$h_S = \operatorname{diam} S, \quad \rho_S = \sup\{\operatorname{diam} U : U \text{ is a } n\text{-ball contained in } S\}, \quad h = \max_{S\in\mathcal{S}_h} h_S.$$

In the sequel we shall consider a *regular* family of triangulations $\{\mathcal{S}_h\}_{h\in(0,h_0]}$ with $h_0 > 0$ fixed, in the sense that there exists a constant $\rho > 0$ such that for all $h \in (0, h_0]$ approaching 0 and for all $S \in \mathcal{S}_h$ it holds that

$$\rho\, h_S \;\leq\; \rho_S$$

(cf. the monographs [CIAR78], p. 132, [HACK92], Section 8.4, [HANE96], Definition 1.10, 1.11 and [KRNE90], Definition 4.3).

In order to handle the different types of boundary conditions in problem (4.1), (4.2) we require a consistence condition upon $\mathcal{S}_h$ in a form requiring that the interior of any face of any $S \in \mathcal{S}_h$ be disjoint to Γ_0 (cf. [KRNE90], condition (6), p. 28). Furthermore, we shall use the *notation* σ_h for the set consisting of all (boundary) faces σ of each $S \in \mathcal{S}_h$ such that σ is a subset of the boundary $\Gamma = \partial\Omega$. The *set* $\sigma_{h,2}$ consists of all boundary faces $\sigma \in \sigma_h$ such that $\sigma \subset \overline{\Gamma_N}$. For the sake of simplicity we assume that the set $\Gamma_\kappa \subseteq \overline{\Gamma_N}$ from (4.4) can be represented for $h \in (0, h_0]$ as a union of the boundary faces σ contained in $\sigma_{h,2}$.

Let us introduce some abbreviations for sets/subsets of vertices of $\mathcal{S}_h$ and for subregions of Ω associated with a vertex/node ξ, i.e.,

$$\begin{aligned}
&\overline{\omega} = \{\xi \in \overline{\Omega} : \xi \text{ is a vertex of at least one } S \in \mathcal{S}_h\}, \quad \omega = \{\xi \in \overline{\omega} : \xi \in \Omega\},\\
&\gamma = \overline{\omega}\backslash\omega, \quad \gamma_D = \{\xi \in \overline{\omega} : \xi \in \overline{\Gamma_D}\}, \quad \gamma_N = \{\xi \in \overline{\omega} : \xi \in \Gamma_N\}, \quad \tilde{\omega} = \overline{\omega}\backslash\gamma_D;\\
&\Omega_\xi = \{S \in \mathcal{S}_h : S \text{ has } \xi \text{ as vertex}\}, \quad \omega_\xi = \{\eta \in (\overline{\omega}\backslash\{\xi\}) : \eta \in \Omega_\xi\} \quad \text{for } \xi \in \overline{\omega}.
\end{aligned} \tag{4.8}$$

The (piecewise linear) finite element subspace V_h^G of V is given by

$$\begin{aligned}
V_h^G &= \{w \in X_h^G : w = 0 \text{ on } \Gamma_D\} \subset V,\\
X_h^G &= \{w \in C(\overline{\Omega}) : w_{|S} \in P_1(S) \quad \forall S \in \mathcal{S}_h\},
\end{aligned}$$

where $P_1(S)$ is the space of linear polynomials defined on the simplex S. The spaces V_h^G and X_h^G, respectively are spanned by the nodal basis functions Φ_ξ^G with $\Phi_\xi^G(\eta) = \delta_{\xi\eta}$ (Kronecker symbol) for $\xi, \eta \in \tilde{\omega}$ (resp. $\xi, \eta \in \overline{\omega}$). Here, the superscript G refers to Galerkin. Obviously, an element $w \in V_h^G$ (resp. $w \in X_h^G$) is uniquely determined by its values $w(\xi)$ at the vertices $\xi \in \tilde{\omega}$ (resp. $\xi \in \overline{\omega}$) of the underlying triangulation $\mathcal{S}_h$ such that a corresponding grid function vector $(w_h)_{\xi\in\tilde{\omega}}$ (resp. $(w_h)_{\xi\in\overline{\omega}}$) is given by $w_{h,\xi} = w(\xi)$.

The grid function vectors w_h form a linear space $L_2(\tilde{\omega})$ equipped with the Euclidean scalar product $\langle .,. \rangle$, i.e.,

$$\langle w_h, v_h \rangle = \sum_{\xi \in \tilde{\omega}} w_{h,\xi}\, v_{h,\xi}.$$

The isomorphism between $L_2(\tilde{\omega})$ and V_h^G is expressed by the Galerkin prolongation P^G, i.e.,

$$L_2(\tilde{\omega}) := \mathbb{R}^{\dim V_h^G}, \quad P^G : L_2(\tilde{\omega}) \longrightarrow V_h^G \quad \text{with} \quad P^G w_h = \sum_{\xi \in \tilde{\omega}} w_{h,\xi}\, \Phi_\xi^G.$$

In a similar way, the linear space $L_2(\overline{\omega})$ for grid function vectors v_h defined on $\overline{\omega}$ is isomorph to X_h^G.

After these preliminaries the *finite element approximation* for the obstacle problem (4.1), (4.2) will be formulated in the following way. Owing to the assumption $\varphi \in W_p^1(\Omega)$, $p > n$ (note the embedding $W_p^1(\Omega) \subset C(\overline{\Omega})$ for $n < p$), the obstacle φ can be controlled pointwise in the constraint set K_h. For the sake of brevity we suppress the subscripts φ and g_D in the notation

$$\begin{aligned} K_h &= \{w_h \in L_2(\overline{\omega}) : w_{h,\xi} \geq \varphi(\xi)\ \forall \xi \in \overline{\omega}, \quad w_{h,\xi} = g_D(\xi)\ \forall \xi \in \gamma_D\} \\ &= \{P^G w_h \in X_h^G : (P^G w_h)(\xi) \geq \varphi(\xi)\ \forall \xi \in \overline{\omega}, \quad (P^G w_h)(\xi) = g_D(\xi)\ \forall \xi \in \gamma_D\} \end{aligned}$$

as an approximation of K_{φ,g_D}.

$$\begin{aligned} &\text{Find} \quad P^G u_h \in K_h : \quad a(P^G u_h, P^G v_h - P^G u_h) \\ &\geq (f, P^G v_h - P^G u_h) + (g_N, P^G v_h - P^G u_h)_{\Gamma_N} \quad \forall P^G v_h \in K_h. \end{aligned} \tag{4.9}$$

This can be equivalently written in matrix notation as:

$$\begin{aligned} &\text{Find} \quad u_h \in K_h : \\ &\langle (L_G + M_G + R_G)\, u_h, v_h - u_h \rangle \geq \langle f_{Gh} + g_{Gh}, v_h - y_h \rangle \quad \forall v_h \in K_h \end{aligned} \tag{4.10}$$

with the (stiffness, mass and boundary) matrices and vectors

$$\begin{aligned} &(L_G)_{\xi,\eta} = \int_\Omega k\, \nabla\Phi_\xi^G\, \nabla\Phi_\eta^G\, dx, \quad (M_G)_{\xi,\eta} = \int_\Omega q\, \Phi_\xi^G\, \Phi_\eta^G\, dx, \\ &(R_G)_{\xi,\eta} = \int_{\Gamma_N} \kappa\, \Phi_\xi^G\, \Phi_\eta^G\, ds, \quad f_{Gh,\xi} = \int_\Omega f\, \Phi_\xi^G\, dx, \\ &g_{Gh,\xi} = \int_{\Gamma_N} g_N\, \Phi_\xi^G\, ds \quad \text{for} \quad \xi \in \tilde{\omega},\ \eta \in \overline{\omega}. \end{aligned} \tag{4.11}$$

Obviously, it holds that

$$a(P^G v_h, P^G w_h) = \langle (L_G + M_G + R_G)\, v_h, w_h \rangle \quad \text{and}$$
$$(f, P^G w_h) + (g_N, P^G w_h)_{\Gamma_N} = \langle f_{Gh} + g_{Gh}, w_h \rangle \quad \text{for} \quad v_h \in L_2(\overline{\omega}),\ w_h \in L_2(\tilde{\omega}).$$

Based essentially on the approximation of the convex set K by K_h (in Mosco's sense; in this context see also the proof of Lemma 3.2), the study of the convergence $P^G u_h \longrightarrow u$ can be found e.g., in the monographs [GLT81], Appendix 1, Section 2, [GLOW84], Section II.2 and [HHNL88], Section 1.2.3 (resigning on the convergence rate for nonregular solutions $u \in H^1(\Omega)$).

Moreover, based on results by Falk (see [FALK74]), Mosco (see [MOSC77]) and by Brezzi, Hager, Raviart (see [BHR77]), an error analysis (in the energy norm) of finite element approximations for obstacle problems of the form (4.1), (4.2) is contained in the monographs [CIAR78], Chapter 5, § 5.1, [GLT81], Appendix 1, Section 3, [GLOW84], Chapter 2, Section 2, [HHNL88], Section 1.1.32, 1.2.3 and [GRRO94], Chapter 8.

Assuming, only for the moment, homogeneous Dirichlet conditions $g_D = 0$ on Γ_D or that $\partial\Omega = \Gamma_N$ holds, the essential point consists in an extension of Cea's Lemma (note $K_h \subset K$ does not hold in general). More precisely, one obtains the estimate

$$\begin{aligned} \left\|u - P^G u_h\right\|_V^2 \leq C\, \Big\{ &\inf_{v \in K} \big[a(u, v - P^G u_h) - (f, v - P^G u_h) \\ &- (g_N, v - P^G u_h)_{\Gamma_N}\big] + \inf_{P^G v_h \in K_h} \big[a(u, P^G v_h - u) \\ &- (f, P^G v_h - u) - (g_N, P^G v_h - u)_{\Gamma_N} + \left\|P^G v_h - u\right\|_V^2\big] \Big\}, \end{aligned}$$

where C depends on the bilinear form a (ellipticity and continuity). Choosing $P^G v_h = \Pi_h u$ as the (piecewise linear) X_h^G-interpolant of u and $v = \max\{P^G u_h, \varphi\}$ as well as using corresponding interpolation error bounds in Sobolev spaces (see e.g., [BRSC96], Section 4.4 and [CIAR78], Chapter 3) one gets the convergence statement

$$\left\|u - P^G u_h\right\|_{H^1(\Omega)} \leq C\, h$$

under the assumptions $u \in H^2(\Omega)$ and $l\varphi \leq g$ (the latter implies $lu = g$ on Γ_N). In the case of nonhomogeneous Dirichlet data, the above a priori estimate has to be modified (see [GLT81], Appendix 1, Lemma 3.2 and cf. Lemma 4.15 below). Concerning error estimates in the maximum norm see Remark 4.29[ii].

At this stage let us recall that the symbols $C, c, C_i, c_i,\ i = 1, 2, \dots$ are used as generic positive constants, independent of the triangulation $\mathcal{S}_h$ (at most depending on h_0 and ρ), which may take different values at different places.

4.1.3 Basics of finite volume approximations

This section is aimed at constructing finite volume (box) approximations in two and three space dimensions for the obstacle problem (4.1), (4.2).

Let us start with some basic features of finite volume (box) approximations. The box method applied to second order elliptic problems is characterized by a trial space consisting of continuous piecewise linear polynomials on the primary triangulation, by a test space consisting of piecewise constants on the dual box mesh and, furthermore, by lumping of the mass and boundary matrices. In other words, the key idea for the construction of the box stiffness matrix is to apply Gauss' integration formula to the second order term on the finite control volumes (boxes) such that volume integrals are converted to surface integrals. Surprisingly, the resulting box stiffness matrix coincides with the Galerkin stiffness matrix. Therefore, finite volume methods lie between the finite difference and finite element methods. Indeed, box schemes are analysed in the mathematical literature both in a generalized finite difference and in a finite element context. With respect to second order elliptic problems in plane domains, investigations based on a generalized finite difference framework have often been done by researchers coming from Eastern Europe (see e.g., [SAMA84], [LMW84], [HAWE85], [HEIN87], [SLM87], [MEIS90], [ANG91] and [GRRO94]). The study of box schemes more in a finite element context can be found e.g., in [IKED83], [BARO87], [HACK89], [CAI91], [CMC90], [CMMC91] [KERK96] and [FFLM97].

An important reason for studying finite volume schemes is that, owing to the lumping of the mass and boundary matrices, the corresponding solutions satisfy a discrete analogy to the maximum principle at least for a weakly acute triangulation (cf. Section 4.4.1).
Furthermore, for the treatment of first order terms in convection-diffusion problems, box schemes are well suited, by virtue of the possible transfer of the upwinding strategy for classical difference methods, to multi-dimensional domains triangulated by simplices (see e.g., [IKED83], [ANG91], [ANG95] and for an overview [GRRO94], Section 7.1.3).

Concerning the implementation of box schemes, it should be emphasized that owing to their intermediate position one can easily shift in the assembling process between the finite element generation (simplex by simplex) and the finite difference procedure (equation by equation). This feature can be advantageous for the exploitation and easy implementation of parallel algorithms based on the computer architecture. Furthermore, note that the lumping process of the mass and boundary matrices for barycentric boxes can be done in a finite element code by using numerical integration formulae having integration nodes in the vertices of the simplex.

Owing to the easily understood physical interpretation of the local modelling of balance equations on finite regions, finite volume approximations are often used in various engineering applications. Of course, the application of finite volume approximations is not restricted to second order problems, but they are also frequently used for the discretization of (nonlinear) systems of conservation laws (see [KROE96]) as well as both fluid flow problems (Navier–Stokes equations, see e.g., the monographs [FEPE99], [PATA80] and the paper [MAMU97]) and solid mechanics (solid body stress analysis, see e.g., [DEMA93], [DEMU94] and [DEMU95]). Finally, we refer to the proceedings [BEVH99] concerning recent developments of finite volume approaches in different (complex) application problems.

At the end of these introductory remarks we emphasize that, concerning the different notations used in the literature for finite volume approximations, the discretization method considered in what follows belongs to the so-called vertex-centered group (in contrast to cell-centered) and is sometimes also called CVFEM (control volume finite element method). We mention that these terms are not used in the literature in a unique manner. Perhaps it is better to classify the different finite volume schemes concerning the conservation property, i.e., on the one hand, conservation is enforced on the primary triangulation cells (cell-centered) and, on the other hand, on the dual triangulation (vertex-centered, our approximation scheme). In other words, in both approaches the name (cell-centered or vertex-centered) refers to the location of the solved variable.

In the sequel we derive and analyse finite volume schemes in the Galerkin framework for *obstacle problems with mixed boundary conditions both in two and three space dimensions.* Due to the basic idea of the finite volume (box) method consisting in the integration of the differential equation/inequality over finite control volumes (finite boxes) we require, besides the (primary) triangulation $\mathcal{S}_h$ given in Section 4.1.2, another (dual) partition of Ω.

At first, let us consider the two-dimensional case $n = 2$, cf. Figure 4.1. For each triangle $S \in \mathcal{S}_h$ we select a point p^S either located in the interior of S or coinciding with one of the edge midpoints of S. This point p^S is connected to the edge midpoints of S by straight-line segments such that we obtain a partition of S into three subregions $D^S_{\xi_i}$, where ξ_i, $i = 1, 2, 3$ should be the vertices of S. In other words, by means of the notations (4.8) we have $S = \bigcup_{\xi \in (\overline{\omega} \cap \partial S)} D^S_\xi$. Furthermore, the regions D^S_ξ consist of at most two sub-triangles, i.e., $D^S_\xi = \bigcup_{\eta \in (\omega_\xi \cap \partial S)} D^S_{\xi;\eta}$, where $D^S_{\xi;\eta}$ is the triangle formed by the vertices ξ, $(\xi + \eta)/2$ (edge midpoint) and p^S.

Now, the box $D_\xi \in \mathcal{D}_h$ with $D_\xi \subset \Omega_\xi$ (cf. (4.8)) associated with a vertex $\xi \in \overline{\omega}$ is defined by the union of the subregions D^S_ξ in Ω_ξ having ξ as vertex. Summarizing, we have

$$\overline{\Omega} = \bigcup_{\xi \in \overline{\omega}} D_\xi, \quad D_\xi = \bigcup_{S \in \mathcal{S}_h : S \subset \Omega_\xi} D^S_\xi, \quad D^S_\xi = \bigcup_{\eta \in (\omega_\xi \cap \partial S)} D^S_{\xi;\eta},$$

$$S = \bigcup_{\xi \in (\overline{\omega} \cap \partial S)} \; \bigcup_{\eta \in (\omega_\xi \cap \partial S)} D^S_{\xi;\eta}.$$

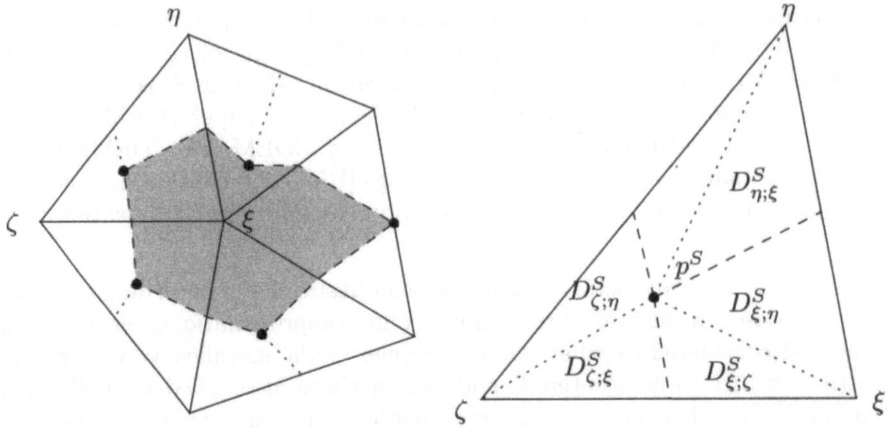

Figure 4.1: *Construction of the boxes in the two-dimensional case (2D). Left: region Ω_ξ, closed box D_ξ (shaded region) with (dashed) boundary lines, points p^S depicted with •. Right: corresponding partition of a triangle S with vertices ξ, η and ζ such that:* $D^S_\xi = D^S_{\xi;\eta} \cup D^S_{\xi;\zeta}$, $D^S_\eta = D^S_{\eta;\xi} \cup D^S_{\eta;\zeta}$, $D^S_\zeta = D^S_{\zeta;\eta} \cup D^S_{\zeta;\xi}$ *and* $S = D^S_\xi \cup D^S_\eta \cup D^S_\zeta$.

The boxes related to a triangulation in $\mathbb{R}^3$ are constructed as follows. Again, for each tetrahedron $S \in \mathcal{S}_h$ we select a point p^S with either $p^S \in \operatorname{int} S$ or coinciding with one of the barycenters of the four triangular faces of S. Now, each $S \in \mathcal{S}_h$ with the vertices ξ_i, $i = 1, \dots, 4$ is decomposed in four subregions $D^S_{\xi_i}$ in such a way that there holds again $S = \bigcup_{\xi \in (\overline{\omega} \cap \partial S)} D^S_\xi$.

Each of these regions D^S_ξ consists of at most six (sub-)tetrahedra $D^S_{\xi;\eta,\zeta}$ in such a way that

$$D^S_\xi = \bigcup_{\eta,\zeta \in (\omega_\xi \cap \partial S), \eta \neq \zeta} D^S_{\xi;\eta,\zeta}$$

holds. Here, $D^S_{\xi;\eta,\zeta}$ is the tedrahedron formed by the vertices ξ, $(\xi + \eta)/2$ (edge midpoint), $(\xi + \eta + \zeta)/3$ (barycenter of a face) and p^S.

Now, the box D_ξ associated with a node $\xi \in \overline{\omega}$ is defined in the same way as above, i.e., as the union of the subboxes D^S_ξ in Ω_ξ having ξ as vertex. Therefore, we summarize in the spatial three-dimensional situation as follows.

$$\overline{\Omega} = \bigcup_{\xi \in \overline{\omega}} D_\xi, \quad D_\xi = \bigcup_{S \in \mathcal{S}_h : S \subset \Omega_\xi} D^S_\xi, \quad D^S_\xi = \bigcup_{\eta,\zeta \in (\omega_\xi \cap \partial S), \eta \neq \zeta} D^S_{\xi;\eta,\zeta},$$

$$S = \bigcup_{\xi \in (\overline{\omega} \cap \partial S)} \bigcup_{\eta,\zeta \in (\omega_\xi \cap \partial S), \eta \neq \zeta} D^S_{\xi;\eta,\zeta}.$$

We want to emphasize that we consider the boxes $D \in \mathcal{D}_h$ as *closed*, i.e., we have $D = \operatorname{int} D \cup \partial D$.

The boxes need not be convex (in contrast to the circumcentric boxes, see below). Furthermore, some of the subboxes $D^S_{\xi;\eta}$ for $n = 2$ ($D^S_{\xi;\eta,\zeta}$ for $n = 3$) can vanish, when p^S coincides with an edge midpoint (a barycenter of one face).

We assume that the boxes D_ξ have a nonzero area. More precisely, there should exist a constant $\delta > 0$, such that for all $S \in \mathcal{S}_h$ it holds that

$$\text{meas}_n(D_\xi \cap S)/\,\text{meas}_n(S) \geq \delta > 0 \quad \forall \xi \in (\overline{\omega} \cap \partial S). \tag{4.12}$$

Actually one needs only $\text{meas}_n(D_\xi)/\,\text{meas}_n(\Omega_\xi) \geq \delta > 0$ (cf. [BARO87]). But, in order to avoid some technical difficulties in the proof of the ellipticity of the box bilinear form a_B (see Theorem 4.2[i]), we shall assume the slightly stronger condition (4.12).

Definition $\mathcal{D}_h(1)$. In what follows, a box partition of Ω as just described for $n = 2, 3$ and satisfying the condition (4.12) will be denoted by $\mathcal{D}_h(1)$.

We observe an essential feature of a box partition $\mathcal{D}_h$ constructed according to $\mathcal{D}_h(1)$, namely

$$\text{meas}_{n-1}(\sigma_i) = n\ \text{meas}_{n-1}(\sigma_i \cap \partial D^S_\xi), \quad \forall \xi \in \overline{\omega},\ S \in \mathcal{S}_h, \tag{4.13}$$

where σ_i, $i = 1, \ldots, n$ denote the boundary faces of the simplex S having ξ as vertex.

At the end of this introductory part of the box method we want to mention two 'classical' choices for the points p^S and furthermore point out briefly the close relation to (optimal) triangulation and mesh generation of two- and three-dimensional regions. A first natural choice consists in

$$p^S = \Big(\sum_{i=1}^{n+1} \xi_i\Big)\Big/(n+1)$$

as barycentre of the simplex S with the vertices ξ_i, $i = 1, \ldots, n+1$. Then, the above subdivision of S leads to barycentric subdomains and can be easily expressed by the barycentric coordinates $\lambda_i(x)$, $i = 1, \ldots, n+1$, i.e.,

$$D^S_{\xi_i} = \{x \in S : \lambda_i(x) \geq \lambda_j(x) \quad \forall j = 1, \ldots, n+1,\ j \neq i\}.$$

By some elementary geometrical considerations one verifies that then

$$\text{meas}_n(D^S_{\xi_i}) = \frac{\text{meas}_n(S)}{n+1} \qquad \forall S \in \mathcal{S}_h \text{ with the vertices } \xi_i,\ i = 1, \ldots, n+1. \tag{4.14}$$

For barycentric boxes the family $\{D_\xi\}_{\xi \in \overline{\omega}}$ is sometimes called a Donald diagram.

Definition $\mathcal{D}_h(2)$. In the sequel let us denote by $\mathcal{D}_h(2)$ a box partition of $\Omega \subset \mathbb{R}^n$, $n = 2, 3$ fulfilling (4.14) and $\mathcal{D}_h(1)$.

The choice of p^S as circumcentre of the simplices S related to circumcentric boxes is the second classical case (at least for $n = 2$, see e.g., [IKED83], [HEIN87], [ANG91], [CMMC91] or [GRRO94]).

Let us briefly point out some connections to the computational geometry relevant to (finite element) mesh generation and to our box partitions $\mathcal{D}_h(1)$ and $\mathcal{D}_h(2)$; for a detailed survey see [FORT92] and [BEEP92].

Let us recall that the so-called Voronoi diagram of $\overline{\omega}$ splits $\mathbb{R}^n$ into regions (Voronoi polyhedra) $\tilde{D}_\xi$ for $\xi \in \overline{\omega}$, such that the points in $\tilde{D}_\xi$ are closer to ξ than to each other $\eta \in \overline{\omega}$. These sets

$$\tilde{D}_\xi = \{x \in \mathbb{R}^n : \|x - \xi\| < \|x - \eta\| \quad \forall \eta \in (\overline{\omega} \backslash \{\xi\})\}$$

are also called Dirichlet regions, Wigner–Seitz cells or Thiessen polyhedra (depending on the application problem). Note, that $\tilde{D}_\xi$ are convex, polyhedral and possibly unbounded sets (for vertices $\xi \in \overline{\omega}$ belonging to the boundary of the convex hull of $\overline{\omega}$). The boxes can be defined by $D_\xi = \overline{\tilde{D}_\xi} \cap \overline{\Omega}$.

Furthermore, each (Voronoi) vertex of the polyhedra generated by the Voronoi diagram is equidistant from at least $(n+1)$ points belonging to $\overline{\omega}$ (in our notation: Voronoi vertex $= p^S$). Then, the Delauney triangulation is dual to the Voronoi diagram in the sense that each pair $\xi, \eta \in \overline{\omega}$ is connected by a straight-line segment, when $\text{meas}_{n-1}(\partial\tilde{D}_\xi \cap \partial\tilde{D}_\eta) > 0$ holds (for an equivalent definition in terms of a so-called cell complex see [FORT92], Section 2.1). These straight-line segments form the edges of a triangulation of the *convex hull* of $\overline{\omega}$. However, there can arise nonsimplicial, but convex Delauney cells (polyhedra) for so-called degenerate Voronoi vertices (if the boundaries of more than $n+1$ Voronoi polyhedra share a Voronoi vertex, then it is called degenerate). Due to the convexity of such cells, they can be subdivided into simplices without introducing new vertices.

For example, such a situation is obtained for a set $\overline{\omega}$ of nodes forming a quadratic or cubic grid. Obviously, the Delauney cells are then quadrangles or cubes and one gets the 'five- (or seven-) point difference star', which gives the connection to the classical finite difference method. Furthermore, one requires some modifications for nonconvex domains Ω 'given' by a point set $\overline{\omega}$ (see [BEEP92], Section 2.2).

An essential property of the Delauney triangulation is that it has empty circumspheres, i.e., the circumsphere of a subset $\hat{\omega} \subset \overline{\omega}$ which forms a Delauney cell contains no nodes of $\overline{\omega} \backslash \hat{\omega}$ in its interior or on its surface. In other words, the n-dimensional ball defined by the (at least) $n+1$ points of $\hat{\omega}$ from which the corresponding (dual) Voronoi vertex is equidistant contains no nodes of $\overline{\omega}$ in its interior. Vice versa, a given triangulation $\mathcal{S}_h$ consisting of simplices is called a

Delauney triangulation if the interior of the circumball of each $S \in \mathcal{S}_h$ does not contain vertices of the triangulation.

From the viewpoint of numerical analysis the just mentioned choice of boxes as circumcentric regions has the disadvantage that a Delauney simplex S does not need to contain its Voronoi vertex (circumcentre) p^S. Note, that $p^S \in S$ can be guaranteed by a weakly acute triangulation for $n = 2$ (no obtuse angle occurs, see Section 4.4.1). This case is included in our defined box partition $\mathcal{D}_h(1)$ (for a right-angled triangle S the circumcentre p^S coincides with one of the edge midpoints). However, for $n = 3$, the assumption $p^S \in S$ for circumcentric boxes is crucial. Moreover, in view of the equality of the stiffness matrices L_G and L_B (see Lemma 4.1[ii] below), the boxes defined according to $\mathcal{D}_h(1)$ and $\mathcal{D}_h(2)$, respectively, are such that (4.13) holds. Obviously, for $n = 2$ this property is also satisfied by the circumcentric boxes, but for $n = 3$ this is in general not fulfilled by circumcentric regions.

Notwithstanding these restrictions, the Delauney triangulation is very valuable because it has some optimality properties. For example, denoting the smallest angle of the triangle $S \in \mathcal{S}_h$ by $\alpha(S)$, the Delauney triangulation maximizes the minimum angle

$$\alpha(\mathcal{S}_h) = \min_{S \in \mathcal{S}_h} \alpha(S)$$

over all triangulations for a given set $\overline{\omega}$ (see [LAWS77] and [FORT92], Lemma 3.1). Recalling Zlamal's condition (cf. [CIAR78], Exercise 3.1.3) equivalent to the assumption of a regular family $\{\mathcal{S}_h\}_{h \in (0,h_0]}$ given in Section 4.1.2, this parameter plays an essential role in the numerical analysis.

For $n = 3$ no extension of this statement is known, but Rajan (see [RAJ94] and [FORT92], Lemma 3.2) gives a criterion independent of the dimension. Let $\operatorname{maxrad}(\mathcal{S}_h)$ be the maximum radius of any min-containment sphere of any simplex $S \in \mathcal{S}_h$, where the min-containment sphere of a simplex S is the smallest sphere containing S. Then, the Delauney triangulation minimizes $\operatorname{maxrad}(\mathcal{S}_h)$ over all triangulations of a set $\overline{\omega} \subset \mathbb{R}^n$.

Summarizing, a suitable way seems to be the generation of a Delauney triangulation by means of the Voronoi diagram and then the use of e.g., barycentric boxes for constructing box schemes for second order (free) boundary value problems.

Let us now come to the definition of finite volume schemes based on box partitions according to $\mathcal{D}_h(1)$. Evidently, there is a correspondence (equal dimension) between the Galerkin subspace V_h^G (resp. X_h^G) and the space of discontinuous piecewise constant functions with respect to the boxes $D \in \mathcal{D}_h$ which will be denoted by

$$V_h^B = \{w \in X_h^B : w = 0 \text{ on } \Gamma_D\}, \quad X_h^B = \{w \in L_\infty(\Omega) : w_{|D} = \text{const } \forall D \in \mathcal{D}_h\}.$$

Furthermore, we introduce the prolongation P^B, the nodal basis Φ_ξ^B of V_h^B (resp. X_h^B), norms and seminorms for grid function vectors $w_h \in L_2(\tilde{\omega})$ (resp. $w_h \in$

$L_2(\overline{\omega})$) by

$$P^B : L_2(\tilde{\omega}) \longrightarrow V_h^B \text{ with } \quad P^B w_h = \sum_{\xi\in\tilde{\omega}} w_{h,\xi}\ \Phi_\xi^B, \quad \Phi_\xi^B(x) = \begin{cases} 1 & x \in D_\xi, \\ 0 & \text{otherwise,} \end{cases}$$

$$\|w_h\|_{0,\overline{\omega}} := \left\|P^B w_h\right\|_{L_2(\Omega)}, \quad |w_h|_{1,\overline{\omega}} := \left|P^G w_h\right|_{H^1(\Omega)},$$
$$\|w_h\|_{1,\overline{\omega}} := \left(|w_h|^2_{1,\overline{\omega}} + \|w_h\|^2_{0,\overline{\omega}}\right)^{1/2} \quad \text{and} \quad \|w_h\|_{0,\gamma} := \left\|P^B w_h\right\|_{L_2(\partial\Omega)}.$$

Now, considering the boundary value problem (4.6), (4.3), the basic idea of the finite volume (box) method consists in integrating the differential equation over the boxes D_ξ, $\xi \in \tilde{\omega}$. In other words, one uses $P^B v_h \in V_h^B$ as test function and integrates the equation over Ω. Then, the volume integrals $\int_{D_\xi} \operatorname{div}(k\nabla w)\, dx$ are transformed to surface integrals by Green's identity and the natural boundary conditions are taken into account. This leads to

$$\begin{aligned} &-\sum_{\xi\in\tilde{\omega}} \int_{\partial D_\xi\setminus\Gamma} (k\ \nabla w\cdot\vec{n})\ P^B v_h\ ds + \int_\Omega (q\ w - f)\ P^B v_h\ dx \\ &+ \int_{\Gamma_N} (\kappa\ w - g_N)\ P^B v_h\ ds = 0. \end{aligned}$$

Now, w is replaced by an element from the Galerkin trial space X_h^G in the first integral, whereas for the zero order and the boundary term the corresponding element from X_h^B (lumping) is used.

4.1.4 Finite volume approximations for the obstacle problem

Performing the same steps as indicated in the previous section for a second order equation (4.6), (4.3), now for the obstacle problem (4.1), (4.2) and exploiting the correspondence between $P^G v_h$ and $P^B v_h$, we arrive at the following finite volume (box) variational inequality.

$$\left|\begin{aligned} &\text{Find} \quad P^G y_h \in K_h : \quad a_B(P^G y_h, P^B v_h - P^B y_h) \\ &\geq\ (f, P^B v_h - P^B y_h) + (g_N, P^B v_h - P^B y_h)_{\Gamma_N} \quad \forall P^G v_h \in K_h \end{aligned}\right. \tag{4.15}$$

with the constraint set

$$K_h = \{P^G w_h \in X_h^G : (P^G w_h)(\xi) \geq \varphi(\xi)\ \forall\xi \in \overline{\omega}, \quad (P^G w_h)(\xi) = g_D(\xi)\ \forall\xi \in \gamma_D\}.$$

This can be equivalently expressed in the following matrix notation.

$$\left|\begin{aligned} &\text{Find} \quad y_h \in K_h : \\ &\left\langle (L_B + M_B + R_B)\ y_h, v_h - y_h\right\rangle \geq \left\langle f_{Bh} + g_{Bh}, v_h - y_h\right\rangle \quad \forall v_h \in K_h \end{aligned}\right. \tag{4.16}$$

with

$$K_h = \{w_h \in L_2(\overline{\omega}) : w_{h,\xi} \geq \varphi(\xi)\ \forall \xi \in \overline{\omega}, \quad w_{h,\xi} = g_D(\xi)\ \forall \xi \in \gamma_D\}.$$

The corresponding matrices and vectors are defined by

$$\begin{aligned}
(L_B)_{\xi,\eta} &= -\int_{\partial D_\xi \setminus \Gamma} (k\ \nabla \Phi_\eta^G \cdot \vec{n})\ ds = -\sum_{D \in \mathcal{D}_h} \int_{\partial D \setminus \Gamma} (k\ \nabla \Phi_\eta^G \cdot \vec{n})\ \Phi_\xi^B\ ds, \\
(M_B)_{\xi,\eta} &= \int_\Omega q\ \Phi_\eta^B\ \Phi_\xi^B\ dx, \quad (R_B)_{\xi,\eta} = \int_{\Gamma_N} \kappa\ \Phi_\eta^B\ \Phi_\xi^B\ ds, \\
f_{Bh,\xi} &= \int_\Omega f\ \Phi_\xi^B\ dx, \quad g_{Bh,\xi} = \int_{\Gamma_N} g_N\ \Phi_\xi^B ds,
\end{aligned} \tag{4.17}$$

where $\xi \in \tilde{\omega}$, $\eta \in \overline{\omega}$, $v_h \in L_2(\overline{\omega})$ and $w_h \in L_2(\tilde{\omega})$. Furthermore, the box bilinear form a_B is given by

$$\begin{aligned}
a_B(P^G v_h, P^B w_h) &= -\sum_{\xi \in \tilde{\omega}} \int_{\partial D_\xi \setminus \Gamma} (k \nabla P^G v_h \cdot \vec{n})\ P^B w_h\ ds \\
&+ \int_\Omega q\ P^B v_h\ P^B w_h\ dx + \int_{\Gamma_N} \kappa\ P^B v_h\ P^B w_h\ ds.
\end{aligned} \tag{4.18}$$

In other words,

$$a_B(P^G v_h, P^B w_h) = \big\langle (L_B + M_B + R_B)\ v_h, w_h \big\rangle.$$

Already at this stage, we refer to Remark 4.4 concerning the definition of the box stiffness matrix L_B.

In the sequel we shall sometimes use the following notations to indicate that grid function vectors are divided (componentwise !) by the area of the boxes and the area of the part of the box boundaries lying on Γ, respectively. More precisely, we denote

$$(v_h/\operatorname{meas}_n(D))_{\xi \in \overline{\omega}} = \frac{v_{h,\xi}}{\operatorname{meas}_n(D_\xi)} \quad \text{for} \quad \xi \in \overline{\omega} \quad \text{and}$$

$$(w_h/\operatorname{meas}_{n-1}(\partial D))_{\xi \in \overline{\omega}} = \begin{cases} \dfrac{w_{h,\xi}}{\operatorname{meas}_{n-1}(\partial D_\xi \cap \Gamma)} & \xi \in \gamma, \\ 0 & \text{otherwise.} \end{cases}$$

We have not scaled the right-hand side vectors f_{Bh}, g_{Bh} and matrices by the box areas in the definition (4.17) which causes utilization of the just introduced notation.

Sometimes we shall use the following abbreviation for the box matrix:

$$\Lambda_B = L_B + M_B + R_B,$$

and furthermore, we will use the (discrete) obstacle grid function vector φ_h with the components $\varphi_{h,\xi} = \varphi(\xi)$ for $\xi \in \overline{\omega}$, i.e., $P^G\varphi_h = \Pi_h\varphi$ is the (piecewise linear) X_h^G-interpolant of $\varphi \in W_p^1(\Omega)$.

Let us now summarize some properties of these finite volume approximations.

Lemma 4.1 *Let the box partitions be constructed according to $\mathcal{D}_h(1)$.*
[i] *For any grid function vector $v_h \in L_2(\overline{\omega})$ there holds*

$$C_1 \left\|P^B v_h\right\|_{L_2(\Omega)} \leq \left\|P^G v_h\right\|_{L_2(\Omega)} \leq C_2 \left\|P^B v_h\right\|_{L_2(\Omega)},$$
$$C_3 \left\|P^B v_h\right\|_{L_2(\partial\Omega)} \leq \left\|P^G v_h\right\|_{L_2(\partial\Omega)} \leq C_4 \left\|P^B v_h\right\|_{L_2(\partial\Omega)}$$

with constants $C_i = C_i(n)$, $i = 1, 3, 4$ and $C_2 = C_2(n, \delta)$ (δ from (4.12)).

[ii] *Assuming a coefficient function*

$$k \in V_h^C = \{w \in L_\infty(\Omega) : w_{|S} = const \quad \forall S \in \mathcal{S}_h\},$$

it holds that

$$\int_\Omega k\ \nabla P^G v_h\ \nabla\Phi_\xi^G\ dx = -\sum_{D\in\mathcal{D}_h} \int_{\partial D\setminus\Gamma} (k\ \nabla P^G v_h \cdot \vec{n})\ \Phi_\xi^B\ ds$$

for each $\xi \in \tilde{\omega}$ and for all $v_h \in L_2(\overline{\omega})$. Hence, the box stiffness matrix L_B coincides with the finite element stiffness matrix L_G for $k \in V_h^C$.

Proof. [i]. Let S be a simplex in $\mathbb{R}^n$ with vertices ξ_i, $n = 2, 3$ which is subdivided into $S = \bigcup_{i=1,\ldots,n+1} D_{\xi_i}^S$ according to the defined box partition $\mathcal{D}_h(1)$. Then there holds

$$\left\|P^B v_h\right\|_{L_2(S)}^2 = \sum_{i=1}^{n+1} v_{h,\xi_i}^2 \operatorname{meas}_n(D_{\xi_i}^S),$$
$$\left\|P^G v_h\right\|_{L_2(S)}^2 = \frac{\operatorname{meas}_n(S)}{(n+1)(n+2)} \Big(\sum_{i=1}^{n+1} v_{h,\xi_i}^2 + (\sum_{i=1}^{n+1} v_{h,\xi_i})^2\Big),$$

that can be verified by (numerical) integration formulae, exact for quadratic polynomials. From here, the claimed inequalities follow by summation over $S \in \mathcal{S}_h$. The statement for the boundary $\Gamma = \partial\Omega$ can be proved in an analogous way by summation over $\sigma \in \sigma_h$ (boundary faces).

[ii]. Owing to $k \in V_h^C$ it is sufficient to show the relation

$$\int_S \nabla P^G v_h\ \nabla\Phi_\xi^G\ dx = -\int_{(\partial D_\xi\setminus\Gamma)\cap S} \nabla P^G v_h \cdot \vec{n}\ ds \quad \forall S \in \mathcal{S}_h,\ \xi \in \tilde{\omega}.$$

Therefore, let us consider an arbitrary simplex $S \in \mathcal{S}_h$. Since $\Delta P^G v_h = 0$ on S, partial integration over S leads to

$$\begin{aligned}\int_S \nabla P^G v_h \, \nabla \Phi_\xi^G \, dx &= \sum_{i=1}^{n} \int_{\sigma_i} \Phi_\xi^G \, (\nabla P^G v_h \cdot \vec{n}) \, ds \\ &= \frac{1}{n} \sum_{i=1}^{n} (\nabla P^G v_h \cdot \vec{n})_{|\sigma_i} \; \text{meas}_{n-1}(\sigma_i),\end{aligned}$$

where σ_i, $i = 1, \ldots, n+1$ are the (boundary) faces of the simplex S, i.e., one has $\partial S = \bigcup_{i=1}^{n+1} \sigma_i$. Furthermore, let σ_i, $i = 1, \ldots, n$ be those faces containing ξ, whereas σ_{n+1} should be the face lying opposite the vertex ξ such that $\Phi_\xi^G = 0$ on σ_{n+1}. By means of Gauss' theorem on $D_\xi^S = D_\xi \cap S$ we obtain

$$\begin{aligned}0 = \int_{D_\xi^S} \Delta P^G v_h \, dx &= \int_{\partial D_\xi^S} \nabla P^G v_h \cdot \vec{n} \, ds \\ &= \int_{\partial D_\xi^S \cap \, \text{int} S} \nabla P^G v_h \cdot \vec{n} \, ds + \int_{\partial D_\xi^S \cap \partial S} \nabla P^G v_h \cdot \vec{n} \, ds.\end{aligned}$$

Now, recalling the essential relation (4.13) in the construction of the boxes according to $\mathcal{D}_h(1)$, we get

$$\begin{aligned}\int_{(\partial D_\xi \setminus \Gamma) \cap S} \nabla P^G v_h \cdot \vec{n} \, ds &= \int_{\partial D_\xi^S \cap \, \text{int} S} \nabla P^G v_h \cdot \vec{n} \, ds \\ &= -\int_{\partial D_\xi^S \cap \partial S} \nabla P^G v_h \cdot \vec{n} \, ds = -\frac{1}{n} \sum_{i=1}^{n} (\nabla P^G v_h \cdot \vec{n})_{|\sigma_i} \; \text{meas}_{n-1}(\sigma_i),\end{aligned}$$

such that after summation over $S \in \mathcal{S}_h$ the proof is completed. ■

Theorem 4.2 *Let the box partitions be constructed according to $\mathcal{D}_h(1)$ and the assumptions (4.4), (4.5) be fulfilled.*
[i] *There exist constants $C_i > 0$, $i = 1, \ldots, 4$, such that*

$$\begin{aligned}a_B(P^G v_h, P^B w_h) &\le C_1 \left\|P^G v_h\right\|_{H^1(\Omega)} \left\|P^G w_h\right\|_{H^1(\Omega)} \le C_2 \; \|v_h\|_{1,\varpi} \; \|w_h\|_{1,\varpi}, \\ a_B(P^G v_h, P^B v_h) &\ge C_3 \; \left\|P^G v_h\right\|^2_{H^1(\Omega)} \ge C_4 \; \|v_h\|^2_{1,\varpi} \quad \forall v_h, w_h \in L_2(\tilde{\omega}).\end{aligned}$$

The problem (4.15), (4.18) possesses a unique solution $P^G y_h \in K_h$ such that

$$\left\|P^G y_h\right\|_{H^1(\Omega)} \le C_5 \; \|f\|_{L_2(\Omega)} + C_6 \; \|g_N\|_{L_2(\Gamma_N)} + C_7 \; \left\|P^G v_h\right\|_{H^1(\Omega)} \tag{4.19}$$

with an arbitrary grid function vector $v_h \in K_h$.

[ii] *The stability estimate*

$$C_8 \ \|y_h - \hat{y}_h\|_{1,\overline{\omega}} \leq \left\|P^G(y_h - \hat{y}_h)\right\|_{H^1(\Omega)} \leq C_9 \left[\|(f_{Bh} - \hat{f}_{Bh})/\operatorname{meas}_n(D)\|_{0,\overline{\omega}}\right.$$
$$\left. + \ \|(g_{Bh} - \hat{g}_{Bh})/\operatorname{meas}_{n-1}(\partial D)\|_{0,\gamma} + \|\varphi_{Wh} - \hat{\varphi}_{Wh}\|_{1,\overline{\omega}}\right],$$

holds for solutions $y_h = y_h(f_{Bh}, g_{Bh}, \varphi_{Wh})$ *and* $\hat{y}_h = \hat{y}_h(\hat{f}_{Bh}, \hat{g}_{Bh}, \hat{\varphi}_{Wh})$ *of (4.15), (4.18). Here, the grid function vectors* φ_{Wh} *and* $\hat{\varphi}_{Wh}$ *are defined by*

$$\varphi_{Wh,\xi} = \begin{cases} \varphi(\xi) & \text{for } \ \xi \in \tilde{\omega} \\ g_D(\xi) & \text{for } \ \xi \in \gamma_D \end{cases} \quad \text{and} \quad \hat{\varphi}_{Wh,\xi} = \begin{cases} \hat{\varphi}(\xi) & \text{for } \ \xi \in \tilde{\omega} \\ \hat{g}_D(\xi) & \text{for } \ \xi \in \gamma_D. \end{cases}$$

[iii] *An element* $P^G y_h \in V_h^G$ *will be a solution of the box obstacle problem (4.15), (4.18) if and only if the grid function vector* $y_h \in L_2(\overline{\omega})$ *satisfies the complementarity problem*

$$\begin{aligned} &\Lambda_B \ y_h \geq f_{Bh} + g_{Bh}, \quad (\Lambda_B \ y_h - f_{Bh} - g_{Bh}) \ (\varphi_h - y_h) = 0, \\ &y_h \geq \varphi_h, \quad \forall \xi \in \tilde{\omega}; \quad y_h = g_D, \quad \xi \in \gamma_D. \end{aligned} \tag{4.20}$$

Proof. [i]. Owing to part [ii] of Lemma 4.1 the box bilinear form a_B can be represented by

$$a_B(P^G v_h, P^B w_h) = \int_\Omega [k \ \nabla(P^G v_h) \ \nabla(P^G w_h) \ + \ q \ (P^B v_h) \ (P^B w_h)] \ dx$$
$$+ \int_{\Gamma_N} \kappa \ (P^B v_h) \ (P^B w_h) \ ds.$$

Consequently, the continuity of a_B is obtained using the results of Lemma 4.1[i]. To show the ellipticity of a_B by means of a Friedrichs–Poincaré type inequality, we choose a subset Ω_q^0 of Ω_q containing all simplices $S \in \mathcal{S}_h$ lying completely in Ω_q. To be more precise, let us define $\Omega_q^0 \subset \bigcap_{0<h\leq h_0} \{ \bigcup_{S\in\mathcal{S}_h : S\subset\Omega_q} S\}$ with $\operatorname{meas}_n(\Omega_q^0) > 0$ if $\operatorname{meas}_n(\Omega_q) > 0$ holds. Hence, by means of Lemma 4.1[i] and (4.4) we deduce

$$\begin{aligned} &a_B(P^G v_h, P^B v_h) \geq k_0 \ |P^G v_h|^2_{H^1(\Omega)} + q_0 \ \left\|P^B v_h\right\|^2_{L_2(\Omega_q^0)} + \kappa_0 \ \left\|P^B v_h\right\|^2_{L_2(\Gamma_\kappa)} \\ &\geq \ k_0 \ |P^G v_h|^2_{H^1(\Omega)} + c_1 \ \left\|P^G v_h\right\|^2_{L_2(\Omega_q^0)} + c_2 \ \left\|P^G v_h\right\|^2_{L_2(\Gamma_\kappa)} \\ &\geq \ c_3 \ \left\|P^G v_h\right\|^2_{H^1(\Omega)} \geq c_4 \ \|v_h\|^2_{1,\overline{\omega}}. \end{aligned}$$

Furthermore, using Lemma 4.1[i] again, we obtain

$$\begin{aligned} &(f, P^B v_h) + (g_N, P^B v_h)_{\Gamma_N} \leq \|f\|_{L_2(\Omega)} \left\|P^B v_h\right\|_{L_2(\Omega)} \\ &+ \ \|g_N\|_{L_2(\Gamma_N)} \left\|P^B v_h\right\|_{L_2(\Gamma_N)} \leq (c_5 \, \|f\|_{L_2(\Omega)} + c_6 \, \|g_N\|_{L_2(\Gamma_N)}) \left\|P^G v_h\right\|_{H^1(\Omega)}. \end{aligned}$$

Therefore, the existence of a unique $P^G y_h \in K_h$ as solution of (4.15), (4.18) is guaranteed by the Lions–Stampacchia theorem.

To show the boundedness of $P^G y_h$ we choose a grid function vector $v_h \in K_h$ in (4.15) such that $P^G(y_h - v_h) \in V_h^G$. This leads to

$$a_B(P^G(y_h - v_h), P^B(y_h - v_h)) \le \int_\Omega f\ (P^B(y_h - v_h))\ dx$$
$$+ \int_{\Gamma_N} g_N\ (P^B(y_h - v_h))\ ds - \ a_B(P^G v_h, P^B(y_h - v_h)).$$

Using the just proved properties of a_B and the triangle inequality we conclude the claimed estimate (4.19).

[ii]. We take $v_h = (\hat{y}_h + \varphi_{Wh} - \hat{\varphi}_{Wh}) \in K_h$ in the inequality (4.16) related to y_h and, vice versa, $v_h = (y_h - \varphi_{Wh} + \hat{\varphi}_{Wh}) \in \hat{K}_h$ in the inequality related to $\hat{y}_h$. Note that the formulation (4.16) is equivalent to (4.15). Adding both inequalities, subtracting $\Lambda_B\ (\varphi_{Wh} - \hat{\varphi}_{Wh})$ and introducing temporarily the notation

$$w_h = (y_h - \hat{y}_h + \hat{\varphi}_{Wh} - \varphi_{Wh}) \in L_2(\tilde{\omega}),$$

we derive

$$a_B(P^G w_h, P^B w_h) = \langle \Lambda_B w_h, w_h \rangle \le \|(f_{Bh} - \hat{f}_{Bh})/\,\mathrm{meas}_n(D)\|_{0,\varpi}\ \|w_h\|_{0,\varpi}$$
$$+\ \|(g_{Bh} - \hat{g}_{Bh})/\,\mathrm{meas}_{n-1}(\partial D)\|_{0,\gamma}\ \|w_h\|_{0,\gamma} + a_B(P^G(\hat{\varphi}_{Wh} - \varphi_{Wh}), P^B w_h).$$

The application of Lemma 4.1[i], part [i] of this theorem, the trace theorem and the triangle inequality complete the proof.

[iii]. Let $P^G y_h \in K_h$ be the solution of the inequality (4.15), (4.16). We define two grid function vectors $v_h^{(\pm)} \in K_h$ by

$$v_{h,\eta}^{(\pm)} = y_{h,\eta} \pm 0.5\ \delta_{\xi\eta}\ (y_{h,\eta} - \varphi_{h,\eta}) \quad \text{for } \eta \in \tilde{\omega}, \quad v_{h,\eta}^{(\pm)} = g_D(\eta) \quad \text{for } \eta \in \gamma_D$$

($\delta_{\xi\eta}$ as Kronecker symbol) for each node $\xi \in \tilde{\omega}$ characterized by $y_{h,\xi} > \varphi_{h,\xi}$. Inserting $v_h^{(\pm)}$ into (4.16), we have

$$0 \le \langle \Lambda_B\ y_h - f_{Bh} - g_{Bh}, v_h^{(\pm)} - y_h \rangle = \pm\ 0.5\ (\Lambda_B\ y_h - f_{Bh} - g_{Bh})_\xi\ (y_h - \varphi_h)_\xi,$$

from which we conclude $\Lambda_B\ y_h = f_{Bh} + g_{Bh}$.

On the other hand, we take grid function vectors $v_h \in K_h$, which are defined by $v_{h,\eta} = y_{h,\eta} + \delta_{\eta\xi}$ in (4.16) for nodes $\xi \in \tilde{\omega}$ characterized by $y_{h,\xi} = \varphi_{h,\xi}$. Then, we derive

$$(\Lambda_B\ y_h - f_{Bh} - g_{Bh})_\xi \ge 0.$$

Thus, y satisfies the conditions in the complementarity problem (4.20). The converse statement is obvious. ■

Remark 4.3 [i] Let us mention that similar proofs of Lemma 4.1[ii] for general boxes according to $\mathcal{D}_h(1)$, but restricted to $n = 2$, are contained in [HACK89], Proposition 3.1.2 and [BARO87], Lemma 3. A modified proof for barycentric boxes in the three-dimensional situation can be found in [IKED83], Lemma 6.1.

[ii] Assuming an arbitrary element $v \in K \cap W_p^1(\Omega)$ with $p > n$, one can take

$$K_h \ni v_h := (\Pi_h v)_{\xi \in \overline{\omega}}$$

in (4.19). Hence, $\Pi_h v = P^G v_h$ holds. Here, Π_h is the (piecewise linear) Lagrange interpolation operator. Note in this context the embedding $W_p^1(\Omega) \subset C(\overline{\Omega})$ for $p > n$. Then, applying interpolation properties in Sobolev spaces one obtains

$$\|P^G v_h\|_{H^1(\Omega)} \leq C_1 \ \|\Pi_h v\|_{W_p^1(\Omega)} \leq C_2 \ \|v\|_{W_p^1(\Omega)} \,.$$

Evidently, if $\partial\Omega = \Gamma_N$ or $\varphi = g_D$ on Γ_D, then $v = \varphi$ is possible.

[iii] Let us emphasize that the grid function norms appearing in Theorem 4.2[ii] can be replaced by the norms of $L_2(\Omega)$ and $L_2(\Gamma_N)$. More precisely, we have

$$\begin{aligned} &\|(f_{Bh} - \hat{f}_{Bh})/\operatorname{meas}_n(D)\|_{0,\overline{\omega}} + \|(g_{Bh} - \hat{g}_{Bh})/\operatorname{meas}_{n-1}(\partial D)\|_{0,\gamma} \\ &\leq \ \|f - \hat{f}\|_{L_2(\Omega)} + \|g_N - \hat{g}_N\|_{L_2(\Gamma_N)} \end{aligned}$$

by means of the Hölder inequality.

[iv] Let us anticipate that the box solution of (4.15), (4.18) fulfills not only the complementarity formulation (4.20) but also Lewy–Stampacchia type inequalities, see Lemma 4.34 in Section 4.5. □

Remark 4.4 The assumption $k \in V_h^C$ which guarantees the equality of the finite element and box stiffness matrices ($L_G = L_B$) does not imply a loss of generality. Let us define $k_C \in V_h^C$ as the mean value

$$(k_C)_{|S} := \Big(\int_S k(x)\ dx \Big) \Big/ \operatorname{meas}_n(S) \quad \text{for} \quad S \in \mathcal{S}_h.$$

Then, we have

$$\begin{aligned} \langle L_G\ v_h, w_h \rangle &= \sum_{S \in \mathcal{S}_h} (\nabla P^G v_h\ \nabla P^G w_h)_{|S} \int_S k(x)\ dx \\ &= \int_\Omega k_C(x)\ \nabla P^G v_h\ \nabla P^G w_h\ dx = \langle L_B\ v_h, w_h \rangle \end{aligned}$$

for $v_h \in L_2(\overline{\omega})$, $w_h \in L_2(\tilde{\omega})$, such that the box stiffness matrices are based on the coefficient k_C.

In general, one uses a quadrature formula for the computation of $\int_S k(x)\ dx$ in the assembly process. This causes an additional error, but, usually, the convergence order of the finite element or box approximation remains unaltered. In this work we shall not discuss the influence of quadrature formulae. □

4.2 Comparison of finite volume and finite element approximations for elliptic variational inequalities

Finite element and finite volume (box) schemes have been compared in [BARO87] and [HACK89] for second order elliptic equations in spatial two-dimensional domains (see also [KERK96]). In this section we will extend the comparison to interior obstacle problems with mixed boundary conditions in two and three dimensions (concerning boundary obstacle problems see Section 4.5). More precisely, we will show that the finite element and finite volume method applied to such variational inequalities lead to (discrete) obstacle problems which have a comparable accuracy.

Therefore, let us now switch over to the derivation of error estimates for the difference $P^G(u_h - y_h)$ in the energy norm with u_h and y_h as solutions of (4.9), (4.10) and (4.15), (4.16), respectively. Choosing $v_h = y_h \in K_h$ in (4.10) and, vice versa, $v_h = u_h \in K_h$ in (4.16) we obtain

$$\begin{aligned}&\langle (L_G + M_G + R_G)\, u_h - (L_B + M_B + R_B)\, y_h, u_h - y_h\rangle \\ &\leq \langle f_{Gh} + g_{Gh} - f_{Bh} - g_{Bh}, u_h - y_h\rangle.\end{aligned}$$

Using the ellipticity of the bilinear form a (denoted by the constant $m > 0$) and the equality of the stiffness matrices $L_B = L_G$ proved in Lemma 4.1[ii], we deduce the following preliminary estimate

$$\left.\begin{aligned}&m \left\|P^G(u_h - y_h)\right\|^2_{H^1(\Omega)} \leq \langle (L_G + M_G + R_G)\; (u_h - y_h), u_h - y_h\rangle \\ &\leq \langle f_{Gh} - f_{Bh}, u_h - y_h\rangle + \langle g_{Gh} - g_{Bh}, u_h - y_h\rangle \\ &+ \langle (M_B - M_G)\; y_h, u_h - y_h\rangle + \langle (R_B - R_G)\; y_h, u_h - y_h\rangle.\end{aligned}\right| \quad (4.21)$$

In order to treat the error terms in the right-hand side of (4.21), we first summarize some auxiliary statements in Lemma 4.6 and Corollary 4.7. The Bramble–Hilbert Lemma in the following form is used for the proof of these statements.

Lemma 4.5 *Let $\Omega \subset \mathbb{R}^n$ be a bounded domain with a Lipschitz boundary. Let Ψ be a linear continuous functional on the Sobolev space $W_p^k(\Omega)$ such that*

$$|\Psi(u)| \leq C_\Psi \; \|u\|_{W_p^k(\Omega)} \quad \forall u \in W_p^k(\Omega) \quad \textit{with} \quad k \geq 1,\ 1 \leq p \leq \infty.$$

[i] *If the functional Ψ fulfills $\Psi(q) = 0$ for all polynomials*

$$q \in P_{k-1}(\Omega) = \{r : r(x) = \sum_\alpha r_\alpha x^\alpha,\ |\alpha| \leq k-1\},$$

then there exists a constant $C_B = C_B(\Omega)$ *such that*

$$|\Psi(u)| \leq C_B\ C_\Psi\ |u|_{W_p^k(\Omega)} \quad \forall u \in W_p^k(\Omega).$$

[ii] *Furthermore, if* $\Psi(q) = 0$ *for all* $q \in P_{k-2}$ *with* $k \geq 2$ *holds, then there exists a constant* $\tilde{C}_B = \tilde{C}_B(\Omega)$ *such that*

$$|\Psi(u)| \leq \tilde{C}_B\ C_\Psi\ \left(\ |u|^2_{W_p^{k-1}(\Omega)} + |u|^2_{W_p^k(\Omega)}\ \right)^{1/2} \quad \forall u \in W_p^k(\Omega).$$

We refer to [CIAR78], Theorem 4.1.3 for the proof of part [i] and to [HEIN87], Appendix ES, Theorem 2 for part [ii].

Lemma 4.6 *Let* $U \subset \mathbb{R}^n$, $n = 1, 2, 3$ *be a (closed) simplex with* $h_U = \operatorname{diam} U$.
[i] *Let* $z \in U$ *be an arbitrary point. Then, the estimate*

$$\|v - v(z)\|_{L_2(U)} \leq C_1\ h_U\ \left(\ |v|^2_{H^1(U)} + h_U^2\ |v|^2_{H^2(U)}\ \right)^{1/2}$$

holds for an arbitrary function $v \in H^2(U)$. *For* $v \in W_p^1(U)$ *with* $p > n$, *there exists a constant* $C_2 > 0$, *such that*

$$\|v - v(z)\|_{L_p(U)} \leq C_2\ h_U\ |v|_{W_p^1(U)}\,.$$

[ii] *Let* $v \in H^1(U)$ *and* v_C *be defined by* $v_C = (\int_U v(x)\ dx)/\operatorname{meas}_n(U)$. *Then, there holds the estimate*

$$\|v - v_C\|_{L_2(U)} \leq C_3\ h_U\ |v|_{H^1(U)}.$$

[iii] *Let* $v \in H^2(U)$ *and* ξ_i, $i = 1, \dots, n+1$ *be the vertices of the simplex* U. *Then*

$$\left| \int_U v(x)\ dx - \frac{\operatorname{meas}_n(U)}{(n+1)} \sum_{i=1}^{n+1} v(\xi_i) \right| \leq C_4\ h_U^2\ (\operatorname{meas}_n(U))^{1/2}\ |v|_{H^2(U)}\,.$$

Proof. [i]. The L_2-norm of $(v - v(z))$ can be rewritten as

$$\|v - v(z)\|_{L_2(U)} = \sup_{w \in L_2(U)} \frac{|\Psi(v,w)|}{\|w\|_{L_2(U)}} \quad \text{with}$$
$$\Psi(v,w) = \int_U (v(x) - v(z))\ w(x)\ dx.$$

Therefore, we consider the functional Ψ. As usual, let $\hat{U}$ be a reference simplex (unit n-simplex) which can be transformed onto the simplex U by an invertible affine mapping $x = F(\hat{x}) = B\hat{x} + b$ such that $U = F(\hat{U})$ (cf. [CIAR78], Section 3.1). Performing such a transformation we obtain

$$\Psi(v,w) = \hat{\Psi}(\hat{v},\hat{w})\ |\det B|, \quad \hat{\Psi}(\hat{v},\hat{w}) = \int_{\hat{U}} (\hat{v} - \hat{v}(\hat{z}))\ \hat{w}\ d\hat{x}$$

with $|\det B| = \operatorname{meas}_n(U)/\operatorname{meas}_n(\hat{U})$. Owing to the embedding $H^2(\hat{U}) \subset C(\overline{\hat{U}})$ we have that $\hat{\Psi}$ is a linear, bounded functional with respect to $\hat{v} \in H^2(\hat{U})$ and $\hat{w} \in L_2(\hat{U})$. Moreover, $\hat{\Psi}(\hat{v}, \hat{w}) = 0$ for $\hat{v} = \text{const}$.
Hence, Lemma 4.5[ii] can be applied, which leads to

$$|\hat{\Psi}(\hat{v}, \hat{w})| \leq C \left(|\hat{v}|^2_{H^2(\hat{U})} + |\hat{v}|^2_{H^1(\hat{U})} \right)^{1/2} \|\hat{w}\|_{L_2(\hat{U})} .$$

Performing now the backward transformation (cf. [CIAR78], Theorems 3.1.2 and 3.1.3), we get

$$|\Psi(v, w)| \leq C \ \|B\| \ \left(|v|^2_{H^2(U)} \ \|B\|^2 + |v|^2_{H^1(U)} \right)^{1/2} \|w\|_{L_2(U)}, \quad \|B\| \leq h_U/\rho_{\hat{U}},$$

where $\rho_{\hat{U}}$ denotes the diameter of the largest n-ball contained in $\hat{U}$.
The second estimate can be demonstrated in a similar way. To be more precise, we consider the norm

$$\|v - v(z)\|_{L_p(U)} = \sup_{w \in L_q(U)} \frac{|\Psi(v, w)|}{\|w\|_{L_q(U)}}, \quad p^{-1} + q^{-1} = 1$$

and use the embedding $W^1_p(\hat{U}) \subset C(\overline{\hat{U}})$ for $p > n$ in the application of the Bramble–Hilbert Lemma 4.5[i].

[ii]. Using the representation

$$\|v - v_C\|_{L_2(U)} = \sup_{w \in L_2(U)} \frac{|\Psi(v, w)|}{\|w\|_{L_2(U)}} \quad \text{with}$$
$$\Psi(v, w) = \int_U (v(x) - v_C) \ w(x) \ dx,$$

this part is proved in a similar way, i.e., by the same transformation and the application of Lemma 4.5[i].

[iii]. We define the following functional Ψ and perform the transformation of U onto the reference simplex $\hat{U}$, i.e.,

$$\Psi(v) = \int_U v(x) \ dx - \frac{\operatorname{meas}_n(U)}{n+1} \sum_{i=1}^{n+1} v(\xi_i) = \hat{\Psi}(\hat{v}) \ \operatorname{meas}_n(U) \quad \text{with}$$
$$\hat{\Psi}(\hat{v}) = \int_{\hat{U}} \hat{v}(\hat{x}) \ d\hat{x} / \operatorname{meas}_n(\hat{U}) - \sum_{i=1}^{n+1} \hat{v}(\hat{\xi}_i)/(n+1).$$

Again, using the embedding $H^2(\hat{U}) \subset C(\overline{\hat{U}})$, we get $|\hat{\Psi}(\hat{v})| \leq C \ \|\hat{v}\|_{H^2(\hat{U})}$. Furthermore, $\hat{\Psi}(\hat{p}) = 0$ holds for all linear polynomials $\hat{p}$. Applying Lemma 4.5[i] and performing the backward transformation, we derive the desired estimate. ■

Corollary 4.7 *Let a box partition be given according to $\mathcal{D}_h(1)$. Then, for grid function vectors $v_h, w_h \in L_2(\overline{\omega})$,*

$$\left\|P^G v_h - P^B v_h\right\|_{L_2(S)} \leq C_1\ h_S\ \left|P^G v_h\right|_{H^1(S)},$$

$$\left\|P^G v_h\ P^G w_h - P^B v_h\ P^B w_h\right\|_{L_1(S)} \leq C_3\ h_S\ \left\|P^G v_h\right\|_{H^1(S)}\ \left\|P^G w_h\right\|_{H^1(S)}$$

and

$$\left\|P^G v_h\ P^G w_h - P^B v_h\ P^B w_h\right\|_{L_2(\sigma)} \leq C_4\ \frac{h_\sigma\ \left\|P^G v_h\right\|_{H^1(\sigma)}\ \left\|P^G w_h\right\|_{H^1(\sigma)}}{(\text{meas}_{n-1}(\sigma))^{1/2}},$$

$$\left\|P^G v_h - P^B v_h\right\|_{L_2(\sigma)} \leq C_2\ h_\sigma \left|P^G v_h\right|_{H^1(\sigma)}$$

for each $S \in \mathcal{S}_h$ and $\sigma \in \sigma_h$, where $h_S = \operatorname{diam} S$ and $h_\sigma = \operatorname{diam} \sigma$.
Assuming a box partition according to $\mathcal{D}_h(2)$, one has

$$\left\|P^G v_h\ P^G w_h - P^B v_h\ P^B w_h\right\|_{L_2(S)} \leq C_5\ \frac{h_S\ \left\|P^G v_h\right\|_{H^1(S)}\ \left\|P^G w_h\right\|_{H^1(S)}}{(\text{meas}_n(S))^{1/2}}$$

for grid function vectors $v_h, w_h \in L_2(\overline{\omega})$. The constants $C_i > 0$, $i = 1, \ldots, 5$ are independent of $S \in \mathcal{S}_h$ and $\sigma \in \sigma_h$.

Proof. To show the first statement for $S \in \mathcal{S}_h$ we make use of the decomposition of a simplex S into subsimplices according to the box partition (cf. Figure 4.1), i.e., corresponding to $n = 2$ (resp. $n = 3$) we have

$$\left\|P^G v_h - P^B v_h\right\|^2_{L_2(S)} = \sum_{\xi \in (\overline{\omega} \cap \partial S)} \begin{cases} \displaystyle\sum_{\eta \in (\omega_\xi \cap \partial S)} \int_{D^S_{\xi;\eta}} (P^G v_h - v_{h,\xi})^2\ dx\,, \\ \displaystyle\sum_{\eta,\zeta \in (\omega_\xi \cap \partial S), \eta \neq \zeta} \int_{D^S_{\xi;\eta,\zeta}} (P^G v_h - v_{h,\xi})^2\ dx. \end{cases}$$

Applying Lemma 4.6[i] to each subsimplex $U := D^S_{\xi;\eta}$ $(n = 2)$ and $U := D^S_{\xi;\eta,\zeta}$ $(n = 3)$ with $v := P^G v_h \in H^2(U)$, $v(z) := (P^G v_h)(\xi) = v_{h,\xi}$, we conclude

$$\begin{aligned} \left\|P^G v_h - v_{h,\xi}\right\|^2_{L_2(U)} &\leq C\ h_U^2\ \left(\left|P^G v_h\right|^2_{H^1(U)} + h_U^2\ \left|P^G v_h\right|^2_{H^2(U)}\right) \\ &= C\ h_U^2\ \left|P^G v_h\right|^2_{H^1(U)}. \end{aligned}$$

Hence, the result is obtained after summation over the subsimplices belonging to the simplex S.
The statement with respect to the L_1-norm is derived by the same decomposition of S into subsimplices. More precisely, the Hölder inequality implies

$$\begin{aligned} \left\|P^G v_h\ P^G w_h - P^B v_h\ P^B w_h\right\|_{L_1(S)} \leq & \sum_{\xi \in (\overline{\omega} \cap \partial S)}\ \sum_{\eta \in (\omega_\xi \cap \partial S)} (\text{meas}_2(D^S_{\xi;\eta}))^{1/2} \\ & * \left\|P^G v_h\ P^G w_h - v_{h,\xi}\ w_{h,\xi}\right\|_{L_2(D^S_{\xi;\eta})} \end{aligned}$$

for $n = 2$ (analogous for $n = 3$). Exploiting Lemma 4.6[i] for each simplex $U := D^S_{\xi;\eta}$ with $v := (P^G v_h\ P^G w_h) \in H^2(U)$, $v(z) := v_{h,\xi}\ w_{h,\xi}$ and using the estimates

$$\left. \begin{aligned} \left|P^G v_h\ P^G w_h\right|^2_{H^1(U)} &\le c_1\ \left\|P^G v_h\right\|^2_{H^1(U)}\ \left\|P^G w_h\right\|^2_{H^1(U)} (\operatorname{meas}_n(U))^{-1}, \\ \left|P^G v_h\ P^G w_h\right|^2_{H^2(U)} &\le c_2\ \left|P^G v_h\right|^2_{H^1(U)}\ \left|P^G w_h\right|^2_{H^1(U)} (\operatorname{meas}_n(U))^{-1}, \end{aligned} \right| \tag{4.22}$$

we obtain

$$\left. \begin{aligned} \left\|P^G v_h\ P^G w_h - v_{h,\xi}\ w_{h,\xi}\right\|^2_{L_2(U)} &\le \frac{C(1+h_U^2)\ h_U^2}{\operatorname{meas}_n(U)}\ \left\|P^G v_h\right\|^2_{H^1(U)} \\ &* \left\|P^G w_h\right\|^2_{H^1(U)}. \end{aligned} \right| \tag{4.23}$$

Thus, the estimate is proved after summation.

We use again the just mentioned decompositions of S into subsimplices for proof of the statement with respect to the L_2-norm of $P^G v_h P^G w_h - P^B v_h P^B w_h$ (box partition according to $\mathcal{D}_h(2)$). Therefore, we obtain

$$\begin{aligned} &\left\|P^G v_h\ P^G w_h - P^B v_h\ P^B w_h\right\|^2_{L_2(S)} \\ &= \sum_{\xi \in (\overline{\omega} \cap \partial S)}\ \sum_{\eta \in (\omega_\xi \cap \partial S)} \left\|P^G v_h\ P^G w_h - v_{h,\xi}\ w_{h,\xi}\right\|^2_{L_2(D^S_{\xi;\eta})} \end{aligned}$$

in the spatial two-dimensional situation $n = 2$ (analogous for $n = 3$). Now, using (4.23) with $U := D^S_{\xi;\eta}$ ($n = 2$) and $U := D^S_{\xi;\eta,\zeta}$ ($n = 3$), noting $h_U < h_S$ and taking into consideration the relation

$$\operatorname{meas}_n(U) = \operatorname{meas}_n(S)/(n+1)!$$

(cf. (4.14)), we deduce the claimed estimate.

The statements for the boundary faces $\sigma \in \sigma_h$ are proved by the application of analogous decompositions of σ into subsimplices and the condition (4.13). ■

Remark 4.8 As a side remark let us briefly mention the well-known lumping operator $\mathcal{L}_h$ (see e.g., [IKED83]), which is defined by

$$\mathcal{L}_h : C(\overline{\Omega}) \longrightarrow L_\infty(\Omega), \qquad (\mathcal{L}_h v)(x) = \sum_{\xi \in \overline{\omega}} v(\xi)\ \chi_{D_\xi}(x),$$

for a general box partition according to $\mathcal{D}_h(1)$. Here, χ is the characteristic function with respect to D_ξ. Evidently, this lumping operator $\mathcal{L}_h$ is closely related to P^B. Taking a function $v \in C(\overline{\Omega})$, one has $\mathcal{L}_h v = P^B v_h$ with the grid function vector

v_h defined by $(v_h)_{\xi\in\overline{\omega}} = \big((\Pi_h v)(\xi)\big)_{\xi\in\overline{\omega}}$ and $\Pi_h v$ is the piecewise linear Lagrange interpolant of v. The estimates

$$\|\mathcal{L}_h v - v\|_{L_2(\Omega)} \le C_1\, h\, \big(|v|^2_{H^1(\Omega)} + h^2\, |v|^2_{H^2(\Omega)} \big)^{1/2} \quad \text{for } v \in H^2(\Omega),\ n = 1,2,3,$$

$$\|\mathcal{L}_h v - v\|_{L_p(\Omega)} \le C_2\, h\, |v|_{W^1_p(\Omega)} \quad \text{for } v \in W^1_p(\Omega),\ n < p$$

can be derived by means of Lemma 4.6[i] as well as the application of the decompositions of $\overline{\Omega}$ into simplices $S \in \mathcal{S}_h$ and of S into subsimplices U as used in Corollary 4.7. □

Now, we establish estimates with respect to the four terms appearing in the right-hand side of inequality (4.21).

Lemma 4.9 [i] *Let a box partition be constructed according to* $\mathcal{D}_h(1)$. *Then,*

$$\langle f_{Gh} - f_{Bh}, v_h \rangle \le C\, h\, \|f\|_{L_2(\Omega)}\, |P^G v_h|_{H^1(\Omega)} \quad \forall v_h \in L_2(\tilde{\omega}).$$

[ii] *Consider a box partition according to* $\mathcal{D}_h(2)$. *Then,*

$$\begin{aligned}\langle f_{Gh} - f_{Bh}, v_h \rangle &\le C\, h^2\, \Big(\sum_{S\in\mathcal{S}_h} |f|^2_{H^1(S)} \Big)^{1/2} |P^G v_h|_{H^1(\Omega)} \\ &\le C\, h^2\, |f|_{H^1(\Omega)}\, |P^G v_h|_{H^1(\Omega)} \quad \forall v_h \in L_2(\tilde{\omega}).\end{aligned}$$

Proof. [i]. Recalling the definition of f_{Gh} in (4.11) and f_{Bh} in (4.17), we obtain

$$\begin{aligned}\langle f_{Gh} - f_{Bh}, v_h \rangle &= \int_\Omega f(x)\, (P^G v_h - P^B v_h)\, dx \\ &\le \sum_{S\in\mathcal{S}_h} \|f\|_{L_2(S)}\, \|P^G v_h - P^B v_h\|_{L_2(S)},\end{aligned}$$

such that the statement is concluded by means of Corollary 4.7.
[ii]. Owing to (4.14) we have $\int_\Omega f_C\, (P^G v_h - P^B v_h)\, dx = 0$ for any piecewise constant function $f_C \in V_h^C$. Consequently, the (optimal) choice

$$(f_C)_{|S} = \int_S f\, dx / \operatorname{meas}_n(S)$$

leads to

$$\begin{aligned}\langle f_{Gh} - f_{Bh}, v_h \rangle &= \int_\Omega (f - f_C)\, (P^G v_h - P^B v_h)\, dx \\ &\le \sum_{S\in\mathcal{S}_h} \|f - f_C\|_{L_2(S)}\, \|P^G v_h - P^B v_h\|_{L_2(S)}.\end{aligned}$$

The claimed statement follows by an application of the results of Lemma 4.6[ii] and Corollary 4.7. ■

The following lemma will be useful in order to treat the differences of the mass matrices, i.e., the term $\langle (M_B - M_G)\, y_h, u_h - y_h \rangle$ in (4.21).

Lemma 4.10 [i] *Let a box partition be constructed according to $\mathcal{D}_h(1)$. Then, for all grid function vectors $v_h, w_h \in L_2(\overline{\omega})$,*

$$\int_\Omega q\, (P^B v_h\; P^B w_h - P^G v_h\; P^G w_h)\, dx$$
$$\leq C\, h\, \|q\|_{L_\infty(\Omega)}\, \|P^G v_h\|_{H^1(\Omega)}\, \|P^G w_h\|_{H^1(\Omega)}.$$

[ii] *Let a box partition be given according to $\mathcal{D}_h(2)$. Then,*

$$\int_\Omega q\, (P^B v_h\; P^B w_h - P^G v_h\; P^G w_h)\, dx \leq C\, h^2\, \left[\|q\|_{L_\infty(\Omega)} + \max_{S \in \mathcal{S}_h}\{|q|_{W^1_\infty(S)}\}\right]$$
$$*\; \|P^G v_h\|_{H^1(\Omega)}\, \|P^G w_h\|_{H^1(\Omega)} \leq C\, h^2\, \|q\|_{W^1_\infty(\Omega)}\, \|P^G v_h\|_{H^1(\Omega)}\, \|P^G w_h\|_{H^1(\Omega)}$$

for all grid function vectors $v_h, w_h \in L_2(\overline{\omega})$.

Proof. The estimate in part [i] follows immediately from Corollary 4.7, so we proceed with part [ii].
Denoting $w = P^B v_h\; P^B w_h - P^G v_h\; P^G w_h$ and using $q_C \in V_h^C$ with

$$(q_C)_{|S} = \int_S q\, dx / \operatorname{meas}_n(S) \quad \text{on} \quad S \in \mathcal{S}_h,$$

we split the integral into

$$\int_\Omega q\, (P^B v_h\; P^B w_h - P^G v_h\; P^G w_h)\, dx = \int_\Omega (q - q_C)\, w\, dx + \int_\Omega q_C\, w\, dx.$$

The first integral on the right-hand side is estimated by means of Lemma 4.6[ii] and Corollary 4.7. Hence, we get

$$\int_\Omega (q - q_C)\, w\, dx \leq \sum_{S \in \mathcal{S}_h} \|q - q_C\|_{L_2(S)}\, \|w\|_{L_2(S)}$$
$$\leq \sum_{S \in \mathcal{S}_h} \frac{C\, (h_S)^2}{(\operatorname{meas}_n(S))^{1/2}}\, |q|_{H^1(S)}\, \|P^G v_h\|_{H^1(S)}\, \|P^G w_h\|_{H^1(S)}$$
$$\leq C \sum_{S \in \mathcal{S}_h} (h_S)^2\, |q|_{W^1_\infty(S)}\, \|P^G v_h\|_{H^1(S)}\, \|P^G w_h\|_{H^1(S)}.$$

Therefore, after summation we obtain

$$\int_\Omega (q - q_C)\, w\, dx \leq C\, \max_{S \in \mathcal{S}_h}\{|q|_{W^1_\infty(S)}\}\, h^2\, \|P^G v_h\|_{H^1(\Omega)}\, \|P^G w_h\|_{H^1(\Omega)}.$$

Exploiting the equilibrium condition (4.14) on the boxes, the second integral from above can be expressed as

$$\int_\Omega q_C\ w\ dx = \sum_{S\in\mathcal{S}_h} (q_C)_{|S} \Big[\frac{\text{meas}_n(S)}{n+1} \sum_{\xi\in(\overline{\omega}\cap\partial S)} v_{h,\xi}\ w_{h,\xi} - \int_S P^G v_h\ P^G w_h\ dx\Big].$$

By the application of Lemma 4.6[iii] with $v := P^G v_h\ P^G w_h,\quad U := S$ and the estimate (4.22) as well as by summation over $S \in \mathcal{S}_h$, we get

$$\begin{aligned}&\int_\Omega q_C\ (P^B v_h\ P^B w_h - P^G v_h\ P^G w_h)\ dx\\ &\le C\ \|q\|_{L_\infty(\Omega)}\ h^2\ \left|P^G v_h\right|_{H^1(\Omega)}\ \left|P^G w_h\right|_{H^1(\Omega)},\end{aligned}$$

which completes the proof of part [ii]. ■

Now, we concentrate on the estimation of the boundary terms occuring in the right-hand side of (4.21).

Lemma 4.11 *Let $v_h \in L_2(\tilde{\omega})$ and a box partition be constructed according to $\mathcal{D}_h(1)$. Then,*

$$\left\langle g_{Gh} - g_{Bh}, v_h\right\rangle \le C\ h^{3/2}\ \Big(\sum_{\sigma\in\sigma_{h,2}} |g_N|^2_{H^1(\sigma)}\Big)^{1/2}\ \left\|P^G v_h\right\|_{H^1(\Omega)}.$$

Furthermore, if $g_N =$ *const. on each* $\sigma \in \sigma_{h,2}$, *then* $\left\langle g_{Gh} - g_{Bh}, v_h\right\rangle = 0$.

Proof. Defining

$$(g_N)_C = \Big(\int_\sigma g_N\ ds\Big)/\ \text{meas}_{n-1}(\sigma),$$

the difference $g_{Gh} - g_{Bh}$ in the right-hand sides of the Neumann/Newton boundary condition can be written as

$$\begin{aligned}\left\langle g_{Gh} - g_{Bh}, v_h\right\rangle &= \int_{\Gamma_N} g_N\ (P^B v_h - P^G v_h)\ ds\\ &= \sum_{\sigma\in\sigma_{h,2}} \int_\sigma (g_N - (g_N)_C)\ (P^B v_h - P^G v_h)\ ds \quad \text{for} \quad \sigma \in \sigma_{h,2}.\end{aligned}$$

Here, we have used that

$$\int_\sigma (P^G v_h - P^B v_h)\ ds = 0,$$

which is a consequence of the equilibrium condition (4.13). The last relation implies also the second statement of the lemma.

Returning to the proof of the first statement, we get

$$\langle g_{Gh} - g_{Bh}, v_h \rangle \leq \sum_{\sigma \in \sigma_{h,2}} \|g_N - (g_N)_C\|_{L_2(\sigma)} \, \left\|P^B v_h - P^G v_h\right\|_{L_2(\sigma)} .$$

Now we apply, on the one hand, Lemma 4.6[ii] and, on the other hand, Corollary 4.7. As a result we obtain

$$\|g_N - (g_N)_C\|_{L_2(\sigma)} \leq C_1 \, h_\sigma \, |g_N|_{H^1(\sigma)}$$

and

$$\left\|P^G v_h - P^B v_h\right\|_{L_2(\sigma)} \leq C_2 \, h_\sigma \, \left|P^G v_h\right|_{H^1(\sigma)} .$$

Considering the seminorm $\left|P^G v_h\right|_{H^1(\sigma)}$, we mention that for a linear function v defined on a simplex S with σ as a boundary face of S,

$$|v|^2_{H^1(\sigma)} \leq C \, h_S^{-1} \, |v|^2_{H^1(S)} \quad \text{with} \quad h_S = \operatorname{diam} S. \tag{4.24}$$

Here, the constant C depends on the parameter ρ (cf. Section 4.1.2). The proof is completed by an application of this property to $P^G v_h$ and by summation over $\sigma \in \sigma_{h,2}$. ∎

Finally, we derive an estimate for the difference between the finite element (boundary) matrix R_G from (4.11) and the lumped diagonal box matrix R_B defined in (4.17).

Lemma 4.12 *Let $v_h, w_h \in L_2(\overline{\omega})$ and the boxes be given according to $\mathcal{D}_h(1)$. Then,*

$$\begin{aligned} &\int_{\Gamma_N} \kappa \, (P^B v_h \, P^B w_h - P^G v_h \, P^G w_h) \, ds \\ &\leq C \, h \left[\|\kappa\|_{L_\infty(\Gamma_N)} + \max_{\sigma \in \sigma_{h,2}} \{|\kappa|_{W^1_\infty(\sigma)}\} \right] \left\|P^G v_h\right\|_{H^1(\Omega)} \left\|P^G w_h\right\|_{H^1(\Omega)} . \end{aligned}$$

Proof. Similar to the proof of Lemma 4.10[ii] we again apply a decomposition of the form $\kappa \, w = (\kappa - \kappa_C) \, w + \kappa_C \, w$ with

$$(\kappa_C)_{|\sigma} = \Big(\int_\sigma \kappa \, ds \Big) / \operatorname{meas}_{n-1}(\sigma) \quad \text{for} \quad \sigma \in \sigma_{h,2}$$

and

$$w = P^B v_h \, P^B w_h - P^G v_h \, P^G w_h .$$

Using on the one hand, Lemma 4.6[ii] with $U := \sigma$ for $\|\kappa - \kappa_C\|_{L_2(\sigma)}$ and, on the other hand, Corollary 4.7, we derive

$$\begin{aligned} &\int_{\Gamma_N} (\kappa - \kappa_C) \, w \, ds \leq C \sum_{\sigma \in \sigma_{h,2}} \frac{h_\sigma^2 \, |\kappa|_{H^1(\sigma)}}{(\operatorname{meas}_{n-1}(\sigma))^{1/2}} \left\|P^G v_h\right\|_{H^1(\sigma)} \left\|P^G v_h\right\|_{H^1(\sigma)} \\ &\leq C \max_{\sigma \in \sigma_{h,2}} \{|\kappa|_{W^1_\infty(\sigma)}\} \Big(\sum_{\sigma \in \sigma_{h,2}} h_\sigma^2 \left\|P^G v_h\right\|^2_{H^1(\sigma)} \Big)^{0.5} \Big(\sum_{\sigma \in \sigma_{h,2}} h_\sigma^2 \left\|P^G w_h\right\|^2_{H^1(\sigma)} \Big)^{0.5} . \end{aligned}$$

Now it is easy to see that

$$\sum_{\sigma\in\sigma_{h,2}} h_\sigma^2 \left\|P^G v_h\right\|^2_{H^1(\sigma)} \leq h^2 \left\|P^G v_h\right\|^2_{L_2(\Gamma)} + \sum_{\sigma\in\sigma_{h,2}} h_\sigma^2 \left|P^G v_h\right|^2_{H^1(\sigma)}$$

holds (analogously for $P^G w_h$). Applying the property (4.24) to $P^G v_h$ and $P^G w_h$ and using the trace theorem, we deduce

$$\int_{\Gamma_N} (\kappa - \kappa_C)\, w\, ds \leq C\, h \max_{\sigma\in\sigma_{h,2}} \{|\kappa|_{W^1_\infty(\sigma)}\} \left\|P^G v_h\right\|_{H^1(\Omega)} \left\|P^G w_h\right\|_{H^1(\Omega)}.$$

Recalling the equilibrium condition (4.13) for boundary faces σ belonging to a simplex $S = S(\sigma) \in \mathcal{S}_h$, the second integral can be represented by

$$\int_{\Gamma_N} \kappa_C\, w\, ds = \sum_{\sigma\in\sigma_{h,2}} \kappa_C|_\sigma \Big(\frac{\text{meas}_{n-1}(\sigma)}{n} \sum_{\xi\in(\overline{\omega}\cap\partial\sigma)} v_{h,\xi}\, w_{h,\xi} - \int_\sigma P^G v_h\, P^G w_h\, ds \Big).$$

Therefore, similar to the proof of Lemma 4.10[ii], we conclude

$$\begin{aligned} \int_\sigma w\, ds &= \int_\sigma (P^B v_h\, P^B w_h - P^G v_h\, P^G w_h)\, ds \\ &\leq C\, h_{S(\sigma)} \left|P^G v_h\right|_{H^1(S(\sigma))} \left|P^G w_h\right|_{H^1(S(\sigma))} \end{aligned}$$

by means of Lemma 4.6[iii], (4.22) (with $U = \sigma$) and additionally (4.24). Consequently, after summation over $\sigma \in \sigma_{h,2}$, we end up with

$$\int_{\Gamma_N} \kappa_C\, w\, ds \leq C\, h \, \|\kappa\|_{L_\infty(\Gamma_N)} \left|P^G v_h\right|_{H^1(\Omega)} \left|P^G w_h\right|_{H^1(\Omega)},$$

which completes the proof. ∎

The previous statements from Lemmas 4.9–4.12 are now summarized in error bounds with respect to the difference of u_h as the finite element solution of (4.9), (4.10) and y_h as the box solution of (4.15), (4.16).

Theorem 4.13 *Let the assumptions (4.4), (4.5) and $k \in V_h^C$ be satisfied.*

[i] *Assume that $g_N \in H^1(\Gamma_N)$, $\kappa \in W^1_\infty(\Gamma_N)$ and a box partition is given according to $\mathcal{D}_h(1)$. Then, there exists a constant $C_1 > 0$ such that the error estimate*

$$\left\|P^G u_h - P^G y_h\right\|_{H^1(\Omega)} \leq C_1\, h$$

holds for the difference of the solutions u_h of (4.9), (4.10) and y_h of (4.15), (4.16).

[ii] *Let a box partition be constructed according to $\mathcal{D}_h(2)$ and furthermore, let $f \in H^1(\Omega)$, $q \in W^1_\infty(\Omega)$, $g_N = const$ on each boundary face $\sigma \in \sigma_{h,2}$ and $\kappa = 0$ be fulfilled. Then,*

$$\left\|P^G u_h - P^G y_h\right\|_{H^1(\Omega)} \leq C_2\, h^2.$$

Proof. Applying the results of Lemmas 4.9[i], 4.10[i], 4.11, 4.12 and the boundedness of $P^G y_h$ according to Theorem 4.2[i] and Remark 4.3[ii], the first order estimate of part [i] is proved by means of (4.21). The second order estimate of part [ii] is obtained using the statements of Lemmas 4.9[ii], 4.10[ii], 4.11 and Theorem 4.2[i]. ■

It should be emphasized that, according to the statements of Lemmas 4.9–4.12, the assumptions on the coefficients f, q, g_N and κ in the previous theorem can be weakened such that only the piecewise (with respect to the triangulation) smoothness/norms are considered (for instance, instead of $f \in H^1(\Omega)$, one requires only $f \in H^1(S)$ for all $S \in \mathcal{S}_h$).

Remark 4.14 We want to stress that, owing to the comparable accuracy of finite element and box schemes, fast solvers, which were developed mainly for (discrete) finite element variational inequalities, are also well applicable to the finite volume (box) discretization (4.15), (4.16). In this context we want to mention adaptive techniques characterized by multilevel iterative solvers and a posteriori estimators (see the research notes [KORN97] and the papers [HAMI83], [HOPP87], [HOKO94], [EFHKW93], [KORN94], [KORN96]); (overlapping) block relaxation methods and domain decomposition methods (see [TARV94], [KNT94], [HOZO92], [HOZO96]). Note, that the algorithms developed in [HOPP87], [HOKO94], [EFHKW93] rely on an outer-inner iterative scheme, the (rigorous) convergence proof of which needs the M-matrix property (see [HOPP87]). This property is also used intensively in [TARV94] and [KNT94]. In contrast to the finite element matrix defined in (4.11), such a discrete maximum principle is fulfilled by the box matrix $\Lambda_B = L_B + M_B + R_B$, assuming a weakly acute triangulation (cf. Section 4.4.1). □

4.3 Error estimates for the finite volume solution

After the study of box schemes in comparison to the finite element schemes in the previous section, a natural question arises concerning the convergence rate of the solution y_h of the box schemes (4.15), (4.16) compared to the exact solution u of the variational inequality problem (4.1), (4.2).

Obviously, a first way consists in the combination of the estimates for the difference between u and the finite element solution $P^G u_h$ of (4.9) with the results of Theorem 4.13. A second possibility is to derive direct error estimates without considering the finite element schemes (4.9), (4.10) explicitly. Such estimates can be found e.g., in [CAI91], [CMC90], [CMMC91], [MEIS90], [ANG91], [ANG95] and also in the monographs [SLM87], [HEIN87] for second order equations in two-dimensional domains.

Here, we derive estimates for variational inequalities of obstacle type in two and three space dimensions. Since we have solutions $u \in H^2(\Omega)$ in mind, Green's formula can be applied and $\Pi_h u$ as the piecewise linear Lagrange interpolant of u is well defined.

Let us start with the derivation of two a priori estimates for the finite volume approximation (4.15), (4.17). Afterwards, in Theorem 4.16, we evaluate the a priori estimates and, as the main result, we obtain a first order convergence statement. This is also the typical convergence rate for finite element approximations; cf. the references given in Section 4.1.2 and Remark 4.20.

Lemma 4.15 *Let the assumptions (4.4) and (4.5) be satisfied. Let u and $P^G y_h$ be the solutions of problem (4.1), (4.2) and its finite volume approximation (4.15), (4.17). Then, the following a priori estimates hold.*

$$\left.\begin{aligned}&a_B(P^G(y_h - v_h), P^B(y_h - v_h)) \le a(u - P^G v_h, P^G(y_h - v_h))\\&+ \Big[a(u, P^G(v_h - y_h)) - (f, P^G(v_h - y_h)) - (g_N, P^G(v_h - y_h))_{\Gamma_N}\Big]\\&+ \big(f, P^B(y_h - v_h) - P^G(y_h - v_h)\big) + \big(g_N, P^B(y_h - v_h) - P^G(y_h - v_h)\big)_{\Gamma_N}\\&+ a(P^G v_h, P^G(y_h - v_h)) - a_B(P^G v_h, P^B(y_h - v_h)) \quad \forall v_h \in K_h\end{aligned}\right| \tag{4.25}$$

and

$$\left.\begin{aligned}&a_B(P^G(y_h - v_h), P^B(y_h - v_h)) \le a(u - P^G v_h, P^G(y_h - v_h))\\&+ \Big[a(u, v - P^G y_h) - (f, v - P^G y_h) - (g_N, v - P^G y_h)_{\Gamma_N}\Big]\\&+ \Big[a(u, P^G v_h - u) - (f, P^G v_h - u) - (g_N, P^G v_h - u)_{\Gamma_N}\Big]\\&+ \big(f, P^B(y_h - v_h) - P^G(y_h - v_h)\big) + \big(g_N, P^B(y_h - v_h) - P^G(y_h - v_h)\big)_{\Gamma_N}\\&+ a(P^G v_h, P^G(y_h - v_h)) - a_B(P^G v_h, P^B(y_h - v_h)) \quad \forall v \in K,\ \forall v_h \in K_h.\end{aligned}\right| \tag{4.26}$$

Proof. The box inequality problem (4.15) can be rewritten as

$$\begin{aligned}a_B(P^G(y_h - v_h), P^B(y_h - v_h)) \quad &\le (f, P^B(y_h - v_h)) + (g_N, P^B(y_h - v_h))_{\Gamma_N}\\&- a_B(P^G v_h, P^B(y_h - v_h)) \quad \forall v_h \in K_h.\end{aligned}$$

Inserting some terms which are zero in the sum, the estimate (4.25) follows immediately.

To prove (4.26) we consider the continuous inequality formulation (4.1), (4.2) and its finite volume approximation (4.15), (4.18) together. For arbitrary $v \in K$, $v_h \in K_h$ we obtain

$$\begin{aligned}&a_B(P^G(y_h - v_h), P^B(y_h - v_h)) \le \big[(f, P^B(y_h - v_h)) + (g_N, P^B(y_h - v_h))_{\Gamma_N}\\&- a_B(P^G v_h, P^B(y_h - v_h))\big] + \big[a(u, v - u) - (f, v - u) + (g_N, v - u)_{\Gamma_N}\big].\end{aligned}$$

In this inequality we insert

$$a(u, v-u) = a(u, v - P^G y_h) + a(u, P^G v_h - u)$$
$$+ a(u - P^G v_h, P^G(y_h - v_h)) + a(P^G v_h, P^G(y_h - v_h))$$

such that the a priori estimate (4.26) is proved. ■

The essence of our above a priori estimates (4.25), (4.26) is that

$$a(u - P^G v_h, P^G(y_h - v_h))$$

and the terms appearing in the square brackets can be treated similar to finite element a priori estimates, whereas the remaining terms on the right-hand sides are caused by our proposed finite volume approximation.

In a sense, Lemma 4.15 can be interpreted as a generalization of Strang's first lemma (see e.g., [CIAR78], Theorem 4.1.1).

We are now able to formulate the main convergence result for $u - P^G y_h$ measured in the H^1-norm. Due to the assumed nonhomogeneous Dirichlet conditions g_D on Γ_D, we prefer to evaluate the a priori estimate (4.25).

Theorem 4.16 *Let the box partitions be constructed according to $\mathcal{D}_h(1)$. Furthermore, let the assumptions (4.4), (4.5), $\varphi \in H^2(\Omega)$, $l\,\varphi \leq g_N$ a.e. on Γ_N, $k \in W^1_\infty(\Omega)$, $\kappa \in W^1_\infty(\Gamma_N)$ and $g_N \in H^1(\Gamma_N)$ be satisfied. Then, provided the solution u of the obstacle problem (4.1), (4.2) belongs to $H^2(\Omega)$, there exists a constant $C > 0$, such that*

$$\left\| u - P^G y_h \right\|_{H^1(\Omega)} \leq C\, h$$

holds for the difference between u and the solution y_h of the finite volume scheme (4.15), (4.18).

Proof. The results derived in Section 4.2 are used for the treatment of the last four terms in the a priori estimate (4.25). The equality of the box and finite element stiffness matrices (cf. Lemma 4.1[ii] and Remark 4.4) as well as the results of Lemmas 4.9[i], 4.10[i], 4.11 and 4.12 lead to

$$\left.\begin{aligned}
&(f, P^B w_h - P^G w_h) + (g_N, P^B w_h - P^G w_h)_{\Gamma_N} \leq \left\| P^G w_h \right\|_{H^1(\Omega)}\, h \\
&\quad * \Big[C_1\, \|f\|_{L_2(\Omega)} + C_2\, h^{1/2} \Big(\sum_{\sigma \in \sigma_{h,2}} |g_N|^2_{H^1(\sigma)} \Big)^{1/2} \Big], \\
&\left|\, a(P^G v_h, P^G w_h) - a_B(P^G v_h, P^B w_h) \,\right| \leq \left\| P^G v_h \right\|_{H^1(\Omega)} \left\| P^G w_h \right\|_{H^1(\Omega)} h \\
&\quad * \Big[C_3\, \|q\|_{L_\infty(\Omega)} + C_4\, \Big(\|\kappa\|_{L_\infty(\Gamma_N)} + \max_{\sigma \in \sigma_{h,2}} \{ |\kappa|_{W^1_\infty(\sigma)} \} \Big) \Big]
\end{aligned}\right| \qquad (4.27)$$

for grid function vectors $v_h \in L_2(\overline{\omega})$, $w_h \in L_2(\tilde{\omega})$.

We apply Green's formula to the terms in the square brackets of the a priori estimate (4.25). Therefore, we have

$$a(u,w)-(f,w)-(g_N,w)_{\Gamma_N}=\int_\Omega (Au-f)w\ dx+\int_{\Gamma_N}(lu-g_N)w\ ds\quad \forall w\in V$$

(see (4.3) for the definition of A and l). Now we take $w = P^G(v_h - y_h) \in V_h^G \subset V$ with $v_h \in K_h$. Furthermore, we use $l\ u = g_N$ a.e. on Γ_N resulting from the assumption $l\ \varphi \leq g_N$ a.e. on Γ_N (cf. the complementarity problem (4.7)).

We derive the following estimate by means of Theorem 4.2[i], the a priori estimate (4.25) and using (4.27).

$$\left.\begin{aligned}
&C_3 \left\|P^G(y_h-v_h)\right\|^2_{H^1(\Omega)} \leq C_2 \left\|u-P^G v_h\right\|_{H^1(\Omega)} \left\|P^G(y_h-v_h)\right\|_{H^1(\Omega)}\\
&+ (Au-f, P^G(v_h-y_h))\\
&+ \left(M_f + M_a\ \left\|P^G v_h\right\|_{H^1(\Omega)}\right) h\ \left\|P^G(y_h-v_h)\right\|_{H^1(\Omega)}
\end{aligned}\right| \tag{4.28}$$

for all $v_h \in K_h$, where M_f and M_a denote the terms in the square brackets of (4.27).

Similar to a corresponding convergence proof for a (piecewise linear) finite element approximation of (4.1), (4.2) (cf. [GLT81], Appendix 1.3 and [HHNL88], Section 1.2.3), we take

$$K_h \ni v_h = \left((\Pi_h u)(\xi)\right)_{\xi\in\overline{\omega}}$$

with $\Pi_h u$ as the piecewise linear interpolant of the function $u \in H^2(\Omega) \subset C(\overline{\Omega})$. Thus, $P^G v_h = \Pi_h u$ holds and interpolation properties (cf. e.g., [CIAR78], Chapter 3 and [BRSC96]) in Sobolev spaces imply

$$\left\|u-P^G v_h\right\|_{H^1(\Omega)} = \left\|u-\Pi_h u\right\|_{H^1(\Omega)} \leq C_1\ h\ |u|_{H^2(\Omega)}$$

and

$$\left\|u-\Pi_h u\right\|_{L_2(\Omega)} \leq C_2\ h^2\ |u|_{H^2(\Omega)}\,.$$

Exploiting the complementarity problem (4.7), $\Pi_h\varphi \leq P^G y_h$ and using

$$\Pi_h u - P^G y_h = \Pi_h u - u + u - \varphi + \varphi - \Pi_h\varphi + \Pi_h\varphi - P^G y_h,$$

we find

$$\begin{aligned}
&(Au-f, \Pi_h u - P^G y_h) \leq (Au-f, \Pi_h u - u) + (Au-f, \varphi-\Pi_h\varphi)\\
&\leq\ \left\|Au-f\right\|_{L_2(\Omega)} C_2\ h^2\ \left(|u|_{H^2(\Omega)} + |\varphi|_{H^2(\Omega)}\right).
\end{aligned}$$

Summarizing, we arrive at

$$\begin{aligned}
C_3 \left\|P^G y_h - \Pi_h u\right\|^2_{H^1(\Omega)} &\le C\, C_1\, h \left\|P^G y_h - \Pi_h u\right\|_{H^1(\Omega)} \\
&+ \|Au - f\|_{L_2(\Omega)}\, C_2\, h^2 \left(|u|_{H^2(\Omega)}\, |\varphi|_{H^2(\Omega)}\right) \\
&+ \left(M_f + M_a\, \|\Pi_h u\|_{H^1(\Omega)}\right) h \left\|P^G y_h - \Pi_h u\right\|_{H^1(\Omega)}.
\end{aligned}$$

Hence, using Young's inequality and

$$\|\Pi_h u\|_{H^1(\Omega)} \le \|\Pi_h u - u\|_{H^1(\Omega)} + \|u\|_{H^1(\Omega)},$$

we obtain

$$\left\|P^G y_h - \Pi_h u\right\|_{H^1(\Omega)} \le C\, h.$$

Finally, applying the triangle inequality to $P^G y_h - u = P^G y_h - \Pi_h u + \Pi_h u - u$, the proof is completed. ■

Remark 4.17 Instead of (4.25), the a priori estimate (4.26) can be evaluated for homogeneous Dirichlet conditions (or more generally for $g_D = \text{const. on } \Gamma_D$) and in the situation $\partial\Omega = \Gamma_N$. Then, one can take

$$v = \max\{P^G y_h, \varphi\} \in K$$

(cf. [HHNL88], Section 1.2.3 and [CIAR78], Section 5.1). □

The error analysis of finite element and finite volume approximations of elliptic variational inequalities should be completed by some additional remarks.

Remark 4.18 Considering the following simple one-dimensional example, we will point out that the obstacle restricts, on the one hand, the regularity of the solution u of (4.1), (4.2) ($u \in C^{1,1}(\Omega)$) and, on the other hand, the convergence rate ($O(h^{1.5})$) of a finite volume/element discretization.

$$\begin{aligned}
&\text{Find}\quad u \in K = \{w \in H^1(-1,1) : w \ge 0 \text{ in } (-1,1),\quad w(x) = 1/9 \text{ for } x = \pm 1\},\\
&(-u'', v - u) \ge (-2, v - u)\quad \forall v \in K.
\end{aligned}$$

This inequality possesses the exact solution

$$u(x) = \left((|x| - 2/3)^+\right)^2.$$

In this context we recall the following definitions of the coincidence set $\Omega^-(u)$, the open no-coincidence set $\Omega^+(u)$ and the free boundary $\Gamma_f = \Gamma_f(u)$.

$$\left.\begin{aligned}
&\Omega^-(u) = \{x \in \Omega : u(x) = \varphi(x)\},\quad \Omega^+(u) = \{x \in \Omega : u(x) > \varphi(x)\},\\
&\Gamma_f = \partial\Omega^-(u) \cap \Omega.
\end{aligned}\right| \tag{4.29}$$

Obviously, in our example we have $\Omega^-(u) = [-2/3, 2/3]$ and the solution u is smooth with respect to the subintervals $\Omega^-(u)$ and $\Omega^+(u)$.

Now we choose a sequence of uniform triangulations consisting of intervals of the length $h = h_k = 10^{-k}$, $k = 1, 2, \ldots$. From (4.11) or (4.17), reduced to the one-dimensional case, we obtain the usual (finite element/difference) tridiagonal stiffness matrix $L_G = L_B$ and the right-hand side vector $f_{Bh} = f_{Gh}$ with the components $-2h$.

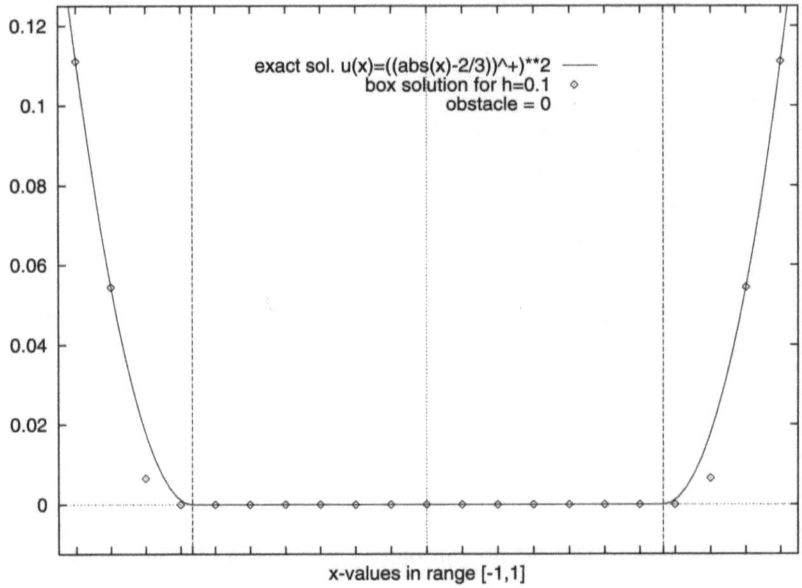

Figure 4.2: *One-dimensional obstacle problem. Solid line: exact solution u. Dots ($\diamond$): discrete (box, finite element) solution y_h for $h = 0.1$.*

Substituting the grid function vectors y_{h_k} defined by

$$y_{h_k,\xi} = \begin{cases} 0 & \text{for} \quad |\xi| \le \xi_0 = \xi_0(h_k) := 2/3 + h_k/3, \\ (|\xi| - \frac{2}{3})^2 + \frac{|\xi| - 1}{3} \frac{h_k^2}{1 - h_k} & \text{for} \quad |\xi| \ge \xi_0 \end{cases}$$

into the equivalent complementarity problem (4.20), we verify that y_{h_k} is the unique solution of the box/finite element inequality. We mention that the points $\xi_0 = \xi_0(h_k)$ are nodes.

Considering now the difference $\Pi_h u - P^G y_h$, depicted in Figure 4.2 for the grid size parameter $h = h_k = 0.1$, we conclude after some calculations that

$$\left|\Pi_h u - P^G y_h\right|_{H^1(\Omega)} > \sqrt{2}/9 \; h^{3/2}$$

for $h = h_k$, $k = 1, 2, \ldots$. □

Remark 4.19 In the report [STEI93] we have studied the convergence behaviour of the box solutions in a *generalized* finite difference context. More precisely, we have chosen circumcentric boxes (cf. Section 4.1.3) for a weakly acute triangulation (cf. Section 4.4.1) in the spatial two-dimensional situation.

The connection to finite differences can be easily seen from

$$\int_{\partial D_\xi \setminus \Gamma} P^G w_h \cdot \vec{n}\, ds = \sum_{\eta \in \omega_\xi} \frac{w_{h,\eta} - w_{h,\xi}}{\text{meas}_1(h_{\xi\eta})}\, \text{meas}_1(s_{\xi\eta}), \quad \omega_\xi \text{ as 'difference star'},$$

where $h_{\xi\eta}$ denotes the face/edge connecting the nodes ξ and η and $s_{\xi\eta}$ is its perpendicular bisector line segment. Assuming a solution $u \in H^2(\Omega)$ we have proved the first order convergence rate $O(h)$ based essentially on local estimates of the interpolation errors for

$$\int_{s_{\xi\eta}} k\, (\nabla u - \Pi_h u) \cdot \vec{n}\, ds \quad \text{and} \quad (\text{meas}_2(D_\xi))^{-1} \int_{D_\xi} u(x)\, dx - u(\xi)$$

as well as on the complementarity problems (4.7), (4.20). But, for this result we need an additional regularity assumption on free boundary Γ_f defined in (4.29). The reason for this is the necessity of a special treatment of the term $\int_{D_\xi}(Au-f)\,dx$ for nodes $\xi \in \tilde{\omega}$ lying in a thin strip along the free boundary Γ_f. Therefore, the general convergence result of Theorem 4.16, including the just mentioned situation of two-dimensional circumcentric boxes, seems to be more valuable.

Let us also mention that in Section 3.4 of the report [STEI93] we have derived the convergence order $O(h^{3/2})$ for finite volume solutions defined by such circumcentric boxes. For this result we need, in addition to the above, a solution u belonging to $H^3(\Omega^+)$, $H^3(\Omega^-)$ and restrictive conditions on the triangulation (similar to those used in [HEIN87] and [CMMC91] for the same convergence order for second order equations).
More precisely, the triangulations have to be uniform in that sense, except that in a subregion $\tilde{\Omega}$ of Ω consisting of a fixed number of strips of the width $O(h)$, the triangulations are formed by equilateral triangles or rectangles of the same shape (cf. Definition 2.9 in [HEIN87] and the definition of so-called essentially symmetric control volumes in [CMMC91]).
The essential point is that for such triangulations the dual circumcentric boxes have a form such that the midpoints of the line segments $s_{\xi\eta}$ and $h_{\xi\eta}$ coincide. In other words, the edge $h_{\xi\eta}$ and its perpendicular bisector segment $s_{\xi\eta}$ form the diagonals of a rhombus which can be mapped onto a reference rhombus by means of an invertible affine transformation. □

Remark 4.20 Using piecewise quadratic finite elements the convergence rate

$$O(h^{3/2-\delta}), \quad \delta > 0$$

is shown in [BHR77]. □

4.4 Penalization methods for solution of the finite volume obstacle problem

The finite volume and finite element approximations (4.15), (4.9) lead to problems of generally nonlinear programming in finite dimension. Therefore, there is a wide variety of techniques for their numerical solution, such as relaxation (SOR with projection), gradient projection, penalization and duality methods (for an overview see [GLT81], Chapter 2 and [GLOW84]). Furthermore, let us recall the close connection of discrete variational inequalities to (finite-dimensional) complementarity problems (cf. (4.20)) which can also be used for the development of algorithms for the numerical solution of obstacle type problems. The theory and application of linear complementarity problems have been studied intensively in [CPS92], including a comprehensive discussion of direct and iterative methods for their solution as well as a historical review.

Based on a discrete maximum principle which is formulated in Section 4.4.1, we will discuss two penalization (regularization) methods for the solution of the finite volume inequality problem. In Section 4.4.2 we study the penalization error and derive comparison results. Combining the penalization error with the convergence rate of the finite volume solution (see Theorem 4.16), we obtain a first order convergence statement for the total error between the solutions of the obstacle problem and the penalization problems (cf. Remark 4.29[i]). Finally, in Section 4.4.3 we are concerned with the iterative solution of the penalization problems.

4.4.1 Discrete maximum principle

An essential tool in the following investigations is a discrete maximum principle which is an important reason to consider box schemes. Hence, besides the lumped matrices M_B and R_B introduced in Section 4.1.3, cf. (4.17), we require that the triangulation be of *weakly acute type*, which means that for $n = 2$ all angles of the triangles $S \in \mathcal{S}_h$ (for $n = 3$: all dihedral angles of faces) are less than or equal to $\pi/2$. In other words, there should be satisfied

$$\nabla\lambda_i \cdot \nabla\lambda_j \leq 0 \quad \text{for any} \quad S \in \mathcal{S}_h, \quad i, j = 1, \ldots, n+1, \ i \neq j,$$

where λ_i are the barycentric coordinates of the simplex S with the vertices ξ_i for $i = 1, \ldots, n+1$. Note in this context that the coefficients of the local element stiffness matrix $l^S = (l^S_{ij})_{i,j=1}^{n+1}$ for a simplex $S \in \mathcal{S}_h$ are given by

$$l^S_{ij} = \text{meas}_n(S)\, |\nabla\lambda_i|\, |\nabla\lambda_j|\, (-\cos\alpha^S_{ij}), \quad i \neq j \quad \text{and} \quad l_{ii} = \text{meas}_n(S)\, |\nabla\lambda_i|^2.$$

Here, α^S_{ij} is the (dihedral) angle between the faces described by $\lambda_i = 0$ and $\lambda_j = 0$. Consequently, all the off-diagonal entries of the matrix $L_B = L_G$ are non-positive for a weakly acute triangulation.

In connection with Theorem 4.2 (ellipticity of the box bilinear form a_B) this property is used to derive a discrete maximum principle of a form that will give

$$\left.\begin{aligned} & C_1 \left\| P^G(w_h^-) \right\|_{H^1(\Omega)}^2 \le C_2 \left\| w_h^- \right\|_{1,\overline{\omega}}^2 \le a_B(P^G(w_h^-), P^B(w_h^-)) \\ & = \langle \Lambda_B \, w_h^-, w_h^- \rangle = \langle \Lambda_B \, (w_h - w_h^+), w_h^- \rangle \\ & \le \langle \Lambda_B \, w_h, w_h^- \rangle = \langle (L_G + M_B + R_B) \, w_h, w_h^- \rangle \end{aligned}\right| \tag{4.30}$$

for grid function vectors $w_h \in L_2(\overline{\omega})$ with $w_h^- \in L_2(\tilde{\omega})$ (i.e., $w_{h,\xi} \ge 0$ for Dirichlet nodes $\xi \in \gamma_D$).

The positive and negative parts of a grid function vector w_h are denoted by

$$w_h^+ = \max\{w_h, 0\} \quad \text{and} \quad w_h^- = \min\{w_h, 0\}$$

(componentwise).

The application of this discrete maximum principle to the box obstacle problem (4.15), (4.16) leads to the following comparison result.

Theorem 4.21 *Let the box partitions be constructed according to $\mathcal{D}_h(1)$ and the assumptions (4.4), (4.5) be fulfilled. Furthermore, let $y_h = y_h(f_{Bh}, g_{Bh}, \varphi_{Wh})$ and $\hat{y}_h = \hat{y}_h(\hat{f}_{Bh}, \hat{g}_{Bh}, \hat{\varphi}_{Wh})$ be the solutions of the box inequality problem (4.16), (4.17), where the grid function vectors φ_{Wh} and $\hat{\varphi}_{Wh}$ are defined as in Theorem 4.2[ii]. Then, assuming a weakly acute triangulation and*

$$f_{Bh} \ge \hat{f}_{Bh}, \quad g_{Bh} \ge \hat{g}_{Bh}, \quad \varphi_{Wh} \ge \hat{\varphi}_{Wh} \quad \forall \xi \in \overline{\omega},$$

a discrete maximum principle holds in the sense that $y_h \ge \hat{y}_h$ for all $\xi \in \overline{\omega}$.

Proof. In order to verify the comparison statement we choose

$$K_h \ni v_h = y_h - (y_h - \hat{y}_h)^- \quad (\text{resp.} \quad \hat{K}_h \ni v_h = \hat{y}_h + (y_h - \hat{y}_h)^-)$$

in the inequality (4.16) related to y_h (resp. to $\hat{y}_h$). Adding both inequalities and using (4.30), we get $\|(y_h - \hat{y}_h)^-\|_{1,\overline{\omega}} = 0$, which implies the relation $y_h \ge \hat{y}_h$. ■

4.4.2 Discussion of penalization techniques

We refer to [GLOW84], [GLT81] and [HANE96] for the general concept of penalization techniques as approximations for variational inequalities. Assuming a symmetric bilinear form, the variational inequality as a minimization problem with constraints is replaced by an unconstrained optimization problem. In other words, if the chosen penalization functional is differentiable, the variational inequality (of first kind) is embedded into a family of nonlinear (penalization) variational equations. However, penalization problems can be formulated without consideration of the associated minimization problem. To be more precise, one introduces a monotone, Lipschitz continuous penalization operator, the kernel of which coincides with the convex set K. Furthermore, having a mixed (Lagrange) formulation for the variational inequality, penalization techniques can be interpreted as regularization of such formulations, cf. e.g., [GRRO94].

Penalty finite element approximations for obstacle problems are studied e.g., in [KIKU81], [SCHO84], [SCHO86] and [STRE89]. In particular, in the last three papers the (original) variational inequality is approximated by penalization problems. Then, these nonlinear differential equations are discretized by finite element penalty problems with a penalization term similar to that in (4.31).

Penalization approximations which are more connected with box schemes are studied in the paper [ADGR95]. Using a two-dimensional weakly acute triangulation with the inverse property (cf. Remark 4.26 below and [CIAR78], p. 140), the authors considered the Poisson problem with homogeneous Dirichlet conditions. The dual partition is given by circumcentric boxes. Considering problems with the penalization functionals

$$j_\varepsilon(v_h) = \varepsilon^{-1}((\varphi_h - v_h)^+)^\alpha, \quad \alpha \geq 1$$

(our approach (4.31) corresponds to $\alpha = 2$) as approximations of a (discrete) Lagrange formulation, the authors derive an L_∞-estimate for the difference between the solutions of the discrete obstacle and penalization problem in this paper. We remark that Theorem 4.27 given below is a generalization of this result (for $\alpha = 2$) obtained by a different argumentation and avoiding the inverse property of the triangulation.

In what follows we want to discuss the approximation of the box inequality (4.15), (4.16) by two different penalization (regularization) methods. We study the relations between the obstacle problem and the penalization equations directly without an explicit consideration of the related minimization and mixed (Lagrange) formulation (see, however, Remark 4.25).

We consider the following problem as an application of the just mentioned general penalization concept to the considered box obstacle problem.

$$\left.\begin{array}{l} \text{Find} \quad y_{h\varepsilon} \in L_2(\overline{\omega}): \quad y_{h\varepsilon,\xi} = g_D(\xi) \quad \text{for } \xi \in \gamma_D, \\ \langle \Lambda_B \, y_{h\varepsilon}, v_h \rangle + \langle \varepsilon^{-1} \, \beta_{1h}(y_{h\varepsilon} - \varphi_h), v_h \rangle = \langle f_{Bh} + g_{Bh}, v_h \rangle \\ \forall v_h \in L_2(\tilde{\omega}) \quad \text{with} \quad \varepsilon > 0 \quad \text{and} \\ \left(\beta_{1h}(w_h)\right)_{\xi \in \tilde{\omega}} = \left(\text{meas}_n(D_\xi) \; (w_{h,\xi})^-\right)_{\xi \in \tilde{\omega}}. \end{array} \right| \tag{4.31}$$

We want to emphasize that for this penalization method we will discuss the usage of penalization parameters $\varepsilon = \varepsilon_\xi$ depending on the nodes $\xi \in \tilde{\omega}$ of the triangulation only at some places (see Remark 4.26, Theorem 4.27 and also Section 5.3.1 below).

Those subsequent statements which, for the sake of brevity, will not directly be formulated for space-dependent penalization parameters, can be transferred to this situation.

In contrast to the penalization equations (4.31), in which the condition that the solution belongs to K_h is replaced by penalization terms $\varepsilon^{-1}\beta_{1h}(y_{h\varepsilon}-\varphi_h)$ becoming progressively larger as the solution moves away from K_h, the solution of the following regularization problem preserves the constraints (see Theorem 4.24).

$$\left|\begin{array}{l} \text{Find} \quad z_{h\varepsilon} \in L_2(\overline{\omega}): \quad z_{h\varepsilon,\xi} = g_D(\xi) \quad \text{for } \xi \in \gamma_D, \\ \langle \Lambda_B\, z_{h\varepsilon}, v_h\rangle + \langle \mu_h\, \beta_{2h}(\frac{z_{h\varepsilon}-\varphi_h}{\varepsilon}), v_h\rangle = \langle f_{Bh} + g_{Bh} + \mu_h, v_h\rangle \\ \forall v_h \in L_2(\tilde{\omega}) \quad \text{with} \quad \varepsilon > 0, \end{array}\right. \tag{4.32}$$

where the grid function vectors $\beta_{2h}(w_h)$ and μ_h are defined by

$$\big(\beta_{2h}(w_h)\big)_{\xi\in\tilde{\omega}} = \big(w_{h,\xi}^+/(1+w_{h,\xi})\big)_{\xi\in\tilde{\omega}} \quad \text{and} \quad \mu_h = \big(\Lambda_B\,\varphi_h - f_{Bh} - g_{Bh}\big)^+.$$

Remark 4.22 In [BELI84], Section 2.5 and [RODR87], Chapter 5, regularization methods of Lewy–Stampacchia type (similar to (4.32)) are used for the derivation of regularity results for the solution of obstacle problems.

Concerning an interpretation of problem (4.32) we refer to Remark 3.7. We stress that other penalization terms $\beta_{2h}(w_h)$ can be constructed based on a nondecreasing Lipschitz function $\beta : \mathbb{R} \longrightarrow [0,1]$ with $\beta(r) = 0$ for $r \le 0$, $\beta(+\infty) = 1$ and $(1-\beta(r))\, r \le C$ (cf. Remark 3.7). □

We emphasize that throughout this and the next Section 4.4.3 we will work with the assumptions made in Theorem 4.21, i.e., we assume a weakly acute triangulation with a box partition according to $\mathcal{D}_h(1)$ and the standard conditions (4.4), (4.5).

Taking into consideration the monotonicity of the penalization terms $\beta_{1h}(w_h)$ and $\beta_{2h}(w_h)$,

$$\begin{array}{l} \langle \beta_{1h}(v_h) - \beta_{1h}(w_h), v_h - w_h\rangle \ge \|(v_h)^- - (w_h)^-\|_{0,\overline{\omega}}^2 \ge 0 \quad \text{and} \\ \langle \beta_{2h}(v_h) - \beta_{2h}(w_h), v_h - w_h\rangle \ge 0 \quad \forall v_h, w_h \in L_2(\tilde{\omega}), \end{array}$$

we obtain stability and comparison statements for the solutions of the box penalization problems (4.31), (4.32) similar to those of Theorems 4.2[i], [ii] and 4.21.

Considering the variational inequality (4.16) together with the penalization problems (4.31) and (4.32), we state the following comparison results.

Theorem 4.23 *Under the assumptions of Theorem 4.21 the solutions of (4.16), (4.31) and (4.32) fulfill*

$$y_{h\varepsilon} \leq y_h \leq z_{h\varepsilon}.$$

The problem (4.31) defines a monotone increasing sequence $y_{h\varepsilon}$*, whereas problem (4.32) defines a monotone decreasing sequence* $z_{h\varepsilon}$ *with respect to* $\varepsilon \longrightarrow 0$.

Proof. To verify the first inequality we take $K_h \ni v_h = y_h - (y_h - y_{h\varepsilon})^-$ in (4.16) and $v_h = (y_h - y_{h\varepsilon})^- \in L_2(\tilde{\omega})$ in (4.31). Adding both and using (4.30), we obtain

$$C \;\big\|(y_h - y_{h\varepsilon})^-\big\|^2_{1,\overline{\omega}} \leq \varepsilon^{-1}\big\langle \beta_{1h}(y_{h\varepsilon} - \varphi_h), (y_h - y_{h\varepsilon})^-\big\rangle = 0.$$

Hence, $y_h \geq y_{h\varepsilon}$ is shown.

Taking $K_h \ni v_h = y_h + (z_{h\varepsilon} - y_h)^-$ in (4.16) and $v_h = -(z_{h\varepsilon} - y_h)^- \in L_2(\tilde{\omega})$ in (4.32), the second inequality will be concluded. Note that we use $z_{h\varepsilon} \in K_h$ already here. It will be proved below in Theorem 4.24. Then, we derive

$$\big\langle \Lambda_B(z_{h\varepsilon} - y_h), (z_{h\varepsilon} - y_h)^-\big\rangle \leq \big\langle \mu_h\,\big[1 - \beta_{2h}((z_{h\varepsilon} - \varphi_h)/\varepsilon)\big], (z_{h\varepsilon} - y_h)^-\big\rangle \leq 0,$$

which shows the relation $y_h \leq z_{h\varepsilon}$ after the application of (4.30).

Let $y_{h\varepsilon}$ and $y_{h\hat{\varepsilon}}$ be the solutions of (4.31) for $\varepsilon > \hat{\varepsilon}$. Then, we have

$$\Lambda_B(y_{h\hat{\varepsilon}} - y_{h\varepsilon}) + \varepsilon^{-1}(\beta_{1h}(y_{h\hat{\varepsilon}} - \varphi_h) - \beta_{1h}(y_{h\varepsilon} - \varphi_h)) = (\varepsilon^{-1} - \hat{\varepsilon}^{-1})\,\beta_{1h}(y_{h\hat{\varepsilon}} - \varphi_h) \geq 0.$$

Therefore, after the multiplication by $(y_{h\hat{\varepsilon}} - y_{h\varepsilon})^- \leq 0$, we verify $y_{h\hat{\varepsilon}} \geq y_{h\varepsilon}$ for $\hat{\varepsilon} < \varepsilon$.

Finally, let $z_{h\varepsilon}$ and $z_{h\hat{\varepsilon}}$ be the solutions of (4.32) for $\varepsilon > \hat{\varepsilon}$. By subtraction we get

$$\begin{aligned}
&\Lambda_B(z_{h\varepsilon} - z_{h\hat{\varepsilon}}) + \big[\beta_{2h}(\frac{z_{h\varepsilon} - \varphi_h}{\varepsilon}) - \beta_{2h}(\frac{z_{h\hat{\varepsilon}} - \varphi_h}{\varepsilon})\big] \\
&= \mu_h\big[\beta_{2h}(\frac{z_{h\hat{\varepsilon}} - \varphi_h}{\hat{\varepsilon}}) - \beta_{2h}(\frac{z_{h\hat{\varepsilon}} - \varphi_h}{\varepsilon})\big] \geq 0
\end{aligned}$$

which, multiplied by $(z_{h\varepsilon} - z_{h\hat{\varepsilon}})^- \leq 0$, gives the last statement $z_{h\hat{\varepsilon}} \leq z_{h\varepsilon}$ for $\hat{\varepsilon} < \varepsilon$. ■

In the next theorem we shall investigate the passage to the limit as $\varepsilon \longrightarrow 0$ in the penalization problems (4.31), (4.32).

Theorem 4.24 *Let the assumptions of Theorem 4.21 be satisfied.*
[i] *The solutions* y_h *of (4.16) and* $y_{h\varepsilon}$ *of (4.31) fulfill the estimates*

$$\begin{aligned}
&\big\|(y_{h\varepsilon} - \varphi_h)^-\big\|_{0,\overline{\omega}} \leq M_{\varphi_h}\varepsilon, \quad \big\|(y_{h\varepsilon} - \varphi_h)^-\big\|_{1,\overline{\omega}} \leq C_1\; M_{\varphi_h}\sqrt{\varepsilon}, \\
&\|y_h - y_{h\varepsilon}\|_{1,\overline{\omega}} \leq C_2\; M_{\varphi_h}\sqrt{\varepsilon}
\end{aligned}$$

with

$$\begin{aligned} M_{\varphi_h} &= \|(\Lambda_B\, \varphi_h - f_{Bh} - g_{Bh})/\operatorname{meas}_n(D)\|_{0,\overline{\omega}} \\ &= \left\|P^B\big((\Lambda_B\, \varphi_h - f_{Bh} - g_{Bh})/\operatorname{meas}_n(D)\big)\right\|_{L_2(\Omega)}. \end{aligned}$$

[ii] *The solutions* y_h *of (4.16) and* $z_{h\varepsilon}$ *of (4.32) satisfy* $z_{h\varepsilon} \in K_h$ *and*

$$\|y_h - z_{h\varepsilon}\|_{1,\overline{\omega}} \leq C_3\, \sqrt{\varepsilon}\, \left[\langle \mu_h, 1\rangle\right]^{1/2} \leq C_4\, \sqrt{\varepsilon}\, \|\mu_h/\operatorname{meas}_n(D)\|_{0,\overline{\omega}}^{1/2}.$$

Proof. *[i].* Taking $v_h = (y_{h\varepsilon} - \varphi_h)^-$ in (4.31) we get

$$\begin{aligned} &\langle \Lambda_B(y_{h\varepsilon} - \varphi_h), (y_{h\varepsilon} - \varphi_h)^-\rangle + \varepsilon^{-1} \left\|(y_{h\varepsilon} - \varphi_h)^-\right\|_{0,\overline{\omega}}^2 \\ &= \langle f_{Bh} + g_{Bh} - \Lambda_B \varphi_h, (y_{h\varepsilon} - \varphi_h)^-\rangle. \end{aligned}$$

Hence, owing to (4.30) the first two statements are obtained.
Now, choosing the representation

$$y_h - y_{h\varepsilon} = r_{h\varepsilon} - (y_{h\varepsilon} - \varphi_h)^-,$$

the proof of the last assertion of part [i] amounts to the estimation of the term $r_{h\varepsilon}$. For this we take

$$v_h = y_h - r_{h\varepsilon} = (y_{h\varepsilon} - (y_{h\varepsilon} - \varphi_h)^-) \in K_h \quad \text{in (4.16) and} \quad v_h = r_{h\varepsilon} \quad \text{in (4.31)},$$

which leads, by subtraction, to

$$\langle \Lambda_B\, (y_{h\varepsilon} - y_h), r_{h\varepsilon}\rangle + \varepsilon^{-1}\langle \beta_{1h}(y_{h\varepsilon} - \varphi_h), r_{h\varepsilon}\rangle \geq 0.$$

Now, taking into account $\langle \beta_{1h}(y_{h\varepsilon} - \varphi_h), r_{h\varepsilon}\rangle \leq 0$ and recalling Theorem 4.2[i], we deduce

$$\begin{aligned} &\langle \Lambda_B\, (y_h - y_{h\varepsilon}), y_h - y_{h\varepsilon}\rangle \leq \langle \Lambda_B(y_{h\varepsilon} - y_h), (y_{h\varepsilon} - \varphi_h)^-\rangle \\ &\leq C\, \|y_h - y_{h\varepsilon}\|_{1,\overline{\omega}}\, \left\|(y_{h\varepsilon} - \varphi_h)^-\right\|_{1,\overline{\omega}}. \end{aligned}$$

[ii]. At first we verify $z_{h\varepsilon} \geq \varphi_h$ for all $\xi \in \overline{\omega}$. Taking $v_h = (z_{h\varepsilon} - \varphi_h)^-$ in (4.32) we get

$$\begin{aligned} &\langle \Lambda_B\, (z_{h\varepsilon} - \varphi_h), (z_{h\varepsilon} - \varphi_h)^-\rangle + \langle \mu_h\, \frac{(z_{h\varepsilon} - \varphi_h)^+}{z_{h\varepsilon} - \varphi_h + \varepsilon},\, (z_{h\varepsilon} - \varphi_h)^-\rangle \\ &= \langle f_{Bh} + g_{Bh} - \Lambda_B\, \varphi_h + \mu_h,\, (z_{h\varepsilon} - \varphi_h)^-\rangle \leq 0. \end{aligned}$$

Owing to $(z_{h\varepsilon} - \varphi_h)^+(z_{h\varepsilon} - \varphi_h)^- = 0$ and (4.30), the relation $z_{h\varepsilon} \geq \varphi_h$ is guaranteed.

In order to conclude the estimate of the penalization error, we put $v_h = y_h - z_{h\varepsilon}$ in (4.32) and $v_h = z_{h\varepsilon} \in K_h$ in (4.16). Consequently, adding both we obtain

$$\langle \Lambda_B\ (z_{h\varepsilon} - y_h), y_h - z_{h\varepsilon} \rangle + \langle \mu_h\ \beta_{2h}((z_{h\varepsilon} - \varphi_h)/\varepsilon),\ y_h - z_{h\varepsilon} \rangle \geq \langle \mu_h, y_h - z_{h\varepsilon} \rangle.$$

Now, using $(z_{h\varepsilon} - \varphi_h)^+ = z_{h\varepsilon} - \varphi_h$, we derive

$$\begin{aligned} C\ \|y_h - z_{h\varepsilon}\|_{1,\overline{\omega}}^2 &\leq \langle \mu_h\ [\beta_{2h}(\frac{z_{h\varepsilon} - \varphi_h}{\varepsilon}) - 1]\ ,\ y_h - z_{h\varepsilon} \rangle \\ &= \varepsilon\ \langle \mu_h,\ \frac{z_{h\varepsilon} - y_h}{z_{h\varepsilon} - \varphi_h + \varepsilon} \rangle \leq \varepsilon\ \langle \mu_h,\ \frac{z_{h\varepsilon} - \varphi_h}{z_{h\varepsilon} - \varphi_h + \varepsilon} \rangle \leq\ \varepsilon\ \langle \mu_h, 1 \rangle \\ &\leq\ \varepsilon\ (\operatorname{meas}_n(\Omega))^{1/2}\ \|\mu_h / \operatorname{meas}_n(D)\|_{0,\overline{\omega}}\,. \end{aligned}$$

Hence, all statements are proved. ■

Remark 4.25 Let us briefly mention the close connection of the penalization terms in (4.31), (4.32) to the Lagrange multiplier of a (primal) saddle point (mixed) formulation of the box obstacle problem (4.16). Just for the moment let us assume homogeneous Dirichlet conditions on γ_D (after a transformation). Then, recalling the complementarity relations (4.20), it can be shown that the pair (y_h, p_h) forms the unique saddle point of the function

$$\mathcal{L}_{Bh}(v_h, q_h) = 0.5\ \langle \Lambda_B\ v_h, v_h \rangle - \langle f_{Bh} + g_{Bh}, v_h \rangle + \langle d_h(q_h), \varphi_h - v_h \rangle$$

('Lagrangian') with $(d_h(q_h))_{\xi\in\tilde{\omega}} = (\operatorname{meas}_n(D_\xi)\ q_{h,\xi})_{\xi\in\tilde{\omega}}$. Here, y_h is the solution of (4.16) and the grid function vector (Lagrange multiplier) p_h is given by

$$0 \leq p_h = (\Lambda_B\ y_h - f_{Bh} - g_{Bh}) / \operatorname{meas}_n(D).$$

In other words, the pair (y_h, p_h) solves the corresponding Kuhn–Tucker system. Moreover, the penalization terms $\varepsilon^{-1}\beta_{1h}(y_{h\varepsilon} - \varphi_h)$ and $\mu_h(\beta_{2h}((z_{h\varepsilon} - \varphi_h)/\varepsilon) - 1)$, respectively, converge for $\varepsilon \longrightarrow 0$ to $d_h(p_h)$ (Lagrange multipliers weighted with the box area). □

Let us discuss some modifications of the error estimates given in Theorem 4.24[i] for the penalization problem (4.31).

Remark 4.26 Using an inverse estimate of the form

$$|P^G v_h|_{H^1(\Omega)}^2 \leq C\ h^{-2}\ \|P^G v_h\|_{L_2(\Omega)}^2$$

(cf. [CIAR78], Section 3.2 and [BRSC96], Section 4.5) in the proof of part [i] of Theorem 4.24, the penalization error can be expressed by

$$\|(y_{h\varepsilon} - \varphi_h)^-\|_{1,\overline{\omega}} \leq c_1\ M_{\varphi_h}\ \frac{\varepsilon}{h} \quad \text{and} \quad \|y_h - y_{h\varepsilon}\|_{1,\overline{\omega}} \leq c_2\ M_{\varphi_h}\ \frac{\varepsilon}{h}$$

(note Lemma 4.1[i]). But we recall that the application of such a global (with respect to Ω) inverse estimate requires an inverse assumption on the triangulation of the form

$$h/h_S \leq \nu \quad \forall S \in \mathcal{S}_h$$

(see [CIAR78], p. 140).

In order to work without this global assumption and, instead of this, to exploit a local inverse relation of the form

$$\left|P^G v_h\right|^2_{H^1(S)} \leq C\ h_S^{-2}\ \left\|P^G v_h\right\|^2_{L_2(S)} \quad \forall S \in \mathcal{S}_h,$$

we propose to use penalization parameters $\varepsilon = \varepsilon_\xi$ depending on the nodes $\xi \in \tilde{\omega}$, e.g.,

$$\varepsilon_\xi = \breve{\varepsilon} \min_{S \in \mathcal{S}_h : S \subset \Omega_\xi} h_S^2.$$

Using $\|(y_{h\varepsilon} - \varphi_h)^-\|_{0,\overline{\omega}} \leq M_{\varphi_h} \max_{\xi \in \tilde{\omega}} \varepsilon_\xi$ (cf. the proof of Theorem 4.24[i]) and the results of Lemma 4.1, this choice leads, after some calculations, to

$$\left|(y_{h\varepsilon} - \varphi_h)^-\right|^2_{1,\overline{\omega}} = \left|P^G((y_{h\varepsilon} - \varphi_h)^-)\right|^2_{H^1(\Omega)} \leq C\ M^2_{\varphi_h}\ \breve{\varepsilon}\ \max_{\xi \in \tilde{\omega}} \varepsilon_\xi.$$

Moreover, we obtain

$$\left\|(y_{h\varepsilon} - \varphi_h)^-\right\|_{1,\overline{\omega}} \leq c_3\ M_{\varphi_h}\ \breve{\varepsilon}\ h \quad \text{and} \quad \|y_h - y_{h\varepsilon}\|_{1,\overline{\omega}} \leq c_4\ M_{\varphi_h}\ \breve{\varepsilon}\ h.$$

Let us anticipate that we return to such a choice of the penalization parameters for the solution of the evolutionary box inequalities in Section 5.3.1 (see Theorem 5.21[ii]). □

In the following theorem we consider the penalization error measured in the maximum norm.

Theorem 4.27 *Under the assumptions of Theorem 4.21 the estimate*

$$\begin{aligned}
&\left\|P^G(y_h - y_{h\varepsilon})\right\|_{L_\infty(\Omega)} \leq \max_{\xi \in \overline{\omega}} (y_{h,\xi} - y_{h\varepsilon,\xi}) \\
&\leq \varepsilon \max_{\xi \in \tilde{\omega}} \frac{|(\Lambda_B\ \varphi_h - f_{Bh} - g_{Bh})_\xi|}{\text{meas}_n(D_\xi)} \\
&\leq \varepsilon\ \left\|P^B((\Lambda_B\ \varphi_h - f_{Bh} - g_{Bh})/\,\text{meas}_n(D))\right\|_{L_\infty(\Omega)}
\end{aligned}$$

holds for the solutions y_h *of (4.16) and* $y_{h\varepsilon}$ *of (4.31).*

Proof. In order to verify the L_∞-error bound we use the notation $w_h = y_h - y_{h\varepsilon}$ and assume that $w_h \not\equiv 0$. From Theorem 4.23 we have $w_h \geq 0$. By means of the (discrete) complementarity problem (4.20) we conclude

$$(\Lambda_B\ w_h)_\xi = \varepsilon^{-1}\ \beta_{1h}(y_{h\varepsilon,\xi} - \varphi_{h,\xi}) \leq 0 \quad \text{for nodes } \xi \in \tilde{\omega} \quad \text{with} \quad y_{h,\xi} > \varphi_{h,\xi}.$$

Let the set $\omega_{\max}$ of nodes be defined by

$$\omega_{\max} = \{\xi \in \tilde{\omega} : w_{h,\xi} = \max_{\eta \in \tilde{\omega}} w_{h,\eta}\}.$$

There exists at least one node $\eta \in \omega_{\max}$ with $(\Lambda_B w_h)_\eta > 0$. Hence, from above it follows that $y_{h,\eta} = \varphi_{h,\eta}$ for this node $\eta \in \omega_{\max}$ and furthermore

$$0 < (\Lambda_B\, w_h)_\eta = (\Lambda_B\, y_h - f_{Bh} - g_{Bh})_\eta - \varepsilon^{-1} \operatorname{meas}_n(D_\eta)\ (y_h - y_{h\varepsilon})_\eta.$$

Using $y_{h,\xi} - \varphi_{h,\xi} \geq 0$ for all $\xi \in \omega_\eta$ and the non-positivity of the off-diagonal elements of Λ_B, we obtain

$$(y_h - y_{h\varepsilon})_\eta \leq \varepsilon\ (\operatorname{meas}_n(D_\eta))^{-1}\ (\Lambda_B\, \varphi_h - f_{Bh} - g_{Bh})_\eta.$$

Consequently, the statement is proved. ∎

We emphasize that the just established L_∞-bound for the penalization error implies an error bound of the form

$$\max_{\xi \in \overline{\omega}} (y_{h,\xi} - y_{h\varepsilon,\xi}) \leq \breve{\varepsilon} \quad \text{for} \quad 0 \leq \varepsilon^{-1} := \varepsilon_\xi^{-1} = \breve{\varepsilon}^{-1}\, \frac{|(\Lambda_B\, \varphi_h - f_{Bh} - g_{Bh})_\xi|}{\operatorname{meas}_n(D_\xi)},$$

such that the penalization error can be controlled pointwise.

Remark 4.28 Let us consider the terms M_{φ_h} and μ_h appearing in Theorem 4.24 and the maximum norm

$$\overline{M}_{\varphi_h} := \left\| P^B\big((\Lambda_B \varphi_h - f_{Bh} - g_{Bh})/\operatorname{meas}_n(D)\big)\right\|_{L_\infty(\Omega)}$$

occuring in Theorem 4.27. We consider an obstacle

$$\varphi \in H^2(\Omega) \quad \text{such that} \quad l\varphi \in L_2(\partial\Omega).$$

In general, one can *not* assume that the obstacle φ fulfills $A\varphi = f$ in Ω and $l\varphi = g_N$ on Γ_N (cf. (4.3)). Therefore, applying Green's formula and recalling the representation

$$a_B(\Pi_h\varphi, P^B z_h) - (f, P^B z_h) - (g_N, P^B z_h)_{\Gamma_N} = \langle \Lambda_B\, \varphi_h - f_{Bh} - g_{Bh}, z_h \rangle$$

for grid function vectors $z_h \in L_2(\tilde{\omega})$, the following error scheme is obtained.

$$\left.\begin{aligned} &(A\varphi - f, P^G z_h) + (l\varphi - g_N, P^G z_h)_{\Gamma_N} \\ &= \sum_{i=1}^{3} \Psi_i(z_h) + \langle \Lambda_B \varphi_h - f_{Bh} - g_{Bh}, z_h \rangle. \end{aligned}\right| \tag{4.33}$$

Here, the error functionals $\Psi_i(z_h)$ are given by

$$\begin{aligned}\Psi_1(z_h) &= a(\varphi - \Pi_h\varphi, P^G z_h),\\ \Psi_2(z_h) &= a(\Pi_h\varphi, P^G z_h) - a_B(\Pi_h\varphi, P^B z_h),\\ \Psi_3(z_h) &= (f, P^B z_h - P^G z_h) + (g_N, P^B z_h - P^G z_h)_{\Gamma_N}.\end{aligned} \tag{4.34}$$

For the estimation of $\overline{M}_{\varphi_h}$ and the terms

$$M_{\varphi_h}^2 = \sum_{\xi\in\tilde\omega} \frac{\left((\Lambda_B\, \varphi_h - f_{Bh} - g_{Bh})_\xi\right)^2}{\mathrm{meas}_n(D_\xi)}, \qquad \langle \mu_h, 1\rangle \le \sum_{\xi\in\tilde\omega} |(\Lambda_B\, \varphi_h - f_{Bh} - g_{Bh})_\xi|,$$

we require local estimates for $\Psi_i(z_h)$, $i = 1,2,3$. Therefore, we introduce the notation e_h^ξ as unit grid function vector associated with a node $\xi \in \tilde\omega$, i.e., the components of the vector $(e_h^\xi)_{\eta\in\tilde\omega}$ are defined by $e_{h,\eta}^\xi = \delta_{\xi\eta}$ (Kronecker symbol). The support of $P^G e_h^\xi = \Phi_\xi^G$ (nodal basis function) coincides with Ω_ξ, whereas the support of $P^B e_h^\xi = \Phi_\xi^B$ is the box $D_\xi \subset \Omega_\xi$. Then, we derive the following local estimates.

$$\begin{aligned}\Psi_1(e_h^\xi) \;&\le \Big[C_1 \,\max\{\|k\|_{L_\infty(\Omega)}, \|q\|_{L_\infty(\Omega)}\}\\ &\quad * \Big(\sum_{S\in\mathcal{S}_h:S\subset\Omega_\xi} h_S^2\,|\Phi_\xi^G|^2_{H^1(S)} + h_S^4\,\|\Phi_\xi^G\|^2_{L_2(S)}\Big)^{1/2}\\ &\quad + C_2\,\|\kappa\|_{L_\infty(\Gamma_N)}\Big(\sum_{S\in\mathcal{S}_h:S\subset\Omega_\xi} h_S^2\,\|\Phi_\xi^G\|^2_{H^1(S)}\Big)^{1/2}\Big]\, |\varphi|_{H^2(\Omega_\xi)},\\ \Psi_2(e_h^\xi) \;&\le (C_3\,\|q\|_{L_\infty(\Omega)} + C_4\,\|\kappa\|_{L_\infty(\Gamma_N)} + C_5 \max_{\sigma\in\sigma_{h,2}:\sigma\subset\partial\Omega_\xi}\{|\kappa|_{W^1_\infty(\sigma)}\})\\ &\quad * \Big(\sum_{S\in\mathcal{S}_h:S\subset\Omega_\xi} h_S^2\,\|\Phi_\xi^G\|^2_{H^1(S)}\Big)^{1/2} \|\Pi_h\varphi\|_{H^1(\Omega_\xi)},\\ \Psi_3(e_h^\xi) \;&\le C_6\,\|f\|_{L_2(\Omega_\xi)}\Big(\sum_{S\in\mathcal{S}_h:S\subset\Omega_\xi} h_S^2\,|\Phi_\xi^G|^2_{H^1(S)}\Big)^{1/2}\\ &\quad + C_7\,|g_N|_{H^1(\partial\Omega_\xi\cap\Gamma_N)}\Big(\sum_{S\in\mathcal{S}_h:S\subset\Omega_\xi} h_S^3\,|\Phi_\xi^G|^2_{H^1(S)}\Big)^{1/2}.\end{aligned} \tag{4.35}$$

We mention that these estimates hold for nodes $\xi \in \gamma_N$, whereas for interior nodes $\xi \in \omega$ the terms regarding to Γ_N do not appear. Furthermore, we note that the relevant norms of $P^G e_h^\xi = \Phi_\xi^G$ occuring in (4.35) can be estimated by

$$\sum_{S\in\mathcal{S}_h:S\subset\Omega_\xi} h_S^2\,\|\Phi_\xi^G\|^2_{H^1(S)} \le C\,\mathrm{meas}_n(\Omega_\xi).$$

The estimate for $\Psi_1(e_h^\xi)$ is obtained by the application of local estimates for interpolation errors (see e.g., [CIAR78], Theorems 3.1.5, 3.1.6) and the relation

$$\int_{\partial S} v^2 \, ds \le C \, h_S^{-1} \left(\|v\|_{L_2(S)}^2 + h_S^2 \, |v|_{H^1(S)}^2 \right)$$

(for $v = \Phi_\xi^G$ and $v = \varphi - \Pi_h\varphi$). This estimate is a consequence of the trace theorem (exploited on the unit simplex $\hat{S}$). Using this estimate and similar tools as in the proofs of Lemma 4.10[i], Corollary 4.7 and Lemma 4.12, we deduce the bound for $\Psi_2(e_h^\xi)$. Finally, the application of similar tools as exploited in the proofs of Lemmas 4.9[i] and 4.11 leads to the estimate for $\Psi_3(e_h^\xi)$.

Returning to the error scheme (4.33), (4.34), now with $z_h = e_h^\xi$ for $\xi \in \tilde{\omega}$, it follows that

$$M_{\varphi h}^2 = \sum_{\xi \in \tilde{\omega}} \text{meas}_n(D_\xi)^{-1} \Big[(A\varphi - f, \Phi_\xi^G) + (l\varphi - g_N, \Phi_\xi^G)_{\Gamma_N} - \sum_{i=1}^3 \Psi_i(e_h^\xi) \Big]^2 .$$

Thus, applying the Hölder inequality and recalling the geometrical condition (nonzero area) (4.12) on the boxes, we derive

$$\begin{aligned} M_{\varphi h}^2 \; &\le C_1 \, \|A\varphi - f\|_{L_2(\Omega)}^2 + C_2 \max_{\xi \in \gamma_N} \frac{\text{meas}_{n-1}(\partial\Omega_\xi \cap \Gamma_N)}{\text{meas}_n(\Omega_\xi)} \, \|l\varphi - g_N\|_{L_2(\Gamma_N)}^2 \\ &+ C_3 \sum_{\xi \in \tilde{\omega}} \Big(\sum_{i=1}^3 \Psi_i(e_h^\xi) \Big)^2 \Big/ \, \text{meas}_n(D_\xi). \end{aligned}$$

Now, we take into account the local error estimates summarized in (4.35). By means of these estimates we deduce that

$$\begin{aligned} \Big(\sum_{i=1}^3 \Psi_i(e_h^\xi) \Big)^2 \; &\le \text{meas}_n(\Omega_\xi) \, \big[C_4 \; |\varphi|_{H^2(\Omega_\xi)}^2 + C_5 \; \|\Pi_h\varphi\|_{H^1(\Omega_\xi)}^2 \\ &+ C_6 \; \|f\|_{L_2(\Omega_\xi)}^2 + C_7 \; |g_N|_{H^1(\partial\Omega_\xi \cap \Gamma_N)}^2 \big]. \end{aligned}$$

Therefore, summing over $\xi \in \tilde{\omega}$, taking into consideration the condition (4.12) and using an interpolation result for $\Pi_h\varphi$, we end up with

$$\left. \begin{aligned} &M_{\varphi h} \le C \quad \text{if} \quad \Gamma_D = \partial\Omega \quad \text{or} \quad l\varphi = g_N \quad \text{a.e. on } \Gamma_N \qquad \text{and} \\ &M_{\varphi h} \le (C_1 + C_2 \; h_{\Gamma_N}^{-1})^{1/2} \quad \text{if} \quad l\varphi \ne g_N \text{ on } \Gamma_N \quad \text{with} \\ &h_{\Gamma_N} := \min_{S \in S_h : \text{meas}_{n-1}(\partial S \cap \Gamma_N) > 0} h_S . \end{aligned} \right| \tag{4.36}$$

Considering the term $\langle \mu_h, 1 \rangle$ appearing in Theorem 4.24[ii], we get

$$\langle \mu_h, 1 \rangle \le \sum_{\xi \in \tilde{\omega}} \Big| (A\varphi - f, \Phi_\xi^G) + (l\varphi - g_N, \Phi_\xi^G)_{\Gamma_N} - \sum_{i=1}^3 \Psi_i(e_h^\xi) \Big|$$

from the error scheme (4.33), (4.34). Hence, by means of the local estimates (4.35) we deduce

$$\begin{aligned}\langle \mu_h, 1\rangle &\leq \sum_{\xi\in\tilde{\omega}} \int_{\Omega_\xi} |A\varphi - f|dx + \sum_{\xi\in\gamma_N} \int_{\Gamma_N\cap\partial\Omega_\xi} |l\varphi - g_N|ds + \sum_{\xi\in\tilde{\omega}}\sum_{i=1}^{3} |\Psi_i(e_h^\xi)| \\ &\leq C_1 \ \|A\varphi - f\|_{L_1(\Omega)} + C_2 \ \|l\varphi - g_N\|_{L_1(\Gamma_N)} + C_3 \ |\varphi|_{H^2(\Omega)} \\ &\quad + C_4 \ \|\Pi_h\varphi\|_{H^1(\Omega)} + C_5 \ \|f\|_{L_2(\Omega)} + C_6 \ |g_N|_{H^1(\Gamma_N)} \,.\end{aligned}$$

Therefore, we obtain the bound

$$\langle \mu_h, 1\rangle \leq C \tag{4.37}$$

for the penalization error of the second method (4.32) given in Theorem 4.24[ii].

Finally, it remains to consider the term $\overline{M}_{\varphi_h}$ characterizing the uniform estimate stated in Theorem 4.27. Assuming additionally $\varphi \in W^2_\infty(\Omega)$, $f \in L_\infty(\Omega)$ and $g_N \in W^1_\infty(\Gamma_N)$, we derive the estimates

$$\left.\begin{array}{l}\overline{M}_{\varphi_h} \leq C \ \text{ if } \ \Gamma_D = \partial\Omega \ \text{ or } \ l\varphi = g_N \ \text{ a.e. on } \Gamma_N \quad \text{and} \\ \overline{M}_{\varphi_h} \leq C_1 + C_2 \ h_{\Gamma_N}^{-1} \ \text{ if } \ l\varphi \neq g_N \text{ on } \Gamma_N, \ \ h_{\Gamma_N} \text{ from (4.36)}\end{array}\right| \tag{4.38}$$

by means of the error scheme (4.33), (4.34) and the local estimates (4.35). □

To estimate the overall error between the solution u of the variational inequality (4.1) and the solutions of the (discrete) penalization problems, we discuss in the following remark coupling of the space discretization parameter with the penalization parameters.

Remark 4.29 [i] Based on (4.36), (4.37), Theorem 4.24, Remark 4.26 and on the error bound for $u - P^G y_h$ (see Theorem 4.16), we obtain a first order convergence estimate $O(h)$ in the energy norm between the solution u of (4.1) and the solutions $y_{h\varepsilon}$, $z_{h\varepsilon}$ of the penalization problems (4.31), (4.32) by coupling the penalization and discretization parameters ε and h.

[ii] To obtain a L_∞-bound for $u - P^G y_{h\varepsilon}$ one needs, besides the result of Theorem 4.27, an estimate for $\left\|u - P^G y_h\right\|_{L_\infty(\Omega)}$. This will not be discussed in detail here. We refer to [BAIO77] and [NIT77] concerning such an estimate (of the order $O(h^{2-\delta})$, $\delta > 0$ arbitrarily small and $O(|\log h|\ h^2)$, respectively) for a finite element solution of the obstacle problem related to the Laplace operator with Dirichlet conditions (then finite element and finite volume method coincide); see also [ADGR95] mentioned at the beginning of this section. Concerning the consideration of more general differential operators we refer to [CORT85].

Furthermore, using the splitting $u - u_{\varepsilon h} = u - u_\varepsilon + u_\varepsilon - R_h u_\varepsilon + R_h u_\varepsilon - u_{\varepsilon h}$, Strehlau (see [STRE89]) derived an L_∞-bound for $u - u_{\varepsilon h}$ for this problem with an upper obstacle φ. Here, u_ε is the solution of the (continuous) penalization problem with $\varepsilon^{-1}(u_\varepsilon - \varphi)^+$ as penalization term, $R_h u_\varepsilon$ denotes the Ritz projection and $u_{\varepsilon h}$ is the solution of the penalization problem with the penalization term $\varepsilon^{-1}(u_{\varepsilon h} - R_h\varphi)^+ \notin V_h^G$.

We mention that for such estimates, one usually requires the inverse property of the triangulation (cf. Remark 4.26), a discrete maximum principle and, in particular, a solution u of (4.1) belonging to $W^2_\infty(\Omega)$. Note that this is the 'limit regularity' for the obstacle problem (cf. Remark 4.18) and can be only guaranteed under restrictive conditions (see [RODR87], Section 5.6 and [CHIP84], Section 3.5). □

4.4.3 Iterative solution of the penalization problems

In the previous section the discrete inequalities (4.15), (4.16) were approximated by discrete penalization problems (cf. (4.31) and (4.32)) and the convergence for the penalization parameter $\varepsilon \longrightarrow 0$ was proved. Now, it remains to indicate how the nonlinear equations (4.31) and (4.32) can be solved effectively by means of iteration processes.
Within this section a superscript m refers to the iteration step.

In order to solve (4.31) we propose the following iteration scheme.

$$\left.\begin{aligned} &\Lambda_B\, y_h^{m+1} + \varepsilon^{-1}\, r_{1h}(y_h^m)\, (y_h^{m+1} - \varphi_h) = f_{Bh} + g_{Bh} \quad \text{for} \quad \xi \in \tilde{\omega}, \\ &(y_h^{m+1})_\xi = g_D(\xi) \quad \text{for} \quad \xi \in \gamma_D; \quad m \geq 0. \end{aligned}\right| \tag{4.39}$$

An arbitrary initial vector $y_h^0 \in L_2(\overline{\omega})$ fulfilling the Dirichlet conditions

$$y^0_{h,\xi} = g_D(\xi) \quad \text{for} \quad \xi \in \gamma_D$$

should be given.
The components of the grid function vector $r_{1h}(v_h) \in L_2(\tilde{\omega})$ are defined by

$$(r_{1h}(v_h))_\xi = \begin{cases} 0 & \text{if} \quad v_{h,\xi} > \varphi_{h,\xi}, \\ \operatorname{meas}_n(D_\xi) & \text{if} \quad v_{h,\xi} \leq \varphi_{h,\xi}. \end{cases}$$

We remark that the iteration scheme (4.39) can be written equivalently as

$$(\Lambda_B + \varepsilon^{-1}\, r_{1h}(y_h^m))\, (y_h^{m+1} - y_h^m) = f_{Bh} + g_{Bh} - \Lambda_B\, y_h^m - \varepsilon^{-1}\, \beta_{1h}(y_h^m - \varphi_h)$$

and $r_{1h}(v_h)\, (v_h - \varphi_h) = \beta_{1h}(v_h - \varphi_h)$ holds.

Defining the grid function vector $r_{2h}(v_h) \in L_2(\tilde{\omega})$ by

$$(r_{2h}(v_h))_\xi = \begin{cases} \dfrac{\varepsilon}{(\varepsilon + v_{h,\xi} - \varphi_{h,\xi})^2} & \text{if} \quad v_{h,\xi} \geq \varphi_{h,\xi}\,, \\ 0 & \text{if} \quad v_{h,\xi} < \varphi_{h,\xi}, \end{cases}$$

we propose the following iteration scheme for the solution of (4.32).

$$\left.\begin{aligned} &(\Lambda_B + \mu_h\, r_{2h}(z_h^m))\,(z_h^{m+1} - z_h^m) = f_{Bh} + g_{Bh} - \Lambda_B\, z_h^m \\ &\quad + \mu_h\big[1 - \beta_{2h}(\frac{z_h^m - \varphi_h}{\varepsilon})\big] \quad \text{for} \quad \xi \in \tilde{\omega}, \\ &(z_h^{m+1})_\xi = g_D(\xi) \quad \text{for} \quad \xi \in \gamma_D; \quad m \geq 0. \end{aligned}\right| \tag{4.40}$$

The initial vector $z_h^0 \in L_2(\overline{\omega})$ is given by $z_{h,\xi}^0 = \varphi_{h,\xi}$ for $\xi \in \tilde{\omega}$ and $z_{h,\xi}^0 = g_D(\xi)$ for Dirichlet nodes $\xi \in \gamma_D$.
We mention that this scheme can be expressed equivalently by

$$\begin{aligned} &\Lambda_B z_h^{m+1} + \mu_h\, r_{2h}(z_h^m)\,(z_h^{m+1} - \varphi_h) \\ &= f_{Bh} + g_{Bh} + \mu_h\,\big[1 + r_{2h}(z_h^m)(z_h^m - \varphi_h) - \beta_{2h}(\frac{z_h^m - \varphi_h}{\varepsilon})\big]. \end{aligned}$$

Remark 4.30 Concerning an overview of methods for the solution of such types of (mildly) nonlinear elliptic equations like (4.31) and (4.32) we refer to [GLOW84], Section V.2. We stress that the proposed iteration schemes (4.39) and (4.40) are Newton-type methods. □

Mainly based on the discrete maximum principle (4.30), the following convergence results (global convergence) for the iteration methods (4.39) and (4.40) are shown.

Theorem 4.31 [i] *Let the assumptions of Theorem 4.21 be satisfied. The solutions $y_{h\varepsilon}$ of (4.31) and y_h^m of (4.39) satisfy*

$$y_h^m \leq y_h^{m+1} \leq y_{h\varepsilon} \quad \textit{for} \quad m \geq 1.$$

Moreover, $y_h^M = y_{h\varepsilon}$ holds after a finite number M of iterations (for fixed triangulation).

[ii] *The solutions $z_{h\varepsilon}$ of (4.32) and z_h^m of (4.40) fulfill*

$$z_h^m \in K_h \quad \textit{and} \quad \varphi_h \leq z_h^m \leq z_h^{m+1} \leq z_{h\varepsilon} \quad \textit{for} \quad m \geq 0.$$

Furthermore, r_{2h} is simplified to $r_{2h}(z_h^m) = \varepsilon/(\varepsilon + z_h^m - \varphi_h)^2$, $m \geq 0$ and the sequence $\{z_h^m\}_{m\in\mathbb{N}}$ converges towards $z_{h\varepsilon}$.

Proof. [i]. Considering the iteration process (4.39) we have

$$(\Lambda_B + \varepsilon^{-1}\, r_{1h}(y_h^m))\, (y_h^{m+1} - y_h^m) = \varepsilon^{-1}\, [r_{1h}(y_h^{m-1}) - r_{1h}(y_h^m)]\, (y_h^m - \varphi_h) \geq 0.$$

Hence, multiplying this relation by $(y_h^{m+1} - y_h^m)^- \leq 0$ we derive, owing to the maximum principle (4.30), that $(y_h^{m+1} - y_h^m)^- = 0$ holds and therefore $y_h^{m+1} \geq y_h^m$. Comparing the solution $y_{h\varepsilon}$ of problem (4.31) with the equation for the iteration step m in (4.39), we obtain

$$\begin{aligned}
&\Lambda_B\, (y_{h\varepsilon} - y_h^m) + \varepsilon^{-1}\, r_{1h}(y_h^{m-1})\, (y_{h\varepsilon} - y_h^m) \\
&= \varepsilon^{-1}\, [r_{1h}(y_h^{m-1}) - r_{1h}(y_{h\varepsilon})]\, (y_{h\varepsilon} - \varphi_h) \geq 0.
\end{aligned}$$

Multiplying this inequality by $(y_{h\varepsilon} - y_h^m)^- \leq 0$ and applying (4.30), we deduce that $(y_{h\varepsilon} - y_h^m)^- = 0$ holds. Consequently, $y_{h\varepsilon} \geq y_h^m$ is proved.

Let us introduce the discrete coincidence set

$$\omega^-(v_h) = \{\xi \in \tilde{\omega} : v_{h,\xi} \leq \varphi_{h,\xi}\}. \tag{4.41}$$

Owing to $y_{h\varepsilon} \geq y_h^{m+1} \geq y_h^m$ for $m \geq 1$, we conclude

$$\omega^-(y_h^1) \supseteq \omega^-(y_h^2) \supseteq \cdots \supseteq \omega^-(y_h^m) \supseteq \cdots \supseteq \omega^-(y_{h\varepsilon}).$$

Recalling the definition of $r_{1h}(v_h)$ in (4.39) we obtain $z_h^{m+1} = z_h^{m+2}$ from $\omega^-(z_h^m) = \omega^-(z_h^{m+1})$. Then, the equation systems for the computation of z_h^{m+1} and z_h^{m+2} coincide.

[ii]. Starting the iteration method (4.40) with $z_{h,\xi}^0 = \varphi_{h,\xi}$ for nodes $\xi \in \tilde{\omega}$ and $z_{h,\xi}^0 = g_D(\xi)$ for Dirichlet nodes $\xi \in \gamma_D$ as well as taking into consideration the relation

$$f_{Bh} + g_{Bh} - \Lambda_B \varphi_h + \mu_h \geq 0,$$

we get

$$(\Lambda_B + \mu_h\, r_{2h}(z_h^0))\, (z_h^1 - z_h^0) \geq 0.$$

Therefore, owing to the maximum principle (4.30), $z_h^1 \geq z_h^0 = \varphi_h$ is guaranteed for nodes $\xi \in \tilde{\omega}$.

Now we shall show that $z_h^{m+1} \geq z_h^m \geq \varphi_h$ holds. Assuming $z_h^m \geq z_h^{m-1} \geq \varphi_h$, the iteration process (4.40) implies

$$\begin{aligned}
&(\Lambda_B + \mu_h\, r_{2h}(z_h^m))\, (z_h^{m+1} - z_h^m) \\
&= \mu_h \left[\beta_{2h}\Big(\frac{z_h^{m-1} - \varphi_h}{\varepsilon}\Big) - \beta_{2h}\Big(\frac{z_h^m - \varphi_h}{\varepsilon}\Big) + r_{2h}(z_h^{m-1})(z_h^m - z_h^{m-1})\right] \\
&= \mu_h \left[\frac{z_h^{m-1} - \varphi_h}{\varepsilon + z_h^{m-1} - \varphi_h} - \frac{z_h^m - \varphi_h}{\varepsilon + z_h^m - \varphi_h} + \frac{\varepsilon\, (z_h^m - z_h^{m-1})}{(\varepsilon + z_h^{m-1} - \varphi_h)^2}\right] \\
&= \frac{\varepsilon\, \mu_h\, (z_h^{m-1} - z_h^m)^2}{(\varepsilon + z_h^{m-1} - \varphi_h)^2\, (\varepsilon + z_h^m - \varphi_h)} \geq 0.
\end{aligned}$$

In other words, the non-negativity of the term in square brackets is a consequence of the fact that $\beta_{2h}(v_h)$ is a concave function. Again by application of the discrete maximum principle (4.30) we verify the desired relation

$$z_h^{m+1} \geq z_h^m \geq \varphi_h.$$

It remains to prove the inequality $z_h^m \leq z_{h\varepsilon}$ for $m \geq 0$. Owing to Theorem 4.24[ii] we have $z_h^0 = \varphi_h \leq z_{h\varepsilon}$ for $\xi \in \tilde{\omega}$. Subtracting the equations (4.40) from the penalization equations in (4.32) as well as taking into account the relations $z_h^m \geq \varphi_h$ and $z_h^{m+1} \geq \varphi_h$, we obtain

$$\begin{aligned}
&\Lambda_B\,(z_{h\varepsilon} - z_h^{m+1}) + \mu_h \left[\beta_{2h}(\frac{z_{h\varepsilon} - \varphi_h}{\varepsilon}) - \beta_{2h}(\frac{z_h^{m+1} - \varphi_h}{\varepsilon})\right] \\
&= \mu_h \left[r_{2h}(z_h^m)\,(z_h^{m+1} - z_h^m) + \beta_{2h}(\frac{z_h^m - \varphi_h}{\varepsilon}) - \beta_{2h}(\frac{z_h^{m+1} - \varphi_h}{\varepsilon})\right] \\
&= \mu_h \left[\frac{\varepsilon\,(z_h^{m+1} - z_h^m)}{(\varepsilon + z_h^m - \varphi_h)^2} + \frac{z_h^m - \varphi_h}{\varepsilon + z_h^m - \varphi_h} - \frac{z_h^{m+1} - \varphi_h}{\varepsilon + z_h^{m+1} - \varphi_h}\right] \geq 0.
\end{aligned}$$

Owing to the monotonicity of $\beta_{2h}(v_h)$ and the maximum principle (4.30), the claimed relation $z_h^{m+1} \leq z_{h\varepsilon}$ is guaranteed. Hence, $\{z_h^m\}_{m\in\mathbb{N}}$ is a monotone and bounded sequence which converges towards $z_{h\varepsilon}$. ■

Remark 4.32 [i] Summarizing the results of Theorems 4.23 and 4.31, we have proved the following chains of inequalities

$$y_h^1 \leq y_h^2 \leq \cdots \leq y_h^m \leq \cdots \leq y_h^M = y_{h\varepsilon} \leq y_h \leq z_{h\varepsilon} \quad \text{and}$$

$$\varphi_h = z_h^0 \leq z_h^1 \leq \cdots \leq z_h^m \leq \cdots \leq z_{h\varepsilon}$$

for the solutions y_h, $y_{h\varepsilon}$, $z_{h\varepsilon}$, y_h^m and z_h^m of the problems (4.16), (4.31), (4.32), (4.39) and (4.40)

[ii] The realization of one iteration step in (4.39) (resp. (4.40)) means to solve (iteratively) a large sparse system of equations. In this context we refer to the monographs [HACK94] and [SAAD96] which are devoted to modern effective algorithms for the solution of sparse systems arising from the discretization of partial differential equations. We mention that at each iteration step in (4.39) (resp. (4.40)) only the diagonal of the system matrix has to be changed. Therefore, no reassembling of the system matrix is necessary in performimg the penalty-iterations.

With respect to the penalty-iteration method (4.31), (4.39) we have observed from a large number of numerical experiments (see Sections 4.6, 5.4) that the number of necessary iterations to arrive at $y_h^M = y_{h\varepsilon}$ depends essentially on the quality of the initial coincidence set $\omega^-(y_h^0)$, i.e.,

$$\omega^-(y_h^0) \approx \omega^-(y_{h\varepsilon})$$

(cf. (4.41)) is desirable. In 'real-life' problems it is often not difficult to choose a suitable initial set $\omega^-(y_h^0)$. Then, only a small number of iterations is necessary to obtain $y_h^M = y_{h\varepsilon}$ (for the injection, compression moulding applications, in general, not more than 5; cf. also Remark 5.26).
If no information is available, then it can be useful to choose $\omega^-(y_h^0) = \omega^-(\tilde{y}_h^0)$, where $\tilde{y}_h^0$ is the interpolation of a coarse grid solution. Compared to (4.31), (4.39), the second penalty-iteration method (4.32), (4.40) is not so sensitive with respect to the initial coincidence set. □

4.5 The Signorini problem as a boundary obstacle problem

The obstacle problem with a unilateral constraint restricted to the boundary $\partial\Omega$ or a part of it is often called the Signorini problem. There are several branches in mathematical physics in which such *boundary* obstacle problems arise. We refer to the monographs [DULI76], [GLT81], [RODR87], [HHNL88] and [KIOD88] for a detailed description.

In what follows we will briefly discuss how results can be derived analogous to those obtained for interior obstacle problems in the foregoing sections (i.e., comparison of finite element, finite volume and exact solution, application of penalization techniques).

The problem we have in mind can be formulated as follows.

$$\left.\begin{array}{l} \text{Find} \quad u \in K_S, \\ a(u, v-u) \geq (f, v-u) + (g_N, v-u)_{\Gamma_N} \quad \forall\, v \in K_S, \end{array}\right| \qquad (4.42)$$

where the bilinear form a is defined in (4.2) and the convex constraint set K_S is given by

$$K_S = K_{\varphi_S, g_D} = \{w \in H^1(\Omega) : \gamma w \geq \varphi_S \text{ on } \Gamma_N, \quad w = g_D \text{ on } \Gamma_D\} \neq \emptyset.$$

The trace of w on $\Gamma = \partial\Omega$ is denoted by γw. In the sequel we omit the notation γ.

The analytical investigation (existence, regularity of the solution) of problem (4.42) is the subject of Section 8.6 in the monograph [RODR87]. Supposing the conditions (4.4) and an obstacle $\varphi_S \in H^{1/2}(\partial\Omega)$ with $\varphi_S \leq g_D$ on Γ_D, the existence of a unique solution of (4.42) is guaranteed. Furthermore, supposing $\partial\Omega = \Gamma_N \in C^{1,1}$, $k \in C^{0,1}(\overline{\Omega})$, $\kappa \in C^{0,1}(\overline{\partial\Omega})$, $g_N \in H^{1/2}(\partial\Omega)$ and $\varphi_S \in H^{3/2}(\partial\Omega)$, the solution u fulfills the regularity $u \in H^2(\Omega)$.

Now we assume a continuous obstacle, e.g., there exists a function $\Phi \in H^2(\Omega)$ (or $\Phi \in W_p^1(\Omega)$, $p > n$) such that $\Phi = \varphi_S$ on $\partial\Omega$ (cf. our assumption (4.5) and

[HHNL88], Section 1.1.7). Then, in agreement with the notation used for interior obstacle problems, the finite element and finite volume approximations of (4.42) read as follows.

$$\left.\begin{array}{l} \text{Find} \quad P^G u_h \in K_{Sh}: \qquad a(P^G u_h, P^G v_h - P^G u_h) \\ \geq (f, P^G v_h - P^G u_h) + (g_N, P^G v_h - P^G u_h)_{\Gamma_N} \quad \forall P^G v_h \in K_{Sh} \end{array}\right| \tag{4.43}$$

and

$$\left.\begin{array}{l} \text{Find} \quad P^G y_h \in K_{Sh}: \qquad a_B(P^G y_h, P^B v_h - P^B y_h) \\ \geq (f, P^B v_h - P^B y_h) + (g_N, P^B v_h - P^B y_h)_{\Gamma_N} \quad \forall P^G v_h \in K_{Sh}. \end{array}\right| \tag{4.44}$$

The essential difference from the (discrete) interior obstacle problems (4.9) and (4.15) consists in the definition of the convex set K_{Sh} instead of K_h, i.e.,

$$\begin{aligned} K_{Sh} &= \{P^G w_h \in X_h^G : (P^G w_h)(\xi) \geq \varphi_S(\xi) \ \forall \xi \in \gamma_N, \\ &\qquad (P^G w_h)(\xi) = g_D(\xi) \ \forall \xi \in \gamma_D\}. \end{aligned}$$

Now, the main result concerning the relation between the finite element solution $P^G u_h$ of (4.43) and the box solution $P^G y_h$ of (4.44) can be obtained by transfer of the convergence property between the finite element and box solutions for the interior obstacle problems stated in Theorem 4.13. Based on equality of the stiffness matrices $L_G = L_B$, we deduce again the estimate (4.21). Then, applying the corresponding statements derived in Section 4.2, we are led to results analogous to those given in Theorem 4.13.

Theorem 4.33 *Let the assumptions (4.4), (4.5) and $k \in V_h^C$ be satisfied. Suppose $g_N \in H^1(\Gamma_N)$, $\kappa \in W^1_\infty(\Gamma_N)$ and a box partition according to $\mathcal{D}_h(1)$, it holds that*

$$\|P^G u_h - P^G y_h\|_{H^1(\Omega)} \leq C_1 \ h$$

for the solutions u_h of (4.43) and y_h of (4.44).
Under the stronger conditions of a box partition according to $\mathcal{D}_h(2)$ and $f \in H^1(\Omega)$, $q \in W^1_\infty(\Omega)$, $g_N = const$ on each boundary face $\sigma \in \sigma_{h,2}$ and $\kappa = 0$, one has

$$\|P^G u_h - P^G y_h\|_{H^1(\Omega)} \leq C_2 \ h^2.$$

Error estimates for the solution $P^G u_h$ of the finite element problem (4.43) compared to the solution u of (4.42) can be found in the paper [BHR77] and furthermore in some monographs, e.g., [GLT81], Section 4.3 and Appendix 4.2, [GLOW84], Section II.4 and [HHNL88], Section 1.1.71. Although the situation is more complicated than for interior obstacle problems (note that $lu = g_N$ can not be assumed on Γ_N), a first order estimate in the energy norm can be established by evaluation of corresponding a priori estimates assuming a reasonable smoothness for u, φ_S and on the free boundary (see below the conditions which guarantee (4.45)).

In order to derive an analogous convergence rate for the box solution $P^G y_h$ of problem (4.44), we study the difference $u - P^G y_h$ in the same way as done in Section 4.3 for the interior obstacle problem. The a priori estimates (4.25), (4.26) from Lemma 4.15 hold equally, only the convex subsets K_h, K have to be replaced by K_{Sh} and K_S. Evaluating the a priori estimate (4.25) as in Theorem 4.16 it can be easily seen that the error representation (4.27) is not influenced by the fact whether we consider interior or boundary obstacle problems. The major difference to the proof of Theorem 4.16 is, that we have now to consider the term

$$(lu - g_N, P^G(v_h - y_h))_{\Gamma_N} \quad \text{instead of} \quad (Au - f, P^G(v_h - y_h))$$

in the estimate (4.28) (see (4.3) for the definition of A and l). Choosing again

$$K_{Sh} \ni v_h = \big((\Pi_h u)(\xi)\big)_{\xi \in \overline{\omega}}$$

and exploiting the complementarity problem equivalent to the Signorini problem (4.42), we arrive at

$$(lu - g_N, \Pi_h u - P^G y_h)_{\Gamma_N} \leq (lu - g_N, \Pi_h u - u + \varphi_S - \Pi_h \varphi_S)_{\Gamma_N}.$$

Based on this representation the proof of Theorem 4.16 shows, that it will suffice to establish the relation

$$(lu - g_N, \Pi_h u - P^G y_h)_{\Gamma_N} \leq C\, h^2 \tag{4.45}$$

to obtain the first order convergence

$$\|u - P^G y_h\|_{H^1(\Omega)} \leq C\, h$$

between u and y_h as the solutions of the Signorini problem (4.42) and its finite volume approximation (4.44). The crucial estimate (4.45) can be derived in the same way as in the above mentioned error analysis for the finite element solution. In particular, we refer to [GLT81], Appendix 4.2, Theorem 2.1 (spatial two-dimensional situation). The additional assumptions are of the form

$$u \in H^2(\Omega) \cap K, \quad \varphi_S \in W^1_\infty(\Gamma_N), \quad g_N \in L_\infty(\Gamma_N),$$
$$u \in W^1_\infty(\Gamma_N) \quad \text{satisfying} \quad \partial u/\partial n \in L_\infty(\Gamma_N).$$

In particular, the set of points from Γ_N at which u changes from $u = \varphi_S$ to $u > \varphi_S$ has to be finite.

The application of solution methods known for interior obstacle problems (see the introduction to Section 4.4) has to be only slightly modified with respect to the Signorini problem (4.44), because the solution y_h of the problem (4.44) is also the solution of an *interior* obstacle problem. To derive the associated interior obstacle problem, we establish at first the Lewy–Stampacchia inequalities fulfilled by the finite volume solution of the interior obstacle problem (4.15), (4.16). In this context we also refer to the complementarity formulation (4.20).

Lemma 4.34 *Suppose a weakly acute triangulation with its box partition according to $\mathcal{D}_h(1)$ and the standard assumptions (4.4), (4.5), the solution $P^G y_h$ of the box inequality problem (4.15), (4.16) satisfies the Lewy–Stampacchia type inequalities*

$$\begin{aligned} f_{Bh} + g_{Bh} \le \Lambda_B y_h &\le \max\{f_{Bh} + g_{Bh}, \Lambda_B \varphi_h\} \\ &= f_{Bh} + g_{Bh} - (f_{Bh} + g_{Bh} - \Lambda_B \varphi_h)^- \quad \forall \xi \in \tilde{\omega}. \end{aligned} \tag{4.46}$$

Here, (4.46) has to be understood componentwise.

Proof. In order to establish the upper bound in the estimate (4.46) let us introduce the notation

$$F_{Bh} := f_{Bh} + g_{Bh} - (f_{Bh} + g_{Bh} - \Lambda_B\, \varphi_h)^-.$$

We will show that the unique solution z_h of the following auxiliary inequality problem with an upper obstacle is equal to the solution y_h of the original problem (4.15), (4.16).

$$\begin{aligned} &\text{Find} \quad P^G z_h \in X_h^G : \quad z_{h,\xi} \le y_{h,\xi}\ \forall \xi \in \tilde{\omega}, \quad z_{h,\xi} = y_{h,\xi}\ \forall \xi \in \gamma_D\,, \\ &\langle \Lambda_B\, z_h, w_h - z_h \rangle \ge \langle F_{Bh}, w_h - z_h \rangle \quad \forall P^G w_h \in X_h^G \quad \text{with} \\ &w_{h,\xi} \le y_{h,\xi}\ \forall \xi \in \tilde{\omega}, \quad w_{h,\xi} = y_{h,\xi}\ \forall \xi \in \gamma_D. \end{aligned} \tag{4.47}$$

Realizing this, the discrete maximum principle (4.30) will be exploited such that a weakly acute triangulation is assumed. At first, let us take

$$w_h = z_h - (z_h - \varphi_h)^- = \max\{z_h, \varphi_h\} \le y_h$$

in (4.47) in order to show $z_h \ge \varphi_h$. Owing to $F_{Bh} \ge \Lambda_B\, \varphi_h$ and $-(z_h - \varphi_h)^- \ge 0$, we get

$$\langle \Lambda_B\, z_h, -(z_h - \varphi_h)^- \rangle \ge \langle F_{Bh}, -(z_h - \varphi_h)^- \rangle \ge \langle \Lambda_B\, \varphi_h, -(z_h - \varphi_h)^- \rangle.$$

Consequently, we have $\langle \Lambda_B (z_h - \varphi_h), (z_h - \varphi_h)^- \rangle \le 0$, such that the application of the discrete maximum principle (4.30) gives $z_h \ge \varphi_h$.
As second step we prove $z_h \ge y_h$. Taking $w_h = z_h - (z_h - y_h)^- = \max\{z_h, y_h\} \le y_h$ in (4.47) and $v_h = y_h + (z_h - y_h)^- = \min\{z_h, y_h\} \ge \varphi_h$ in (4.16) and adding both inequalities, we obtain

$$\langle \Lambda_B\, (z_h - y_h), (z_h - y_h)^- \rangle \le \langle -(f_{Bh} + g_{Bh} - \Lambda_B\, \varphi_h)^-, (z_h - y_h)^- \rangle \le 0.$$

Therefore, by means of (4.30) we verify $z_h \geq y_h$. Hence, we arrive at $z_h = y_h$.
Now, returning to (4.47), we have

$$\big\langle \Lambda_B\, z_h, w_h - z_h \big\rangle = \big\langle \Lambda_B\, y_h, w_h - y_h \big\rangle \geq \big\langle F_{Bh}, w_h - y_h \big\rangle \quad \text{for all} \quad w_h \leq y_h.$$

Taking $w_h = y_h - v_h$ with an arbitrary grid function vector $v_h \geq 0$, we conclude the upper bound $\Lambda_B\, z_h = \Lambda_B\, y_h \leq F_{Bh}$ for all nodes $\xi \in \tilde{\omega}$. ■

Now, we construct a (discrete) interior obstacle φ_h related to the boundary obstacle φ_{S_h} with $\varphi_{S_{h,\xi}} = \varphi_S(\xi)$ for $\xi \in \gamma = \gamma_D \cup \gamma_N$. Then, by means of the results of Lemma 4.34, the (finite volume) Signorini problem (4.44) can be led back to a (box) interior obstacle problem of the form (4.15), (4.16).

Therefore, let a (discrete) interior obstacle φ_h be constructed by solving the following finite volume problem with Dirichlet conditions (given by φ_{S_h}) on the whole (discrete) boundary $\gamma = \gamma_D \cup \gamma_N$.

$$(L_B + M_B)\, \varphi_h = f_{Bh} \quad \text{for } \xi \in \omega, \qquad \varphi_h = \varphi_{S_h} \quad \text{for } \xi \in \gamma. \tag{4.48}$$

Note, that we can also take $\varphi_{h,\xi} = g_D(\xi)$ in (4.48) for nodes $\xi \in \gamma_D$.

Considering now the interior obstacle problem

$$\left.\begin{array}{l} \text{Find} \quad P^G \tilde{y}_h \in K_h : \\ \big\langle \Lambda_B\, \tilde{y}_h, v_h - \tilde{y}_h \big\rangle \geq \big\langle f_{Bh} + g_{Bh}, v_h - \tilde{y}_h \big\rangle \quad \forall P^G v_h \in K_h \quad \text{with} \\ K_h = \{P^G w_h \in X_h^G : w_{h,\xi} \geq \varphi_{h,\xi}\ \forall \xi \in \overline{\omega}, \quad w_{h,\xi} \geq g_D(\xi)\ \forall \xi \in \gamma_D\}, \end{array}\right| \tag{4.49}$$

the subject of the next theorem is to show that $\tilde{y}_h = y_h$ holds with $\tilde{y}_h$ and y_h as solutions of (4.49) and (4.44), respectively.

Theorem 4.35 *Let us assume (4.4), (4.5) and let be given a weakly acute triangulation and its box partition according to $\mathcal{D}_h(1)$. Then, the solution $\tilde{y}_h$ of the interior obstacle problem (4.49) with the convex set K_h characterized by the obstacle φ_h according to (4.48) is equal to the solution y_h of the boundary obstacle problem (4.44).*

Proof. Due to $\tilde{y}_h \in K_h \subset K_{Sh}$ and the uniqueness of the solution of (4.44) it remains to verify that $\tilde{y}_h$ satisfies the inequality (4.44) for any $v_h \in K_{Sh}$. Owing to Lemma 4.34 the solution $\tilde{y}_h \in K_h$ of problem (4.49) is characterized by

$$f_{Bh} + g_{Bh} \leq \Lambda_B\, \tilde{y}_h \leq \max\{f_{Bh} + g_{Bh}, \Lambda_B\, \varphi_h\} \quad \forall \xi \in \tilde{\omega}.$$

By virtue of equation (4.48) we even have $\Lambda_B \tilde{y}_h = \Lambda_B \varphi_h = f_{Bh}$ for interior nodes $\xi \in \omega$.

Now, taking an arbitrary $v_h \in K_{Sh}$ and an associated $w_h \in K_h$ with $w_{h,\xi} = v_{h,\xi}$ for $\xi \in \gamma$, it follows

$$\begin{aligned}\langle \Lambda_B \tilde{y}_h - f_{Bh} - g_{Bh}, v_h - \tilde{y}_h \rangle &= \sum_{\xi \in \gamma_N} (\Lambda_B \tilde{y}_h - f_{Bh} - g_{Bh})_\xi \, (v_h - \tilde{y}_h)_\xi \\ &= \langle \Lambda_B \tilde{y}_h - f_{Bh} - g_{Bh}, w_h - \tilde{y}_h \rangle \geq 0,\end{aligned}$$

which completes the proof. ■

Remark 4.36 Under some assumptions on the data, in particular on the obstacle, the transformation of the Signorini problem to an interior obstacle problem can also be performed on the continuous level. The problem (4.42) is assigned an obstacle problem of the form (4.1) with an obstacle 'extended' by solving a second order equation. For details we refer to [RODR87], Section 8.6. □

Now, the penalization methods investigated in Sections 4.4.2, 4.4.3 can immediately be applied to problem (4.49) and hence to the Signorini problem (4.44). The term M_{φ_h} occuring in the penalization error estimates for problem (4.31) (see Theorem 4.24 and Remarks 4.26, 4.28), is given by

$$M_{\varphi_h} = \Big[\sum_{\xi \in \gamma_N} (f_{Bh} - g_{Bh} - \Lambda_B \, \varphi_h)^2 / \operatorname{meas}_n(D_\xi) \Big]^{1/2}.$$

The regularization method (4.32) is characterized by the auxiliary grid function vector μ_h the components of which are now given by

$$\mu_{h,\xi} = 0 \quad \text{for } \xi \in \omega, \quad \mu_{h,\xi} = (\Lambda_B \, \varphi_h - f_{Bh} - g_{Bh})_\xi^+ \quad \text{for } \xi \in \gamma_N.$$

4.6 Results from numerical experiments for elliptic obstacle problems

In this section we discuss results from numerical experiments for three obstacle problems of the form (4.1), (4.2) with known exact solution u.

Firstly, according to Section 4.2, we compare the finite element solution u_h of problem (4.9), (4.10) with the box solution y_h of problem (4.15), (4.16). Secondly, the difference between $\Pi_h u$ (piecewise linear interpolant of u) and $P^G y_h$ will be evaluated. Thirdly, a number of numerical experiments is performed to observe the error behaviour between the exact solution u and the solutions of the iteration schemes (4.39) and (4.40). This corresponds to our proposed numerical solution procedure for the obstacle problem (4.1), (4.2), which comprises

- the discretization of (4.1) by the box inequality (4.15), (4.16),

- the replacement of these box inequalities by the penalization problems (4.31), (4.32) and, finally,
- the iterative solution of the penalization equations by means of (4.39), (4.40).

The boxes are chosen exclusively as barycentric domains. This is motivated by the good convergence of the box method expected in this case and by the easy implementation of the barycentric box method in a finite element code (cf. Section 4.1.3).

The first example is an obstacle problem on the unit square characterized by simple coefficients in the differential operator and by homogeneous Dirichlet boundary conditions, but also by a strongly changing right-hand side. The second example contains more complicated coefficients and also mixed boundary conditions. The third example is a (slightly) modified elasto-plastic torsion inequality on a circle.

We have chosen a sequence of mesh-size parameters

$$h = h_i = (20\, i)^{-1}, \quad i = 1, \ldots, 10$$

for these examples.

The iteration schemes (4.39), (4.40) have been solved with different penalization parameters ε coupled with the discretization parameter h by $\varepsilon = h^{\delta}$, $\delta \geq 2$. Furthermore, a standard relaxation procedure was used (successive over-relaxation (SOR) with projection as described in [GLT81] and [GLOW84]).

We emphasize that in this section our main interest consists in the (numerical) examination of the predicted convergence rates between exact, finite element, finite volume and (finite volume) penalization solutions. Therefore, besides the penalty-iterations (4.31), (4.39) and (4.32), (4.40), we have applied the SOR method with projection which requires an enormous cpu-time compared to the penalty-iterations (at least for 'fine discretizations'). We refer to Section 4.4.3 (see, in particular, Remark 4.32 and also Remark 5.26 in Section 5.3.2) for computational aspects of the penalty-iteration methods.

The computations were carried out in the framework of the C++ finite element toolbox *FeliCs* (see [BAUE99]) developed at the Chair for Applied Mathematics of the Technical University Munich.

4.6.1 Examples with known exact solution

Example 4.37 Our first obstacle problem is characterized by a free boundary Γ_f (for the definition see (4.29)) which is composed of four circumferences touching each other. To be more precise, we consider the variational inequality (4.1), (4.2) on the unit square Ω with $k = q = 1$, $\Gamma = \partial\Omega = \Gamma_D$ and $\varphi = g_D = 0$. The right-hand side f is chosen according to

$$f(x) = 10^4 \left\{ 8\, \alpha^2\, (\rho(x))^{\alpha-1} \left[0.5\, d^{\alpha} - (\rho(x))^{\alpha}\right] + 0.5\, \left[\left(d^{\alpha} - (\rho(x))^{\alpha}\right)^{+}\right]^2 \right\}$$

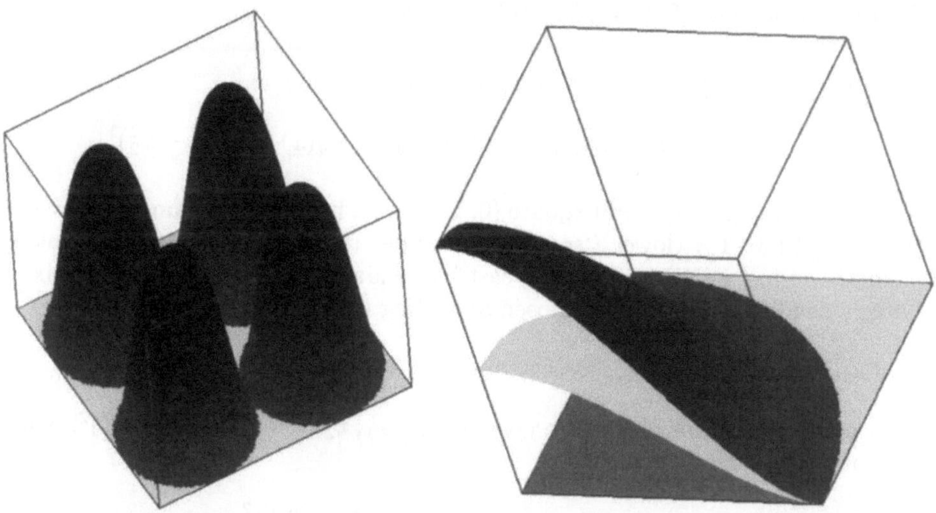

Figure 4.3: *Numerical results of the finite volume approximation for two obstacle problems. Left: example 4.37. Right: example 4.38 for* $h = 0.01$. *Obstacles are represented by light grey.*

with $\alpha = 1.5$, $d = 16^{-1}$ and

$$\rho(x) = \rho(x_1, x_2) = (x_1 - 0.25)^2 + (x_2 - 0.25)^2, \quad 0 \le x_1 \le 0.5, \quad 0 \le x_2 \le 0.5,$$

where $\rho(x_1 + 0.5, x_2) = \rho(x_1 + 0.5, x_2 + 0.5) = \rho(x_1, x_2 + 0.5) = \rho(x_1, x_2)$ for $x_1 \in [0, 0.5]$, $x_2 \in [0, 0.5]$.
Then, the exact solution u turns out to be

$$u(x) = 5000\ \left[\left(d^\alpha - (\rho(x))^\alpha\right)^+\right]^2 = \begin{cases} 5000\ \left[d^\alpha - (\rho(x))^\alpha\right]^2 & \text{for} \quad \rho(x) \le d, \\ 0 & \text{for} \quad \rho(x) \ge d. \end{cases}$$

The free boundary Γ_f is described by the circumferences of the four circles with centres $(0.25, 0.25)$, $(0.75, 0.25)$, $(0.25, 0.75)$, $(0.75, 0.75)$ and the radius $r = 0.25$. Obviously, this test problem is symmetric with respect to $(0.5, 0.5)$, cf. also Figure 4.3. □

Example 4.38 The second example is characterized by mixed boundary conditions. We consider the inequality (4.1), (4.2) with

$$k(x) = k(x_1, x_2) = 1 + (x_1\ x_2)^{3/2},$$

a zero-order term

$$q(x) = q(x_1, x_2) = |x_1 - 0.5| + |x_2 - 0.5|$$

and a right-hand side $f(x) = f(x_1, x_2)$ given by

$$\begin{aligned} f(x) &= -8\,(x_1^2 + x_2^2)\,\left(1 + (x_1\,x_2)^{3/2}\right) \\ &\quad + (1 - x_1^2 - x_2^2)\,\left[8 + 20(x_1\,x_2)^{3/2} + q(x)\,(1 - x_1^2 - x_2^2)\right]. \end{aligned}$$

Again, the domain Ω is the unit square $(0,1)\times(0,1)$, but now its boundary $\Gamma = \partial\Omega$ is partitioned into Γ_D (lower ('south') and upper ('north') boundary segment of Ω) and Γ_N (left ('west') and right ('east') boundary segment of Ω). The mixed boundary conditions on Γ_N are chosen according to

$$\kappa(x_1, x_2) = (1 - x_1)\,|x_2 - 0.75| \quad \text{and} \quad g_N(x_1, x_2) = \kappa(x_1, x_2)\,(1 - x_2^2)^2.$$

The Dirichlet conditions on Γ_D are given by $g_D(x_1, x_2) = (1 - x_2)\,(1 - x_1^2)^2$. Now, let the obstacle φ be defined by

$$\varphi(x) = \varphi(x_1, x_2) = 0.5\,\left((1 - x_1^2 - x_2^2)^+\right)^2.$$

Then, the exact solution u turns out to be

$$u(x) = u(x_1, x_2) = 2\,\varphi(x) = \begin{cases} (1 - x_1^2 - x_2^2)^2 & \text{for} \quad x_1^2 + x_2^2 \le 1, \\ 0 & \text{for} \quad x_1^2 + x_2^2 \ge 1. \end{cases}$$

The free boundary Γ_f is identical with the circular arc $x_1^2 + x_2^2 = 1$, $(x_1, x_2) \in \Omega$, cf. Figure 4.3. □

Example 4.39 Our third example is based on the well-known elliptic variational inequality describing the elasto-plastic torsion. A cylindrical bar with a cross-section $\Omega \subset \mathbb{R}^2$ is subjected to a torsion applied at the upper end around the longitudinal axis in such a way that the lateral boundary remains stress-free. Assuming the 'von Mises' criterion for an elastic perfectly plastic material, it has been shown (cf. [RODR87], Section 1.6 and 8.4, 8.5) that for a positive twist angle θ per unit length the stress potential u is given as the solution of the elliptic variational inequality (4.1), (4.2) with $k = 1$, $q = 0$, $f = -2\theta$, $\Gamma = \partial\Omega = \Gamma_D$, $g_D = 0$ and the obstacle $\varphi(x) = -\mathrm{dist}(x, \Gamma)$. It should be mentioned that the mathematical modelling originally leads to an elliptic inequality with a gradient constraint, but Brézis and Sibony (see [BRSI71]) proved the equivalence to the obstacle problem studied here. A theoretical analysis of the problem can be found in [RODR87], Section 8.4 and 8.5. From the numerical point of view it is considered by many authors (see e.g., [GLT81], [HOKO94] and [EFHKW93]).

In the case of radial symmetry, i.e., $\Omega = \{x = (x_1, x_2) \in \mathbb{R}^2 : x_1^2 + x_2^2 < R^2\}$, one can find the explicit solution. Assuming $\theta \ge R^{-1}$, an elementary calculation shows, that the torsion is elasto-plastic and the stress potential u is given by

$$u = u(\rho) = \begin{cases} \rho - R & \text{in } P = \{\rho : \theta^{-1} \leq \rho \leq R\}, \\ 0.5\,\theta\,[\rho^2 - R^2 + (R - \theta^{-1})^2] & \text{in } E = \{\rho : 0 \leq \rho \leq \theta^{-1}\}, \end{cases}$$

where $\rho = \rho(x_1, x_2) = \sqrt{x_1^2 + x_2^2}$ and the obstacle is now $\varphi = \varphi(\rho) = \rho - R$. Due to the fact that we are specially interested in comparing finite element and finite volume schemes for the approximation of obstacle problems, we consider the following modified problem, characterized by the same obstacle φ and exact solution u as above, but with

$$k = 1, \quad q(x_1, x_2) = \max\{x_1^2 - x_2^2, 0\}, \quad f(x_1, x_2) = -2\,\theta + q(x_1, x_2)\, u(x_1, x_2),$$
$$\kappa(x_1, x_2) = \max\{x_2^2 - x_1^2, 0\}, \quad g_N(x_1, x_2) = 1 + \kappa(x_1, x_2)\, u(x_1, x_2).$$

The domain Ω is chosen as a circle with the radius $R = 0.5$. Instead of Dirichlet conditions we use natural (Neumann/Newton) conditions on the boundary $\partial\Omega = \Gamma_N$. Furthermore, the twist angle is taken as $\theta = 10/3$, such that the free boundary coincides with the circumference for $\rho(x_1, x_2) = (x_1^2 + x_2^2)^{1/2} = 0.3$. □

	Example 4.37		Example 4.38		Example 4.39	
h	error	error/h^2	error	error/h^2	error	error/h^2
1/20	8.819E-1	3.528E+2	4.397E-3	1.759	9.957E-7	3.983E-4
1/40	2.147E-1	3.435E+2	1.107E-3	1.772	2.471E-7	3.954E-4
1/80	5.221E-2	3.342E+2	2.781E-4	1.780	5.951E-8	3.809E-4
1/120	2.350E-2	3.384E+2	1.239E-4	1.785	2.625E-8	3.780E-4
1/160	1.336E-2	3.420E+2	6.972E-5	1.785	1.478E-8	3.785E-4
1/200	8.523E-3	3.409E+2	4.464E-5	1.786	9.566E-9	3.826E-4

Table 4.1: *Error behaviour* $\left\|P^G u_h - P^G y_h\right\|_{H^1(\Omega)}$ *between finite element* (u_h) *and finite volume* (y_h) *solution (SOR with projection).*

4.6.2 Numerical results for the error between the finite element and the finite volume solution

In order to compare the solutions of the finite element and the finite volume inequalities, we have applied the successive over-relaxation (SOR) method with projection, see [GLT81], [GLOW84]. The numerical results obtained for the difference $P^G u_h - P^G y_h$ in the H^1-norm are contained in Table 4.1 and depicted in Figure 4.4.

On the other hand, we have compared the finite element and box solutions of the corresponding penalization problems. To solve the finite element inequalities (4.9), (4.10), we use a penalty-iteration procedure analogous to (4.31), (4.39). To obtain $u_{h\varepsilon}^{(1)}$ as the corresponding solution, we apply a numerical integration technique proposed by Kikuchi (see [KIKU81]) for treatment of the corresponding penalization term

$$\beta_1(P^G w_h) = (P^G w_h)^- \quad \text{with} \quad V_h^G \ni P^G w_h = (P^G u_h - P^G \varphi_h).$$

Note that one has $\beta_1(P^G w_h) \in V$, but in general *not* $\beta_1(P^G w_h) \in V_h^G$. More precisely, the integral $(\beta_1(P^G w_h), P^G v_h)$ is replaced by a numerical quadrature scheme $Q_h(\beta_1(P^G w_h)\ P^G v_h)$, where Q_h is defined by

$$Q_h(v) = \sum_{S \in \mathcal{S}_h} \sum_{l=1}^{L} \omega_l^S \ v(x_l^S)$$

with x_l^S as the local coordinate of the l-th integration point within the element S and ω_l^S as the corresponding weights.
For a simplicial triangulation a natural choice consists in ω_l^S = const and x_l^S as the coordinate of the l-th nodal point (vertex) of the simplex S. In other words, using barycentric boxes, this quadrature scheme leads exactly to the term $\langle \beta_{1h}(w_h - \varphi_h), v_h \rangle$ in the penalization scheme (4.31).

In a similar way, we have computed a finite element solution $u_{h\varepsilon}^{(2)}$ by a penalty-iteration method analogous to (4.32), (4.40). The auxiliary grid function vector μ_{Gh} is defined by $\mu_{Gh} = ((L_G + M_G + R_G)\ \varphi_h - f_{Gh} - g_{Gh})^+$. The penalization term $\beta_{2h}(P^G w_h)$ is evaluated at the nodal points $\xi \in \tilde{\omega}$.

We have performed computations for different couplings between the penalization and discretization parameters (e.g., $\varepsilon = h^2$, $\varepsilon = h^3$). But, in all these cases the error behaves nearly like $O(h^2)$ measured in the H^1-norm and, moreover, also the absolute errors are not very sensitive about the coupling $\varepsilon = h^\delta$. Some of the results obtained by means of the penalty-iteration procedure (4.32), (4.40) with $\varepsilon = h^2$ are shown in Table 4.2.

For all numerical examples we have computed the exponent δ in $\|.\|_{H^1(\Omega)} \approx C\, h^\delta$ by averaging the single exponents

$$\delta_{ij} = (\log e_i - \log e_j)\ /\ (\log h_i - \log h_j), \quad i,j = 1, 10, \quad i > j.$$

Here, e_i denotes the error in the H^1-norm obtained for the mesh-size parameter h_i. In those columns of Tables 4.1, 4.2 in which we have taken h^2, this numerically determined exponent δ is actually between 1.99 and 2.01 (cf. Figure 4.4 for the Examples 4.38 and 4.39).

Summarizing, the numerical results presented in Tables 4.1, 4.2 and Figure 4.4 confirm the convergence order $O(h^2)$ in the H^1-norm theoretically derived in Theorem 4.13. In particular, we get the second order convergence also for Examples 4.38 and 4.39 with Neumann/Newton boundary conditions.

4.6.3 Error behaviour of the finite volume and the penalization solutions

In order to examine the error behaviour between the piecewise linear interpolant $\Pi_h u$ of the exact solution u and the finite volume solution $P^G y_h$, the finite

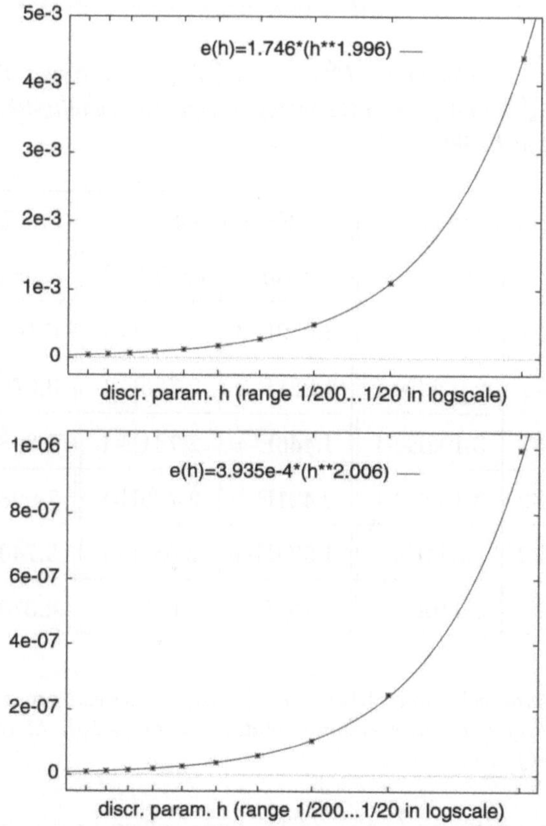

Figure 4.4: *Error behaviour between finite element* (u_h) *and finite volume* (y_h) *solution (SOR with projection). Solid lines: comparison curves* $e(h)$. *Stars* ($*$): *error* $\|P^G u_h - P^G y_h\|_{H^1(\Omega)}$ *for* $h = 1/200, \ldots, 1/20$. *Top: Example 4.38. Bottom: Example 4.39.*

	Example 4.37		Example 4.38		Example 4.39	
h	error	error$/h^2$	error	er.$/h^{1.976}$	error	err.$/h^{1.95}$
1/20	8.474E-1	3.390E+2	4.123E-3	1.535	8.987E-7	3.095E-4
1/40	2.108E-1	3.373E+2	1.070E-3	1.567	2.032E-7	2.703E-4
1/80	5.217E-2	3.339E+2	2.736E-4	1.576	5.322E-8	2.736E-4
1/120	2.342E-2	3.373E+2	1.226E-4	1.574	2.444E-8	2.770E-4
1/160	1.323E-2	3.387E+2	6.926E-5	1.570	1.399E-8	2.778E-4
1/200	8.478E-3	3.391E+2	4.446E-5	1.566	9.067E-9	2.783E-4

Table 4.2: *Error behaviour $\|P^G(z_{h\varepsilon} - u_{h\varepsilon}^{(2)})\|_{H^1(\Omega)}$ between finite element ($u_{h\varepsilon}^{(2)}$) and finite volume ($z_{h\varepsilon}$) solution (penalty-iteration (4.32), (4.40) with $\varepsilon = h^2$).*

	Example 4.37		Example 4.38		Example 4.39	
h	error	er.$/h^{1.447}$	error	er.$/h^{1.466}$	error	er.$/h^{1.176}$
1/20	4.579E-1	3.495E+1	3.059E-3	2.471E-1	6.163E-3	2.088E-1
1/40	1.396E-1	2.905E+1	1.221E-3	2.723E-1	3.070E-3	2.350E-1
1/80	5.637E-2	3.198E+1	4.446E-4	2.741E-1	1.298E-3	2.245E-1
1/120	3.128E-2	3.190E+1	2.421E-4	2.705E-1	7.998E-4	2.229E-1
1/160	1.956E-2	3.024E+1	1.574E-4	2.681E-1	5.740E-4	2.244E-1
1/200	1.459E-2	3.116E+1	1.137E-4	2.687E-1	4.370E-4	2.221E-1

Table 4.3: *Error behaviour $\|\Pi_h u - P^G y_h\|_{H^1(\Omega)}$ between exact solution u of (4.1) and finite volume solution y_h of (4.15), (4.16) (SOR with projection).*

volume inequality problems (4.15), (4.16) were solved by the SOR method with projection. Besides this, we have applied the penalty-iteration schemes (4.31), (4.39) and (4.32), (4.40) with 'small' penalization parameters ε. We have chosen different couplings of penalization and discretization parameters to keep the influence of the penalization small compared to the error $\left\|\Pi_h u - P^G y_h\right\|_{H^1(\Omega)}$.

Then, we looked for 'nearly minimal' errors $\left\|\Pi_h u - P^G y_{h\varepsilon}\right\|_{H^1(\Omega)}$ in (4.31), (4.39) (resp. $\left\|\Pi_h u - P^G z_{h\varepsilon}\right\|_{H^1(\Omega)}$ in (4.32), (4.40)).

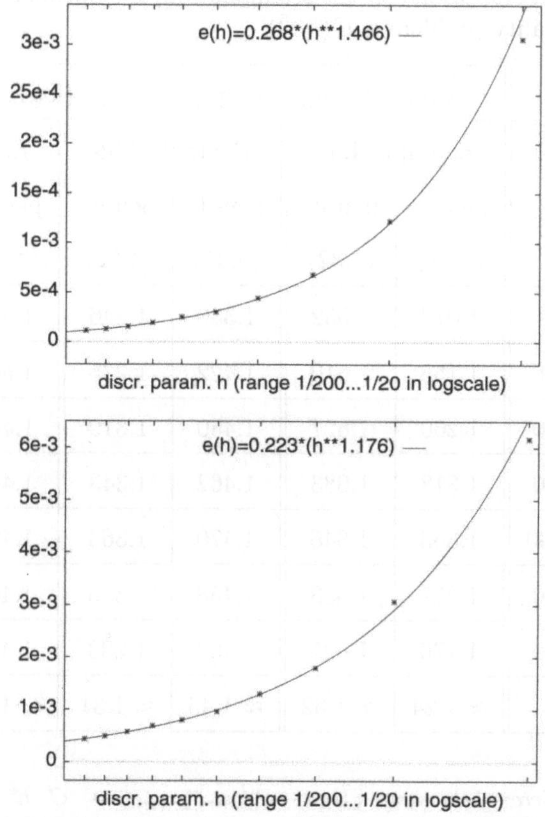

Figure 4.5: *Error behaviour between exact (u) and finite volume (y_h) solution (SOR with projection). Solid lines: comparison curves e(h). Stars (*): error* $\left\|\Pi_h u - P^G y_h\right\|_{H^1(\Omega)}$ *for* $h = 1/200, \ldots, 1/20$. *Top: Example 4.38. Bottom: Example 4.39.*

The absolute errors for these three methods are almost identical so that the numerical results are shown in Table 4.3 and Figure 4.5 for only the SOR method with projection. The exponents δ in $\left\|\Pi_h u - P^G y_h\right\|_{H^1(\Omega)} \approx C\, h^{\delta}$ are again computed by averaging the single exponents δ_{ij} (cf. Section 4.6.2).

Summarizing, the results presented in Table 4.3 and Figure 4.5 point out a convergence order of about $O(h^{1.45})$ for Examples 4.37, 4.38 and $O(h^{1.15})$ for the elasto-plastic torsion problem Example 4.39. The better convergence order obtained in Examples 4.37, 4.38 seems to be caused by using a (structured) mesh consisting of right-angled triangles for the unit square. Hence, the interior nodes $\xi \in \omega$ are the barycentres of their associated boxes. In contrast to this, an unstructured mesh consisting of general triangles was used for the circle in the elasto-plastic torsion Example 4.39. In this context we recall Remark 4.19 where we have discussed that, under restrictive assumptions on the triangulation, the optimal convergence order for the box inequality problem is $O(h^{1.5})$.

	exponents $\delta_{ij} = (\log e_i - \log e_j)/(\log h_i - \log h_j)$					
	Example 4.37		Example 4.38		Example 4.39	
$h_i : h_j$	pen-it. (4.31)	pen-it. (4.32)	pen-it. (4.31)	pen-it. (4.32)	pen-it. (4.31)	pen-it. (4.32)
1/20 : 1/40	1.072	1.552	1.330	1.246	1.483	1.173
1/40 : 1/80	1.156	1.610	1.422	1.275	1.560	1.197
1/60 : 1/120	1.260	1.627	1.450	1.319	1.499	1.250
1/80 : 1/160	1.318	1.683	1.462	1.343	1.440	1.276
1/100 : 1/200	1.353	1.646	1.470	1.363	1.401	1.286
1/60 : 1/180	1.297	1.633	1.458	1.336	1.465	1.268
1/20 : 1/140	1.170	1.625	1.400	1.283	1.485	1.203
average	≈ 1.24	≈ 1.62	≈ 1.43	≈ 1.31	≈ 1.47	≈ 1.24

Table 4.4: *Error behaviour* $\|\Pi_h u - P^G y_{h\varepsilon}\|_{H^1(\Omega)} \approx C\ h^\delta$ *between exact solution* u *of (4.1) and finite volume solution* $y_{h\varepsilon}$ *of the penalty-iteration (pen-it.) (4.31), (4.39)* ($\varepsilon = h^2$). *Error* $\|\Pi_h u - P^G z_{h\varepsilon}\|_{H^1(\Omega)} \approx C\ h^\delta$ *with* $z_{h\varepsilon}$ *as solution of the penalty-iteration (pen-it.) (4.32), (4.40)* ($\varepsilon = h^2$).

Finally, in Table 4.4 some numerical results are presented for the penalty-iteration methods (4.31), (4.39) and (4.32), (4.40), where the coupling $\varepsilon = h^2$ is taken according to Remark 4.29[i]. The exponents δ_{ij} are computed by

$$\delta_{ij} = (\log e_i - \log e_j)/(\log h_i - \log h_j),$$

where e_i denotes the error measured in the H^1-norm between the solution of (4.39) (respectively (4.40)) and $\Pi_h u$ obtained for the mesh-size parameter h_i.

In all examples the error behaviour is (somewhat) better than the theoretically predicted first order convergence (cf. Remark 4.29[i]). In particular, for Example 4.37

the absolute values of the errors obtained by (4.31), (4.39) with $\varepsilon = h^2$ are greater than the corresponding values of (4.32), (4.40), which seems to be caused by the 'great' value of the term M_{φ_h} in this example. Therefore, from the practical point of view, we recommend the mesh-dependent choice of the penalization parameter for the penalty-iteration scheme (4.31), (4.39) resulting from Theorem 4.27.

International Series of Numerical Mathematics
Vol. 136, © 2002 Birkhäuser Verlag Basel/Switzerland

Chapter 5

Numerical Analysis of the Evolutionary Variational Inequalities

We will now derive and analyse finite element and finite volume approximations for evolutionary variational inequalities studied in Chapter 3 from the analytical point of view. For this purpose we often exploit results obtained in Chapter 4 for the numerical treatment of elliptic variational inequalities.

Both the finite element and box schemes will be investigated in a variational (Galerkin) framework. The numerical treatment of the evolutionary variational inequalities with mixed boundary conditions will be performed for spatial two- and three-dimensional domains and includes the same topics as in Chapter 4, i.e., solvability and stability of discrete problems, maximum principle, comparison of finite element and box schemes, error estimates based exactly on regularity of the solution gained in Chapter 3 and different penalization (regularization) methods.

Let us anticipate an essential property (advantage) of our memory term discretization. It does *not* require us to store the (solution-) grid function vectors of preceding time instants. The computational effort for treatment of the memory term comprises only two matrix-vector multiplications at each time instant (cf. Remark 5.2).

Before we give the outline of the present chapter, we refer to the comprehensive paper [MCTH93] by McLean and Thomée dealing with the numerical treatment of an evolution *equation* with a *positive-type* memory term. In Section 5.1 we will formulate the finite element and box schemes including an explanation of the key idea for discretizing the memory term in the evolutionary inequalities (3.1), (3.2). Furthermore, some basic properties of the proposed (discrete) inequality formulations will be derived (boundedness of the solutions, several stability estimates, equivalent complementarity problem, time evolution of the solution).

In Section 5.2 we are concerned with the accuracy of the finite element and finite volume approximations. On the one hand, the finite element is compared with the finite volume approximation (cf. also Section 4.2) and, on the other hand, we analyse the convergence rate of both approximations compared to the solution u of the original evolutionary inequality formulation (3.1), (3.2). It should be mentioned that our error analysis is based on the regularity of the solution u obtained in Chapter 3 (see, in particular, Sections 3.3, 3.4 and 3.5). As the main result we will obtain the convergence order

$$O(h) + O(\tau)$$

in the norm of the space $L_\infty(0,T;H^1(\Omega))$ (τ as time discretization parameter).

Similar to Section 4.4, in Section 5.3 we propose and analyse penalization and regularization methods for solution of the discrete evolutionary inequalities. Our investigations are based again on the discrete maximum principle (cf. Section 4.4.1), so we concentrate on the finite volume problems. Coupling the space- and time-discretization parameters with the penalization parameters the overall error can be reduced (cf. Remark 5.24). In Section 5.3.2 we discuss the iterative solution of (discrete) penalization problems.

Section 5.4 contains results from numerical experiments for two problems of the form (3.1), (3.2) with known exact solution. These two model problems are also closely related to the degenerate free boundary problem introduced in Section 2.1.

5.1 Finite element and finite volume approximations for the evolutionary variational inequalities

Restricting ourselves again to bounded *polygonal (polyhedral)* domains Ω in $\mathbb{R}^2$ ($\mathbb{R}^3$) we retain all the notations used in Chapter 4 for the numerical treatment of elliptic inequalities.

Besides the spatial discretization introduced in Sections 4.1.2 and 4.1.3 we require a discretization of the time interval $[0,T]$ of form such that

$$\tau = T/N$$

denotes the time step and

$$t^j = j\,\tau, \quad j = 0,\ldots,N$$

are the time instants.
In the sequel, the superscripts i and j should indicate the values of grid function vectors and matrices at the time instants t^i and t^j, $i,j = 0,\ldots,N$.

Let us recall the standard assumptions (3.4), (3.3) from Section 3.1 on which our analytical investigations are based. For the numerical treatment we will basically assume the same conditions. However, to control the obstacle and the Dirichlet conditions pointwise we require slightly stronger assumptions which can be summarized as follows.

$$\left.\begin{array}{l} k,\ q \in W^1_\infty(0,T;L_\infty(\Omega)), \quad \kappa \in L_\infty(\Gamma_N), \quad f \in C([0,T];L_2(\Omega)), \\ g_N \in C([0,T];L_2(\Gamma_N)), \quad g_D \in \big(C([0,T];H^1(\Omega)) \cap C(\overline{\Gamma_D} \times [0,T])\big), \\ k(x,t) \geq k_0 > 0, \quad q(x,t) \geq 0 \quad \text{for } x \in \Omega,\ t \in [0,T], \\ \kappa(x) \geq 0 \ \text{ for } x \in \Gamma_N; \quad \text{meas}_{n-1}(\Gamma_D) + \text{meas}_{n-1}(\Gamma_\kappa) + \text{meas}_n(\Omega_q) > 0 \\ \text{with} \quad \Omega_q = \{x \in \Omega : q(x,t) \geq q_0 > 0\ \forall t \in [0,T]\} \quad \text{and} \\ \Gamma_\kappa = \{x \in \overline{\Gamma_N} : \kappa(x) \geq \kappa_0 > 0\} \end{array}\right| \tag{5.1}$$

and with respect to the obstacle $\varphi = \varphi(x)$

$$\varphi \in W^1_p(\Omega), \quad p > n; \quad \varphi \leq g_D(t) \quad \text{on } \Gamma_D, \quad \forall t \in [0,T]. \tag{5.2}$$

5.1.1 Formulation of the finite element and finite volume approximations

For the numerical solution of the inequality problem (3.1), (3.2) we propose a finite element approximation which reads in matrix notation as follows.

$$\left.\begin{array}{l} \text{Find} \quad u^j_h \in K^j_h : \qquad \langle \Lambda^j_G\, u^j_h, v_h - u^j_h \rangle \\ \geq \langle f^j_{Gh} + g^j_{Gh}, v_h - u^j_h \rangle + \displaystyle\sum_{i=0}^{j-1} \langle (\Lambda^{i+1}_G - \Lambda^i_G)\, u^i_h, v_h - u^j_h \rangle \\ \forall v_h \in K^j_h, \quad j = 0, \ldots, N \end{array}\right| \tag{5.3}$$

with the time-dependent constraint sets

$$\begin{aligned} K^i_h &= \{w_h \in L_2(\overline{\omega}) : w_{h,\xi} \geq \varphi(\xi)\ \forall \xi \in \overline{\omega}, \quad w_{h,\xi} = g_D(\xi, t^i)\ \forall \xi \in \gamma_D\} \\ &= \{P^G w_h \in X^G_h : (P^G w_h)(\xi) \geq \varphi(\xi)\ \forall \xi \in \overline{\omega}, \quad (P^G w_h)(\xi) = g_D(\xi, t^i)\ \forall \xi \in \gamma_D\} \end{aligned}$$

for $i = 0, \ldots, N$.

Remark 5.1 For the sake of clarity, dependence of the coefficients on the space variable x and time t will be given in the definition of the matrices and grid function vectors in (5.4) and (5.10) below. In contrast to this, for the sake of brevity we will suppress the spatial dependence of the coefficients at some places in the sequel. □

The matrices and grid function vectors for the finite element formulation are defined in the following way.

$$\left.\begin{aligned}
&\Lambda_G^i = (1-\alpha)\ (L_G^i + M_G^i) + \alpha\ (L_G^{i-1} + M_G^{i-1}) + R_G, \quad \alpha \in [0,1],\\
&(L_G^i)_{\xi,\eta} = \int_\Omega k(x,t^i)\ \nabla\Phi_\xi^G(x)\ \nabla\Phi_\eta^G(x)\ dx,\\
&(M_G^i)_{\xi,\eta} = \int_\Omega q(x,t^i)\ \Phi_\xi^G(x)\ \Phi_\eta^G(x)\ dx,\\
&(R_G)_{\xi,\eta} = \int_{\Gamma_N} \kappa(x)\ \Phi_\xi^G(x)\ \Phi_\eta^G(x)\ ds,\\
&g_{Gh,\xi}^i = \int_{\Gamma_N} g_N(x,t^i)\ \Phi_\xi^G(x)\ ds,\\
&f_{Gh,\xi}^i = \int_\Omega f(x,t^i)\ \Phi_\xi^G(x)\ dx, \quad \xi \in \tilde{\omega},\ \eta \in \overline{\omega}, \quad i = 0,\dots,N.
\end{aligned}\right| \tag{5.4}$$

Additionally, we define

$$L_G^{-1} := L_G^0 \quad \text{and} \quad M_G^{-1} := M_G^0.$$

Obviously, the matrices and grid function vectors just defined (5.4) can be equivalently written as

$$\begin{aligned}
&\langle \Lambda_G^i\, v_h, w_h \rangle = (1-\alpha)\ a(t^i; P^G v_h, P^G w_h) + \alpha\ a(t^{i-1}; P^G v_h, P^G w_h),\\
&\langle f_{Gh}^i + g_{Gh}^i, w_h \rangle = (f(t^i), P^G w_h) + (g_N(t^i), P^G w_h)_{\Gamma_N}
\end{aligned}$$

for $v_h \in L_2(\overline{\omega})$ and $w_h \in L_2(\tilde{\omega})$.
The underlying idea for the just given finite element discretization (5.3) should be explained in the following remark considering, in particular, the memory term in (3.1), (3.2).

Remark 5.2 The discretization scheme (5.3) arises from splitting of the memory term in (3.1), (3.2) into subintegrals, i.e.,

$$\int_0^{t^j} b(t'; u(t'), v)\ dt' = \sum_{i=0}^{j-1} \int_{t^i}^{t^{i+1}} b(t'; u(t'), v)\ dt'.$$

Each of these subintegrals is approximated by a trapezoidal rule weighted with a parameter $\alpha \in [0,1]$. This leads to the finite element inequality problem:

$$\left.\begin{aligned}
&\text{Find } u_h^j \in K_h^j: \ \ a(t^j; P^G u_h^j, P^G v_h - P^G u_h^j) \geq (f(t^j), P^G v_h - P^G u_h^j)\\
&+ (g_N(t^j), P^G v_h - P^G u_h^j)_{\Gamma_N} + \sum_{i=0}^{j-1} b_G^i(P^G u_h, P^G v_h - P^G u_h^j)\\
&\forall v_h \in K_h^j, \quad j = 0,\dots,N.
\end{aligned}\right| \tag{5.5}$$

The memory term in its discretized form is given by

$$
\begin{aligned}
& b^i_G(P^G u_h, P^G v_h) \\
& = \int_\Omega \left(k(t^{i+1}) - k(t^i)\right) \left[\alpha\, \nabla(P^G u_h^{i+1}) + (1-\alpha)\nabla(P^G u_h^i)\right] \nabla P^G v_h \, dx \\
& + \int_\Omega \left(q(t^{i+1}) - q(t^i)\right) \left[\alpha\, P^G u_h^{i+1} + (1-\alpha)\, P^G u_h^i\right] P^G v_h \, dx.
\end{aligned}
\tag{5.6}
$$

Then, at the time instant $t = t^j$, all terms containing the unknown u_h^j are collected in the left-hand side (cf. also Lemma 5.13 below). Consequently, by means of the additional definitions

$$k(t^{-1}) := k(t^0) \quad \text{and} \quad q(t^{-1}) := q(t^0),$$

we arrive at the finite element approximation (5.3), (5.4).

As already indicated at the beginning of the present chapter, we do *not* need explicitly the (solution-) vectors u_h^i, $i = 0, \ldots, j-1$ for the computation of u_h^j at the time instant t^j, $j = 0, \ldots, N$ according to (5.3), (5.4). In other words, it is not necessary to store the preceding (solution-) vectors. At each time instant t^j, two matrix-vector multiplications, i.e.,

$$(\Lambda_G^j - \Lambda_G^{j-1})\, u_h^{j-1},$$

have to be performed. The results of these multiplications are added to *one* vector which represents the necessary additional storage caused by the memory term in its discretized form. The same property which is advantageous for computer implementation is also valid for the finite volume approximation introduced in what follows. □

Recalling the basic principles for the construction of finite volume schemes from Section 4.1.3 and using ideas for the discretization of the memory term similar to those in the finite element approximation (5.5), we obtain the following finite volume inequalities.

$$
\begin{aligned}
& \text{Find } y_h^j \in K_h^j : \; a_B(t^j; P^G y_h^j, P^B v_h - P^B y_h^j) \ge (f(t^j), P^B v_h - P^B y_h^j) \\
& + (g_N(t^j), P^B v_h - P^B y_h^j)_{\Gamma_N} + \sum_{i=0}^{j-1} b_B^i(P^G y_h, P^B v_h - P^B y_h^j) \\
& \forall v_h \in K_h^j, \quad j = 0, \ldots, N
\end{aligned}
\tag{5.7}
$$

with the box bilinear forms a_B and b_B defined by

$$\begin{aligned}
a_B(t^i; P^G y_h, P^B v_h) &= -\sum_{\xi\in\tilde\omega} \int_{\partial D_\xi\setminus\Gamma} k(t^i)\ (\nabla P^G y_h \cdot \vec{n})\ P^B v_h\ ds \\
&+ \int_\Omega q(t^i)\ P^B y_h\ P^B v_h\ dx + \int_{\Gamma_N} \kappa\ P^B y_h\ P^B v_h\ ds, \\
b_B^i(P^G y_h, P^B v_h) &= -\sum_{\xi\in\tilde\omega} \int_{\partial D_\xi\setminus\Gamma} (k(t^{i+1}) - k(t^i))\ \big[\alpha\ \nabla(P^G y_h^{i+1})\cdot\vec{n} \\
&+ (1-\alpha)\ \nabla(P^G y_h^i)\cdot\vec{n}\big]\ P^B v_h\ ds \\
&+ \int_\Omega (q(t^{i+1}) - q(t^i))\ \big[\alpha\ P^B y_h^{i+1} + (1-\alpha)\ P^B y_h^i\big]\ P^B v_h\ dx.
\end{aligned} \tag{5.8}$$

Again, at the time instant $t = t^j$, all terms containing the unknown y_h^j are collected in the left-hand side. Therefore, we end up with the following box inequalities given in matrix notation.

$$\begin{aligned}
&\text{Find } y_h^j \in K_h^j : \quad \langle \Lambda_B^j\ y_h^j, v_h - y_h^j\rangle \\
&\geq \langle f_{Bh}^j + g_{Bh}^j, v_h - y_h^j\rangle + \sum_{i=0}^{j-1} \langle (\Lambda_B^{i+1} - \Lambda_B^i)\ y_h^i, v_h - y_h^j\rangle \\
&\forall v_h \in K_h^j, \quad j = 0,\dots,N
\end{aligned} \tag{5.9}$$

with the time-dependent convex sets

$$K_h^i = \{w_h \in L_2(\overline{\omega}) : w_{h,\xi} \geq \varphi(\xi)\ \forall\xi\in\overline{\omega}, \quad w_{h,\xi} = g_D(\xi, t^i)\ \forall\xi\in\gamma_D\}$$

for $i = 0,\dots,N$.

The matrices and grid function vectors are defined as follows.

$$\begin{aligned}
\Lambda_B^i &= (1-\alpha)\ (L_B^i + M_B^i) + \alpha\ (L_B^{i-1} + M_B^{i-1}) + R_B, \quad \alpha\in[0,1], \\
(L_B^i)_{\xi,\eta} &= -\sum_{D\in\mathcal{D}_h} \int_{\partial D\setminus\Gamma} k(x,t^i)\ (\nabla\Phi_\eta^G(x)\cdot\vec{n})\ \Phi_\xi^B(x)\ ds \\
&= -\int_{\partial D_\xi\setminus\Gamma} k(x,t^i)\ (\nabla\Phi_\eta^G(x)\cdot\vec{n})\ ds, \\
(M_B^i)_{\xi,\eta} &= \int_\Omega q(x,t^i)\ \Phi_\eta^B(x)\ \Phi_\xi^B(x)\ dx, \\
(R_B)_{\xi,\eta} &= \int_{\Gamma_N} \kappa(x)\ \Phi_\eta^B(x)\ \Phi_\xi^B(x)\ ds, \\
g_{Bh,\xi}^i &= \int_{\Gamma_N} g_N(x,t^i)\ \Phi_\xi^B(x)\ ds, \\
f_{Bh,\xi}^i &= \int_\Omega f(x,t^i)\ \Phi_\xi^B(x)\ dx, \quad \xi\in\tilde\omega,\ \eta\in\overline{\omega}, \quad i = 0,\dots,N.
\end{aligned} \tag{5.10}$$

For the sake of completeness we define $L_B^{-1} := L_B^0$ and $M_B^{-1} := M_B^0$. Furthermore, we refer to Remark 5.3 for the definition of the box stiffness matrices Λ_B^i.

The matrices and grid function vectors defined in (5.10) can be equivalently expressed by

$$\langle \Lambda_B^i\, v_h, w_h \rangle = (1-\alpha)\; a_B(t^i; P^G v_h, P^B w_h) + \alpha\; a_B(t^{i-1}; P^G v_h, P^B w_h),$$
$$\langle f_{Bh}^i + g_{Bh}^i, w_h \rangle = (f(t^i), P^B w_h) + (g_N(t^i), P^B w_h)_{\Gamma_N}$$

for $v_h \in L_2(\overline{\omega})$ and $w_h \in L_2(\tilde{\omega})$.

Remark 5.3 Similar to our study of the box schemes applied to elliptic inequalities in Chapter 4, we will now use equality of the finite element and box stiffness matrices. Recalling Lemma 4.1[ii] we require the condition $k(t^i) \in V_h^C$ for $i = 0, \dots, N$ to guarantee $L_G^i = L_B^i$. Remembering Remark 4.4 we note that this assumption does not imply a loss of generality. Therefore, we will make use of the equality

$$L_G^i = L_B^i \quad \text{for} \quad i = 0, \dots, N$$

without repeating this condition in the subsequent lemmas and theorems of this chapter. □

5.1.2 Properties of the discrete inequality problems

In this section we will show the boundedness of solutions of the discrete inequalities (5.3), (5.9) (independent of the discretization parameters) and derive several stability estimates. Furthermore, the equivalent (finite-dimensional) complementarity problem will be formulated. First, let us summarize some basic properties of the proposed approximations in the following lemma.

Lemma 5.4 *Let the box partitions be constructed according to $\mathcal{D}_h(1)$ and the assumptions (5.1) be fulfilled.*

[i] *There exist constants $C_i > 0$, $i = 1, \dots, 8$, such that*

$$\langle \Lambda_G^i\, w_h, w_h \rangle \ge C_1 \left\| P^G w_h \right\|_{H^1(\Omega)}^2 \ge C_2 \left\| w_h \right\|_{1,\overline{\omega}}^2,$$
$$\langle \Lambda_G^i\, v_h, w_h \rangle \le C_3 \left\| P^G v_h \right\|_{H^1(\Omega)} \left\| P^G w_h \right\|_{H^1(\Omega)} \le C_4 \left\| v_h \right\|_{1,\overline{\omega}} \left\| w_h \right\|_{1,\overline{\omega}},$$
$$\langle \Lambda_B^i\, w_h, w_h \rangle \ge C_5 \left\| P^G w_h \right\|_{H^1(\Omega)}^2 \ge C_6 \left\| w_h \right\|_{1,\overline{\omega}}^2,$$
$$\langle \Lambda_B^i\, v_h, w_h \rangle \le C_7 \left\| P^G v_h \right\|_{H^1(\Omega)} \left\| P^G w_h \right\|_{H^1(\Omega)} \le C_8 \left\| v_h \right\|_{1,\overline{\omega}} \left\| w_h \right\|_{1,\overline{\omega}}$$

hold for $i = 0, \dots, N$ and arbitrary grid function vectors $v_h \in L_2(\overline{\omega})$, $w_h \in L_2(\tilde{\omega})$.

[ii] *Let L_k and L_q be the Lipschitz constants of the mappings*

$$[0,T] \ni t \longrightarrow k(t) \in L_\infty(\Omega) \quad \text{and} \quad [0,T] \ni t \longrightarrow q(t) \in L_\infty(\Omega),$$

respectively. Then,

$$\langle (\Lambda_G^{i+1} - \Lambda_G^i)\, v_h, w_h \rangle \leq \max\{L_k, L_q\}\ \tau\ \|P^G v_h\|_{H^1(\Omega)}\ \|P^G w_h\|_{H^1(\Omega)},$$
$$\langle (\Lambda_B^{i+1} - \Lambda_B^i)\, v_h, w_h \rangle \leq \max\{L_k, L_q/C_1^2\}\ \tau\ \|P^G v_h\|_{H^1(\Omega)}\ \|P^G w_h\|_{H^1(\Omega)}$$

for $i = 0, \dots, N-1$ *and grid function vectors* $v_h \in L_2(\overline{\omega})$, $w_h \in L_2(\tilde{\omega})$. *The constant* C_1 *is given in Lemma 4.1[i].*

[iii] *Let* $w_h \in L_2(\tilde{\omega})$. *Then*

$$\langle f_{Gh}^i + g_{Gh}^i, w_h \rangle \leq \|f(t^i)\|_{L_2(\Omega)}\ \|P^G w_h\|_{L_2(\Omega)} + c\, \|g_N(t^i)\|_{L_2(\Gamma_N)}\ \|P^G w_h\|_{H^1(\Omega)}$$

and

$$\begin{aligned}\langle f_{Bh}^i + g_{Bh}^i, w_h \rangle\ &\leq C_1^{-1}\, \|f(t^i)\|_{L_2(\Omega)}\ \|P^G w_h\|_{L_2(\Omega)} \\ &+ \frac{c}{C_3}\, \|g_N(t^i)\|_{L_2(\Gamma_N)}\ \|P^G w_h\|_{H^1(\Omega)}\end{aligned}$$

hold for $i = 0, \dots, N$. *The constants* C_1, C_3 *are given in Lemma 4.1[i] and* c *denotes the constant in the trace theorem.*

Proof. [i]. The statements for the finite element matrix Λ_G^i are an immediate consequence of the V-ellipticity and continuity of the bilinear form a as well as Lemma 4.1[i].
The matrices Λ_B^i can be represented by the box bilinear form a_B, so we refer to Theorem 4.2[i] concerning the claimed properties of Λ_B^i.

[ii]. According to the definition of Λ_G^i in (5.4) we have

$$\begin{aligned}\langle (\Lambda_G^{i+1} - \Lambda_G^i)\, v_h, w_h \rangle\ &= (1-\alpha)\big(a(t^{i+1}; P^G v_h, P^G w_h) - a(t^i; P^G v_h, P^G w_h)\big) \\ &+ \alpha\ \big(a(t^i; P^G v_h, P^G w_h) - a(t^{i-1}; P^G v_h, P^G w_h)\big).\end{aligned}$$

Consequently, the estimate of the finite element approximation of the memory term is obtained by means of properties of the bilinear form a. Using the analogous representation for $\langle (\Lambda_B^{i+1} - \Lambda_B^i)\, v_h, w_h \rangle$ and Lemma 4.1[i] we get the statement for the difference of the box matrices.

[iii]. The boundedness of the (finite element) grid function vectors f_{Gh}^j and g_{Gh}^j is obvious, whereas the analogous estimates for f_{Bh}^j and g_{Bh}^j are obtained using additionally Lemma 4.1[i]. ■

Remark 5.5 Sometimes it will be useful to exploit the boundedness of the right-hand sides in the corresponding grid function vector norms, i.e.,

$$\begin{aligned}
\langle f_{Bh}^i, w_h\rangle &\le \|f_{Bh}^i/\operatorname{meas}_n(D)\|_{0,\varpi}\ \|w_h\|_{0,\varpi} \le \|f(t^i)\|_{L_2(\Omega)}\ \|P^B w_h\|_{L_2(\Omega)},\\
\langle g_{Bh}^i, w_h\rangle &\le \|g_{Bh}^i/\operatorname{meas}_{n-1}(\partial D)\|_{0,\gamma}\ \|w_h\|_{0,\gamma} \le \|g_N(t^i)\|_{L_2(\Gamma_N)}\ \|P^B w_h\|_{L_2(\Gamma)},\\
\langle f_{Gh}^i, w_h\rangle &\le \|f_{Gh}^i/\operatorname{meas}_n(D)\|_{0,\varpi}\ \|w_h\|_{0,\varpi} \le c\ \|f(t^i)\|_{L_2(\Omega)}\ \|P^B w_h\|_{L_2(\Omega)},\\
\langle g_{Gh}^i, w_h\rangle &\le \|g_{Gh}^i/\operatorname{meas}_{n-1}(\partial D)\|_{0,\gamma}\ \|w_h\|_{0,\gamma} \le C\ \|g_N(t^i)\|_{L_2(\Gamma_N)}\ \|P^B w_h\|_{L_2(\Gamma)}
\end{aligned}$$

hold for $i = 0,\ldots,N$ and $w_h \in L_2(\tilde{\omega})$.
These relations can be derived by means of the boxes constructed in Section 4.1.3; cf., in particular, (4.12), (4.13) and Remark 4.3[iii]. □

Now, all auxiliary tools are available to show solvability of the discrete inequality problems (5.3), (5.9) and boundedness of their solutions. This is the subject of the next theorem.

Theorem 5.6 *Let the conditions (5.1), (5.2) be fulfilled. Furthermore, let the box partitions be defined according to $\mathcal{D}_h(1)$. Then, for all $j = 0,\ldots,N$, the finite element inequality problem (5.3), (5.4) (resp. the box inequality problem (5.9), (5.10)) possesses a unique solution*

$$P^G u_h^j \in K_h^j \quad (\textit{resp.}\quad P^G y_h^j \in K_h^j),$$

such that

$$\begin{aligned}
\|P^G u_h^j\|_{H^1(\Omega)} &\le \exp(c_1\, t^j) \max_{i=0,\ldots,j} \{c_2\,\|f(t^i)\|_{L_2(\Omega)} + c_3\,\|g_N(t^i)\|_{L_2(\Gamma_N)}\\
&\quad + c_4\,\|P^G v_h^i\|_{H^1(\Omega)}\}
\end{aligned}$$

and

$$\begin{aligned}
\|P^G y_h^j\|_{H^1(\Omega)} &\le \exp(c_5\, t^j) \max_{i=0,\ldots,j} \{c_6\,\|f(t^i)\|_{L_2(\Omega)} + c_7\,\|g_N(t^i)\|_{L_2(\Gamma_N)}\\
&\quad + c_8\,\|P^G v_h^i\|_{H^1(\Omega)}\}
\end{aligned}$$

hold for all $j = 0,\ldots,N$ and arbitrary grid function vectors $v_h^i \in K_h^i$ with $i = 0,\ldots,N$.

Proof. We conclude *inductively* that, for each time instant t^j with $j = 0,\ldots,N$, the finite element scheme (5.3), (5.4) represents an elliptic variational inequality with a right-hand side F_{Gh}^j defined by

$$F_{Gh}^j = f_{Gh}^j + g_{Gh}^j + \sum_{i=0}^{j-1}(\Lambda_G^{i+1} - \Lambda_G^i)\, u_h^i.$$

Owing to the results of Lemma 5.4 the assumptions of the Lions–Stampacchia theorem (cf. Sections 3.2 and 4.1.1) are satisfied. Hence, a unique solution u_h^j exists for $j = 0, \ldots, N$.

To prove the claimed estimate for the finite element solution u_h^j, we choose grid function vectors $v_h = v_h^j \in K_h^j$ for $j = 0, \ldots, N$, such that $(P^G(u_h^j - v_h^j)) \in V_h^G$ is guaranteed. Consequently, we obtain

$$\begin{aligned}\langle \Lambda_G^j\,(u_h^j - v_h^j), v_h^j - u_h^j\rangle \;&\geq \langle f_{Gh}^j + g_{Gh}^j - \Lambda_G^j\, v_h^j, v_h^j - u_h^j\rangle \\ &\quad + \sum_{i=0}^{j-1} \langle (\Lambda_G^{i+1} - \Lambda_G^i)\, u_h^i, v_h^j - u_h^j\rangle.\end{aligned}$$

Using the results of Lemma 5.4 and the triangle inequality, we derive

$$\begin{aligned}\left\| P^G u_h^j \right\|_{H^1(\Omega)} \;&\leq \max_{i=0,\ldots,j} \{c_2 \left\| f(t^i) \right\|_{L_2(\Omega)} + c_3 \left\| g_N(t^i) \right\|_{L_2(\Gamma_N)} \\ &\quad + c_4 \left\| P^G v_h^i \right\|_{H^1(\Omega)} \} + c_1\,\tau \sum_{i=0}^{j-1} \left\| P^G u_h^i \right\|_{H^1(\Omega)}\end{aligned}$$

for $j = 0, \ldots, N$.

Applying Gronwall's inequality (Lemma 3.1) we deduce the above estimate for $P^G u_h^j$, $j = 0, \ldots, N$.

In an analogous way, the statements for the box inequality problem (5.9), (5.10) can be proved. ∎

Remark 5.7 [i] The constant c_1 in the estimate for the finite element solution $P^G u_h$ stated in Theorem 5.6 is given by the maximum of the Lipschitz constants L_k and L_q divided by the ellipticity constant of the bilinear form a. In the same way, the constant c_5 depends on these Lipschitz constants and the ellipticity of the box bilinear form a_B (cf. Lemma 5.4).

[ii] Assuming a function $v \in L_\infty(0, T; W_p^1(\Omega))$, $p > n$ with

$$v(t) \in (K(t) \cap W_p^1(\Omega)) \quad \text{and} \quad \|v(t)\|_{W_p^1(\Omega)} \leq C \quad \forall t \in [0, T],$$

one can take $K_h^i \ni v_h^i := \left(\Pi_h(v(t^i))\right)_{\xi \in \overline{\omega}}$, $i = 0, \ldots, N$ in the estimates stated in Theorem 5.6 (cf. also Remark 4.3[ii]). Owing to the Sobolev embedding $W_p^1(\Omega) \subset C(\overline{\Omega})$ for $p > n$ and interpolation properties in Sobolev spaces, we get

$$\begin{aligned}\left\| P^G v_h^i \right\|_{H^1(\Omega)} \;&\leq\; C_1\, \left\| \Pi_h(v(t^i)) \right\|_{W_p^1(\Omega)} \leq C_2\, \left\| v(t^i) \right\|_{W_p^1(\Omega)} \\ &\leq\; C_2 \max_{t \in [0,T]} \|v(t)\|_{W_p^1(\Omega)} \leq C \quad \text{for all} \quad i = 0, \ldots, N.\end{aligned}$$

In the absence of Dirichlet conditions or in the situation $\varphi = g_D$ on Γ_D, we can take $v = \varphi$. Furthermore, in Chapter 3 (see, in particular, Section 3.5) we have shown that the solution u of the evolutionary inequality (3.1), (3.2) fulfills $u \in L_\infty(0,T;H^2(\Omega))$ with $u(t) \in H^2(\Omega)$ and $\|u(t)\|_{H^2(\Omega)}$ is bounded for all $t \in [0,T]$ (resp. even in the space $W_p^2(\Omega)$ for $2 \le p < \infty$). Consequently, $v = u$ is possible.

Furthermore, let us introduce $I_\tau^0(P^G v_h)$ as the piecewise constant in time function of $P^G v_h$ with respect to the time discretization

$$[t^0, t^1 - 0.5\tau],\ [t^1 - 0.5\tau, t^1 + 0.5\tau],\ \ldots,\ [t^N - 0.5\tau, t^N].$$

More precisely, we have

$$(I_\tau^0(P^G v_h))(.,t) = \begin{cases} P^G v_h^0(.) & 0 \le t < t^0 + \tau/2, \\ P^G v_h^i(.) & t^i - \tau/2 < t \le t^i + \tau/2, \quad i = 1,\ldots,N-1, \\ P^G v_h^N(.) & t^N - \tau/2 \le t \le T. \end{cases}$$

Consequently, the estimate for the box solution $P^G y_h$ of Theorem 5.6 can be rewritten in the form

$$\begin{aligned} \left\|I_\tau^0(P^G y_h)\right\|_{L_\infty(0,T;H^1(\Omega))} &\le \exp(c_5\, T)\ \big[c_6\ \|f\|_{C([0,T];L_2(\Omega))} \\ &\quad + c_7\ \|g_N\|_{C([0,T];L_2(\Gamma_N))} + c_9 \max_{t\in[0,T]} \|v(t)\|_{W_p^1(\Omega)}\big] \end{aligned}$$

(analogously for the finite element solution $P^G u_h$). □

Our aim is now to derive several a priori estimates which are essential for the stability of the fully discrete schemes (5.3), (5.4) and (5.9), (5.10).

Theorem 5.8 *Let the same assumptions as in Theorem 5.6 be fulfilled.*
[i] *Let $u_h = u_h(f_{Gh}, g_{Gh}, \varphi_{Wh})$ and $\hat{u}_h = \hat{u}_h(\hat{f}_{Gh}, \hat{g}_{Gh}, \hat{\varphi}_{Wh})$ be solutions of the finite element approximation (5.3), where the grid function vectors φ_{Wh}^i and $\hat{\varphi}_{Wh}^i$ are defined by*

$$\varphi_{Wh,\xi}^i = \begin{cases} \varphi(\xi) & \text{for } \xi \in \tilde{\omega}, \\ g_D(\xi, t^i) & \text{for } \xi \in \gamma_D, \end{cases} \quad \text{and} \quad \hat{\varphi}_{Wh,\xi}^i = \begin{cases} \hat{\varphi}(\xi) & \text{for } \xi \in \tilde{\omega}, \\ \hat{g}_D(\xi, t^i) & \text{for } \xi \in \gamma_D, \end{cases}$$

for $i = 0,\ldots,N$. Then,

$$\begin{aligned} &C_1 \left\|P^G(u_h^j - \hat{u}_h^j)\right\|_{H^1(\Omega)} \le \left\|u_h^j - \hat{u}_h^j\right\|_{1,\overline{\omega}} \\ &\le C_2 \exp(C_3\, t^j) \max_{i=0,\ldots,j} \Big[\left\|\varphi_{Wh}^i - \hat{\varphi}_{Wh}^i\right\|_{1,\overline{\omega}} + \left\|(f_{Gh}^i - \hat{f}_{Gh}^i)/\operatorname{meas}_n(D)\right\|_{0,\overline{\omega}} \\ &+ \left\|(g_{Gh}^i - \hat{g}_{Gh}^i)/\operatorname{meas}_{n-1}(\partial D)\right\|_{0,\gamma}\Big] \quad \text{for all} \quad j = 0,\ldots,N. \end{aligned}$$

Analogously, for solutions $y_h = y_h(f_{Bh}, g_{Bh}, \varphi_{Wh})$ *and* $\hat{y}_h = \hat{y}_h(\hat{f}_{Bh}, \hat{g}_{Bh}, \hat{\varphi}_{Wh})$ *of the box inequality problems (5.9), one has*

$$C_1 \left\| P^G (y_h^j - \hat{y}_h^j) \right\|_{H^1(\Omega)} \le \left\| y_h^j - \hat{y}_h^j \right\|_{1,\overline{\omega}}$$
$$\le C_4 \exp(C_5\, t^j) \max_{i=0,\ldots,j} \left[\|\varphi_{Wh}^i - \hat{\varphi}_{Wh}^i\|_{1,\overline{\omega}} + \left\| (f_{Bh}^i - \hat{f}_{Bh}^i)/\,\mathrm{meas}_n(D) \right\|_{0,\overline{\omega}} \right.$$
$$\left. + \|(g_{Bh}^i - \hat{g}_{Bh}^i)/\,\mathrm{meas}_{n-1}(\partial D)\|_{0,\gamma} \right] \quad \textit{for all} \quad j = 0, \ldots, N.$$

[ii] *The estimates*

$$C_1 \left\| P^G (u_h^{j+1} - u_h^j) \right\|_{H^1(\Omega)} \le \left\| u_h^{j+1} - u_h^j \right\|_{1,\overline{\omega}}$$
$$\le C_6 \left\| (f_{Gh}^{j+1} - f_{Gh}^j)/\,\mathrm{meas}_n(D) \right\|_{0,\overline{\omega}} + C_7 \left\| (g_{Gh}^{j+1} - g_{Gh}^j)/\,\mathrm{meas}_{n-1}(\partial D) \right\|_{0,\gamma}$$
$$+ C_8 \left\| \varphi_{Wh}^{j+1} - \varphi_{Wh}^j \right\|_{1,\overline{\omega}}$$

and

$$C_1 \left\| P^G (y_h^{j+1} - y_h^j) \right\|_{H^1(\Omega)} \le \left\| y_h^{j+1} - y_h^j \right\|_{1,\overline{\omega}}$$
$$\le C_6 \left\| (f_{Bh}^{j+1} - f_{Bh}^j)/\,\mathrm{meas}_n(D) \right\|_{0,\overline{\omega}} + C_7 \left\| (g_{Bh}^{j+1} - g_{Bh}^j)/\,\mathrm{meas}_{n-1}(\partial D) \right\|_{0,\gamma}$$
$$+ C_9 \left\| \varphi_{Wh}^{j+1} - \varphi_{Wh}^j \right\|_{1,\overline{\omega}}$$

hold for the difference of the solutions u_h *of (5.3) (resp.* y_h *of (5.9)) at time instants* t^{j+1} *and* t^j, $j = 0, \ldots, N-1$.

Proof. We restrict ourselves to the statements related to the box problem (5.9). The estimates for the finite element approximation can be derived analogously.

[i]. We take

$$v_h = (\hat{y}_h^j + \varphi_{Wh}^j - \hat{\varphi}_{Wh}^j) \in K_h^j$$

in the inequality related to y_h^j and, vice versa,

$$v_h = (y_h^j - \varphi_{Wh}^j + \hat{\varphi}_{Wh}^j) \in \hat{K}_h^j$$

in the inequality related to $\hat{y}_h^j$. We temporarily introduce the notation

$$w_h := (y_h^j - \hat{y}_h^j - \varphi_{Wh}^j + \hat{\varphi}_{Wh}^j) \in L_2(\tilde{\omega}).$$

By addition of both inequalities and subtraction of $\Lambda_B^j\,(\varphi_{Wh}^j - \hat{\varphi}_{Wh}^j)$, we obtain

$$\begin{aligned} \langle \Lambda_B^j\, w_h, w_h \rangle \;&\le \langle f_{Bh}^j - \hat{f}_{Bh}^j + g_{Bh}^j - \hat{g}_{Bh}^j,\ w_h \rangle + \langle \Lambda_B^j\,(\hat{\varphi}_{Wh}^j - \varphi_{Wh}^j), w_h \rangle \\ &+ \sum_{i=0}^{j-1} \langle (\Lambda_B^{i+1} - \Lambda_B^i)\,(y_h^i - \hat{y}_h^i)\ ,\ w_h \rangle. \end{aligned}$$

Applying the results of Lemma 5.4, Remark 5.5 and using the triangle inequality, we deduce

$$\begin{aligned}\left\|y_h^j - \hat{y}_h^j\right\|_{1,\varpi} &\le c_1 \left\|(f_{Bh}^j - \hat{f}_{Bh}^j)/\operatorname{meas}_n(D)\right\|_{0,\varpi} + c_3 \left\|\varphi_{Wh}^j - \hat{\varphi}_{Wh}^j\right\|_{1,\varpi} \\ &+ c_2 \left\|(g_{Bh}^j - \hat{g}_{Bh}^j)/\operatorname{meas}_{n-1}(\partial D)\right\|_{0,\gamma} + c_4\,\tau \sum_{i=0}^{j-1} \left\|y_h^i - \hat{y}_h^i\right\|_{1,\varpi} .\end{aligned}$$

Now, taking into account Gronwall's inequality (Lemma 3.1), the above estimate is proved.

[ii]. We take

$$v_h = (y_h^{j+1} - \varphi_{Wh}^{j+1} + \varphi_{Wh}^j) \in K_h^j$$

in the inequality at the time instant t^j and

$$v_h = (y_h^j - \varphi_{Wh}^j + \varphi_{Wh}^{j+1}) \in K_h^{j+1}$$

in the inequality at the time instant t^{j+1}. By addition of both inequalities, we arrive at

$$\begin{aligned}&\langle \Lambda_B^{j+1}(y_h^{j+1} - y_h^j),\ y_h^{j+1} - y_h^j - (\varphi_{Wh}^{j+1} - \varphi_{Wh}^j)\rangle \\ &\le \langle f_{Bh}^{j+1} - f_{Bh}^j + g_{Bh}^{j+1} - g_{Bh}^j\,,\ y_h^{j+1} - y_h^j - (\varphi_{Wh}^{j+1} - \varphi_{Wh}^j)\rangle.\end{aligned}$$

Subtracting $\Lambda_B^{j+1}(\varphi_{Wh}^{j+1} - \varphi_{Wh}^j)$ and using the results of Lemma 5.4 and Remark 5.5 the claimed estimate is obtained. ■

We mention that the term $\varphi_{Wh}^{j+1} - \varphi_{Wh}^j$ is different from zero only for Dirichlet nodes $\xi \in \gamma_D$.

Furthermore, let us recall that, according to Remark 5.5, the grid function norms appearing in the previous theorem for the right-hand sides and boundary terms can be replaced by the norms of the spaces $L_2(\Omega)$ and $L_2(\Gamma_N)$.

We finish our discussion of some basic properties of the finite element and finite volume approximations with the statement that, similar to the elliptic situation, the solutions of the discrete inequalities (5.3) and (5.9) can be equivalently characterized by a finite-dimensional complementarity problem. In the following theorem we restrict ourselves to this result for the box approximation. The analogous formulation for the finite element approximation is obvious.

Theorem 5.9 *Let the assumptions of Theorem 5.6 be fulfilled. Then the box inequality problem (5.9) is equivalent to the following complementarity problem.*

$$\left.\begin{array}{l} \textit{Find} \quad y_h^j \in L_2(\overline{\omega}) \quad \textit{such that} \\ \Lambda_B^j\, y_h^j - f_{Bh}^j - g_{Bh}^j - \sum\limits_{i=0}^{j-1} (\Lambda_B^{i+1} - \Lambda_B^i)\, y_h^i \geq 0, \quad y_h^j \geq \varphi_h, \\ \left(\Lambda_B^j\, y_h^j - f_{Bh}^j - g_{Bh}^j - \sum\limits_{i=0}^{j-1} (\Lambda_B^{i+1} - \Lambda_B^i)\, y_h^i\right) (y_h^j - \varphi_h) = 0 \quad \forall \xi \in \tilde{\omega}; \\ y_h^j = g_D(\xi, t^j) \quad \forall \xi \in \gamma_D \qquad \textit{for all} \quad j = 0, \ldots, N. \end{array}\right| \quad (5.11)$$

Proof. The key idea is to repeat the proof of Theorem 4.2[iii] with a modified right-hand side

$$F_{Bh}^j = f_{Bh}^j + g_{Bh}^j + \sum_{i=0}^{j-1} (\Lambda_B^{i+1} - \Lambda_B^i)\, y_h^i.$$

■

5.1.3 Time evolution of the finite volume solution

The variational inequality formulation for the general free boundary problem (2.1) was derived in Section 2.3.3 under the assumption of an advancing moving boundary (cf. Lemma 2.1). From the physical point of view we have discussed this important condition in the introduction of the application problems in Section 2.2. Furthermore, the property that the corresponding variational inequality solution is nondecreasing with respect to time played an essential role in Section 3.3 in the study of its regularity with respect to time. Therefore, it is desirable to show that the solutions of the discrete inequality problems introduced in Section 5.1.1 are also nondecreasing with respect to time.
For such a result we require a discrete maximum principle (cf. Section 4.4.1), such that we restrict ourselves to the box inequality formulation (5.9), (5.10).

Theorem 5.10 *Let a weakly acute triangulation together with its box partition according to $\mathcal{D}_h(1)$ be given. Furthermore, let the assumptions (5.1), (5.2) be fulfilled. Then, assuming Dirichlet boundary conditions $g_D = g_D(x,t)$ on Γ_D nondecreasing with respect to time and*

$$(f_{Bh}^{j+1} + g_{Bh}^{j+1})_\xi \geq (f_{Bh}^j + g_{Bh}^j)_\xi \quad \forall \xi \in \tilde{\omega}, \quad j = 0, \ldots, N-1,$$

the solution of the box inequality (5.9) satisfies

$$y_{h,\xi}^{j+1} \geq y_{h,\xi}^j \quad \forall \xi \in \overline{\omega}, \quad j = 0, \ldots, N-1.$$

Proof. For the sake of brevity let us denote

$$w_h = y_h^{j+1} - y_h^j.$$

Owing to the nondecreasing Dirichlet conditions we can take

$$v_h = y_h^{j+1} - w_h^- = \max\,\{y_h^{j+1}, y_h^j\} \in K_h^{j+1}$$

in the inequality (5.9) at the time t^{j+1} and, vice versa,

$$v_h = y_h^j + w_h^- = \min\,\{y_h^{j+1}, y_h^j\} \in K_h^j$$

in the inequality (5.9) at the time t^j. Consequently, we obtain

$$\langle \Lambda_B^{j+1}\, y_h^{j+1}, -(w_h^-)\rangle \geq \langle f_{Bh}^{j+1} + g_{Bh}^{j+1}, -(w_h^-)\rangle + \sum_{i=0}^{j} \langle (\Lambda_B^{i+1} - \Lambda_B^i)\, y_h^i\,, -(w_h^-)\rangle$$

$$\text{and}\quad \langle \Lambda_B^j\, y_h^j, w_h^-\rangle \geq \langle f_{Bh}^j + g_{Bh}^j, w_h^-\rangle + \sum_{i=0}^{j-1} \langle (\Lambda_B^{i+1} - \Lambda_B^i)\, y_h^i\,, w_h^-\rangle.$$

The addition of these two inequalities gives

$$\langle \Lambda_B^{j+1}\, w_h, w_h^-\rangle \leq \langle f_{Bh}^{j+1} - f_{Bh}^j + g_{Bh}^{j+1} - g_{Bh}^j,\ w_h^-\rangle \leq 0.$$

Applying the discrete maximum principle (4.30) we get $w_h^- = 0$. Therefore, we have $w_h = y_h^{j+1} - y_h^j \geq 0$. ■

Note that the monotonicity assumptions of the foregoing theorem are equivalent to

$$\int_{D_\xi} (f(x,t^{j+1}) - f(x,t^j))\, dx + \int_{\partial D_\xi \cap \Gamma_N} (g_N(x,t^{j+1}) - g_N(x,t^j))\, ds \geq 0 \quad \forall \xi \in \tilde{\omega},$$

so that we are in the same situation as in Section 3.3.

5.2 Error estimates for the finite element and finite volume solutions

In this section, as an important part of this work, we are concerned with error analysis for the finite element approximation (5.3) and the finite volume approximation (5.9). More precisely, the purpose of this section is threefold.

In Section 5.2.1 we will compare the finite element and finite volume (box) approximations in a similar way as done for the elliptic variational inequalities in Section 4.2. Furthermore, in Sections 5.2.2 and 5.2.3, there will be derived error estimates *both* for the finite element solution of problem (5.3) and the box solution of (5.9) compared to the solution of the evolutionary variational inequality (3.1), (3.2).

5.2.1 Comparison of the finite element and finite volume approximations

In what follows we will compare the solutions of the finite element (5.3), (5.4) and the box scheme (5.9), (5.10) as approximations for the solution of the evolutionary variational inequality (3.1), (3.2).

As the main result we shall prove that both approximations have a comparable accuracy, i.e., first and second order convergence can be established depending on choice of the boxes and smoothness of the data. We use a similar line of reasoning as in the elliptic case studied in Section 4.2. Therefore, some auxiliary tools derived there will now again be exploited.

Theorem 5.11 *Let the assumptions (5.1), (5.2) be fulfilled.*

[i] Let the boxes be constructed according to $\mathcal{D}_h(1)$*. Then, for all* $j = 0,\ldots,N$*, the first order estimate*

$$\begin{aligned}\left\|P^G(u_h^j - y_h^j)\right\|_{H^1(\Omega)} &\leq h\ \exp(c_1\ t^j)\ \Big[\ c_2 \max_{i=0,\ldots,j} \|f(t^i)\|_{L_2(\Omega)} \\ &+ \max_{i=0,\ldots,j} \Big\{ h^{1/2} \Big(\sum_{\sigma\in\sigma_{h,2}} |g_N(t^i)|^2_{H^1(\sigma)} \Big)^{1/2} \Big\} \\ &+ \Big(c_4 + L_q\ t^{j-1} + c_5\big(\|\kappa\|_{L_\infty(\Gamma_N)} + \max_{\sigma\in\sigma_{h,2}} \{|\kappa|_{W^1_\infty(\sigma)}\}\big)\Big) \max_{i=0,\ldots,j} \|P^G y_h^i\|_{H^1(\Omega)}\Big]\end{aligned}$$

holds for the difference of the finite element solution u_h *of (5.3), (5.4) and the box solution* y_h *of (5.9), (5.10). Here,* L_q *is given in Lemma 5.4[ii] and* c_1 *is the same constant as in Theorem 5.6 (cf. also Remark 5.7[i]).*

[ii] Let the box partitions be given according to $\mathcal{D}_h(2)$*. Furthermore, let* $\kappa \equiv 0$ *and* $g_N(t^i) = const.$ *on each boundary face* $\sigma \in \sigma_{h,2}$ *for all* $i = 0,\ldots,N$ *be fulfilled. Then, for all* $j = 0,\ldots,N$*, a second order estimate holds of the form*

$$\begin{aligned}\left\|P^G(u_h^j - y_h^j)\right\|_{H^1(\Omega)} &\leq h^2\ \exp(c_1\ t^j)\ \Big[\max_{i=0,\ldots,j} \Big\{c_6\ \Big(\sum_{S\in\mathcal{S}_h} |f(t^i)|^2_{H^1(S)} \Big)^{1/2}\Big\} \\ &+ (M_q + (L_q + \tilde{L}_q)\ t^{j-1}) \max_{i=0,\ldots,j} \|P^G y_h^i\|_{H^1(\Omega)}\Big].\end{aligned}$$

Besides L_q *given already in part [i], the constants* M_q *and* $\tilde{L}_q$ *are defined by*

$$M_q = \max_{i=0,\ldots,N} \max_{S\in\mathcal{S}_h} |q(t^i)|_{W^1_\infty(S)} \quad \text{and} \quad \tilde{L}_q = \max_{S\in\mathcal{S}_h} \tilde{L}_q(S)$$

with

$$|q(t) - q(\hat{t})|_{W^1_\infty(S)} \leq \tilde{L}_q(S)\ |t - \hat{t}| \quad \forall\ t,\hat{t} \in [0,T] \quad \text{and} \quad S \in \mathcal{S}_h.$$

Before we prove this theorem, let us make some additional remarks on the stated estimates.

Remark 5.12 The solution $P^G y_h^i$ of the box inequality problem (5.9) appearing in the right-hand side of the above estimates can be bounded according to Theorem 5.6 and Remark 5.7[ii].

Furthermore, assuming a corresponding smoothness of the coefficients, the terms on the right-hand side can be bounded independently of the spatial and time discretizations. For instance, provided $q \in C^{0,1}([0,T]; W_\infty^1(\Omega))$, we have

$$M_q + (L_q + \tilde{L}_q)\, t^{j-1} \leq c_7.$$

Taking this together with the assumption $f \in C([0,T]; H^1(\Omega))$, the estimate stated in part [ii] of the previous theorem can be expressed by

$$\left\| P^G(u_h^j - y_h^j) \right\|_{H^1(\Omega)} \leq c_8\, h^2\, \exp(c_1\, t^j) \max_{i=0,\dots,j} \|P^G y_h^i\|_{H^1(\Omega)} .$$

Recalling $I_\tau^0(P^G v_h)$ as the piecewise constant in time function of $P^G v_h$ (defined in Remark 5.7[ii]), we note that the just mentioned estimate given in Theorem 5.11[ii] can be represented by

$$\|I_\tau^0(P^G(u_h - y_h))\|_{L_\infty(0,T;H^1(\Omega))} \leq c_8\, h^2\, \exp(c_1 T)\, \|I_\tau^0(P^G y_h)\|_{L_\infty(0,T;H^1(\Omega))}$$

for the difference of the finite element u_h and the box solution y_h. For an estimate of the norm $\|I_\tau^0(P^G y_h)\|_{L_\infty(0,T;H^1(\Omega))}$ we refer again to Theorem 5.6 and Remark 5.7[ii]. □

Proof of Theorem 5.11. We take $v_h = y_h^j \in K_h^j$ in the finite element inequality (5.3) and, vice versa, $v_h = u_h^j \in K_h^j$ in the box inequality (5.9). Now, adding the inequalities (5.3) and (5.9) at the time instant t^j, we obtain the following intermediate result.

$$\left.\begin{aligned} &\langle \Lambda_G^j\,(u_h^j - y_h^j),\, u_h^j - y_h^j\rangle \leq \langle f_{Gh}^j - f_{Bh}^j,\, u_h^j - y_h^j\rangle \\ &+ \langle g_{Gh}^j - g_{Bh}^j,\, u_h^j - y_h^j\rangle + \langle(\Lambda_B^j - \Lambda_G^j)\, y_h^j,\, u_h^j - y_h^j\rangle \\ &+ \sum_{i=0}^{j-1}\langle(\Lambda_G^{i+1} - \Lambda_G^i)\,(u_h^i - y_h^i),\, u_h^j - y_h^j\rangle \\ &+ \sum_{i=0}^{j-1}\langle(\Lambda_G^{i+1} - \Lambda_B^{i+1} - \Lambda_G^i + \Lambda_B^i)\, y_h^i\,,\, u_h^j - y_h^j\rangle \end{aligned}\right| \tag{5.12}$$

for all $j = 0, \dots, N$.

In what follows the terms on the right-hand side of this basic relation (5.12) will be treated separately. Depending on the construction of the boxes (either according to $\mathcal{D}_h(1)$ or $\mathcal{D}_h(2)$, see Section 4.1.3) and the smoothness of the right-hand side f, we get the bounds

$$\left.\begin{array}{ll} \langle f^j_{Gh} - f^j_{Bh}, u^j_h - y^j_h\rangle & \leq \left|P^G(u^j_h - y^j_h)\right|_{H^1(\Omega)} C_1\, h\, \|f(t^j)\|_{L_2(\Omega)} \\ \quad \text{for } \mathcal{D}_h(1) \quad \text{and} & \\ \langle f^j_{Gh} - f^j_{Bh}, u^j_h - y^j_h\rangle & \leq \left|P^G(u^j_h - y^j_h)\right|_{H^1(\Omega)} C_2\, h^2 \\ & \quad * \Big(\sum_{S\in S_h} |f(t^j)|^2_{H^1(S)}\Big)^{1/2} \quad \text{for } \mathcal{D}_h(2) \end{array}\right| \tag{5.13}$$

by means of Lemma 4.9.

Exploiting Lemma 4.11, the difference of the right-hand sides in the Neumann/Newton boundary condition can be estimated by

$$\left.\begin{array}{ll} \langle g^j_{Gh} - g^j_{Bh}, u^j_h - y^j_h\rangle & \leq C_3\, h^{3/2} \Big(\sum_{\sigma\in\sigma_{h,2}} |g_N(t^j)|^2_{H^1(\sigma)}\Big)^{1/2} \\ & \quad * \left\|P^G(u^j_h - y^j_h)\right\|_{H^1(\Omega)}, \\ \text{and} & \\ \langle g^j_{Gh} - g^j_{Bh}, u^j_h - y^j_h\rangle & = 0 \quad \text{if} \quad g_N(t^j) = \text{const. on each} \quad \sigma \in \sigma_{h,2}. \end{array}\right| \tag{5.14}$$

Let us now consider the difference

$$\langle (\Lambda^j_B - \Lambda^j_G)\, y^j_h, u^j_h - y^j_h\rangle.$$

Recalling the definitions (5.4) and (5.10) and owing to the equality of the stiffness matrices $L^j_G = L^j_B$ for $j = 0, \ldots, N$, we have

$$\Lambda^j_B - \Lambda^j_G = (1-\alpha)\,(M^j_B - M^j_G) + \alpha\,(M^{j-1}_B - M^{j-1}_G) + (R_B - R_G).$$

Therefore, we obtain

$$\begin{aligned} &\langle (\Lambda^j_B - \Lambda^j_G)\, v_h,\ w_h\rangle \\ &= \int_\Omega \left[(1-\alpha)\, q(t^j) + \alpha\, q(t^{j-1})\right] \left[P^B v_h\, P^B w_h - P^G v_h\, P^G w_h\right] dx \\ &\quad + \int_{\Gamma_N} \kappa\, (P^B v_h\, P^B w_h - P^G v_h\, P^G w_h)\, ds \end{aligned}$$

for $v_h \in L_2(\overline{\omega})$, $w_h \in L_2(\tilde{\omega})$.

Applying the results of Lemmas 4.10 and 4.12 to this representation, we derive the following estimates (cf. also (4.27)).

$$
\begin{aligned}
&\langle (\Lambda_B^j - \Lambda_G^j)\, y_h^j,\ u_h^j - y_h^j \rangle \le h\, \left\| P^G y_h^j \right\|_{H^1(\Omega)} \left\| P^G (u_h^j - y_h^j) \right\|_{H^1(\Omega)} \\
&* \Big[C_4 \left((1-\alpha) \left\| q(t^j) \right\|_{L_\infty(\Omega)} + \alpha \left\| q(t^{j-1}) \right\|_{L_\infty(\Omega)} \right) \\
&+ C_5 \left(\|\kappa\|_{L_\infty(\Gamma_N)} + \max_{\sigma \in \sigma_{h,2}} \{ |\kappa|_{W^1_\infty(\sigma)} \} \right) \Big] \quad \text{for } \mathcal{D}_h(1) \\
&\text{and} \\
&\langle (\Lambda_B^j - \Lambda_G^j)\, y_h^j,\ u_h^j - y_h^j \rangle \le C_6\, h^2 \left\| P^G y_h^j \right\|_{H^1(\Omega)} \left\| P^G (u_h^j - y_h^j) \right\|_{H^1(\Omega)} \\
&* \Big[(1-\alpha) \left(\left\| q(t^j) \right\|_{L_\infty(\Omega)} + \max_{S \in \mathcal{S}_h} \{ |q(t^j)|_{W^1_\infty(S)} \} \right) \\
&+ \alpha \left(\left\| q(t^{j-1}) \right\|_{L_\infty(\Omega)} + \max_{S \in \mathcal{S}_h} \{ |q(t^{j-1})|_{W^1_\infty(S)} \} \right) \Big] \quad \text{for } \ \mathcal{D}_h(2), \quad \kappa \equiv 0.
\end{aligned}
\tag{5.15}
$$

Proceeding with the treatment of the terms in (5.12), we deduce

$$
\begin{aligned}
&\sum_{i=0}^{j-1} \langle (\Lambda_G^{i+1} - \Lambda_G^i)\, (u_h^i - y_h^i), (u_h^j - y_h^j) \rangle \ \le\ \max\{L_k, L_q\}\ \tau \\
&* \left\| P^G (u_h^j - y_h^j) \right\|_{H^1(\Omega)} \sum_{i=0}^{j-1} \left\| P^G (u_h^i - y_h^i) \right\|_{H^1(\Omega)}
\end{aligned}
\tag{5.16}
$$

by means of Lemma 5.4.

It remains to consider the term

$$
\sum_{i=0}^{j-1} \langle (\Lambda_G^{i+1} - \Lambda_B^{i+1} - \Lambda_G^i + \Lambda_B^i)\, y_h^j\ ,\ u_h^j - y_h^j \rangle .
$$

Using again the equality of the finite element and box stiffness matrices, the essential point is to estimate the difference of the mass matrices. More precisely, the representation

$$
\begin{aligned}
&\Lambda_G^{i+1} - \Lambda_B^{i+1} - (\Lambda_G^i - \Lambda_B^i) = (1-\alpha)\ \left(M_G^{i+1} - M_G^i - (M_B^{i+1} - M_B^i) \right) \\
&+ \alpha\ \left(M_G^i - M_G^{i-1} - (M_B^i - M_B^{i-1}) \right)
\end{aligned}
$$

leads to

$$
\begin{aligned}
&\langle (\Lambda_G^{i+1} - \Lambda_B^{i+1} - \Lambda_G^i + \Lambda_B^i)\, v_h,\ w_h \rangle = \int_\Omega \left[(1-\alpha) \left(q(t^{i+1}) - q(t^i) \right) \right. \\
&\left. + \alpha\ \left(q(t^i) - q(t^{i-1}) \right) \right] \left[P^G v_h\ P^G w_h - P^B v_h\ P^B w_h \right] dx
\end{aligned}
$$

for $v_h \in L_2(\overline{\omega}),\ \ w_h \in L_2(\tilde{\omega})$.

We deduce

$$\left.\begin{array}{l}\displaystyle\sum_{i=0}^{j-1}\langle(\Lambda_G^{i+1}-\Lambda_B^{i+1}-\Lambda_G^{i}+\Lambda_B^{i})\,y_h^i\ ,\ u_h^j-y_h^j\rangle\le\left\|P^G(u_h^j-y_h^j)\right\|_{H^1(\Omega)}\\ \displaystyle *\sum_{i=0}^{j-1}\|P^G y_h^i\|_{H^1(\Omega)}\ \tau*\begin{cases}C_7\ L_q\ h & \text{for}\quad \mathcal{D}_h(1),\\ C_8\ (L_q+\tilde{L}_q)\ h^2 & \text{for}\quad \mathcal{D}_h(2),\end{cases}\end{array}\right| \qquad (5.17)$$

by means of Lemma 4.10.

Now, we substitute the estimates (5.13)–(5.17) into the basic relation (5.12) and use the ellipticity of Λ_G^j (cf. Lemma 5.4[i]). As a result we obtain

$$\begin{aligned}&C\left\|P^G(u_h^j-y_h^j)\right\|_{H^1(\Omega)}\le h\max_{i=0,\dots,j}\Big\{C_1\left\|f(t^i)\right\|_{L_2(\Omega)}\\&+C_3\,h^{1/2}\Big(\sum_{\sigma\in\sigma_{h,2}}|g_N(t^i)|^2_{H^1(\sigma)}\Big)^{1/2}\Big\}+h\,\Big[C_4\left(\|q\|_{C([0,T];L_\infty(\Omega))}+L_q\,t^{j-1}\right)\\&+C_5\left(\|\kappa\|_{L_\infty(\Gamma_N)}+\max_{\sigma\in\sigma_{h,2}}\{|\kappa|_{W^1_\infty(\sigma)}\}\right)\Big]\max_{i=0,\dots,j}\{\|P^G y_h^i\|_{H^1(\Omega)}\}\\&+\max\{L_k,L_q\}\,\tau\sum_{i=0}^{j-1}\|P^G(u_h^i-y_h^i)\|_{H^1(\Omega)}\qquad\text{for all}\quad j=0,\dots,N\end{aligned}$$

for box partitions constructed according to $\mathcal{D}_h(1)$. Applying Gronwall's inequality in its discrete form (Lemma 3.1), the assertion [i] is proved.

In the same way, the second order estimate stated in part [ii] is shown by the combination of (5.12)–(5.17) and Gronwall's inequality. ■

5.2.2 A priori estimates for the finite element and finite volume solutions

After the comparison of the finite element and finite volume approximations performed in the previous Section 5.2.1, our aim is now to study the convergence rate of the finite element solution u_h of (5.3) and the finite volume solution y_h of (5.9).

First we will establish general a priori estimates for the errors

$$u-u_h \quad\text{and}\quad u-y_h,$$

with u as the solution of the inequality formulation (3.1), (3.2)). Then, in Section 5.2.3, we will show how these estimates can be evaluated for the finite element solution of (5.3) and the box solution of problem (5.9).

Let us start with two general a priori estimates for the finite element inequality formulation (5.3), (5.4). Already at this point let us refer to Remark 5.15 concerning an interpretation of the following a priori estimates.

Lemma 5.13 *Let the assumptions (5.1) and (5.2) be satisfied. Furthermore, let* u *and* $P^G u_h$ *be the solutions of problem (3.1), (3.2) and its finite element approximation (5.3), (5.4). Then, for all* $j = 0, \ldots, N$, *there holds the following a priori estimate.*

$$
\begin{aligned}
&\langle \Lambda_G^j\,(u_h^j - v_h^j),\ u_h^j - v_h^j\rangle \le a(t^j; u(t^j) - P^G v_h^j,\ P^G(u_h^j - v_h^j)) \\
&+ \sum_{i=0}^{j-1} \langle (\Lambda_G^{i+1} - \Lambda_G^i)\,(u_h^i - v_h^i), u_h^j - v_h^j\rangle \\
&+ \Big[\, a(t^j; u(t^j),\ P^G(v_h^j - u_h^j)) - (f(t^j), P^G(v_h^j - u_h^j)) \\
&\quad - (g_N(t^j), P^G(v_h^j - u_h^j))_{\Gamma_N} - \int_0^{t^j} b(t'; u(t'), P^G(v_h^j - u_h^j))\, dt' \,\Big] \\
&+ \sum_{i=0}^{j-1} \Big[\, b_G^i(P^G v_h, P^G(u_h^j - v_h^j)) - \int_{t^i}^{t^{i+1}} b(t'; u(t'), P^G(u_h^j - v_h^j))\, dt' \,\Big]
\end{aligned}
$$

with arbitrary grid function vectors v_h^i *satisfying*

$$v_h^i \in K_h^i \quad \textit{for} \quad i = 0, \ldots, N.$$

A second a priori estimate holds in the form that, for all $j = 0, \ldots, N$, *one has*

$$
\begin{aligned}
&\langle \Lambda_G^j\,(u_h^j - v_h^j),\ u_h^j - v_h^j\rangle \le a(t^j; u(t^j) - P^G v_h^j,\ P^G(u_h^j - v_h^j)) \\
&+ \sum_{i=0}^{j-1} \langle (\Lambda_G^{i+1} - \Lambda_G^i)\,(u_h^i - v_h^i), u_h^j - v_h^j\rangle \\
&+ \Big[\, a(t^j; u(t^j), v - P^G u_h^j) - (f(t^j), v - P^G u_h^j) \\
&\quad - (g_N(t^j), v - P^G u_h^j)_{\Gamma_N} - \int_0^{t^j} b(t'; u(t'), v - P^G u_h^j)\, dt' \,\Big] \\
&+ \Big[\, a(t^j; u(t^j), P^G v_h^j - u(t^j)) - (f(t^j), P^G v_h^j - u(t^j)) \\
&\quad - (g_N(t^j), P^G v_h^j - u(t^j))_{\Gamma_N} - \int_0^{t^j} b(t'; u(t'), P^G v_h^j - u(t^j))\, dt' \,\Big] \\
&+ \sum_{i=0}^{j-1} \Big[\, b_G^i(P^G v_h, P^G(u_h^j - v_h^j)) - \int_{t^i}^{t^{i+1}} b(t'; u(t'), P^G(u_h^j - v_h^j))\, dt' \,\Big]
\end{aligned}
$$

with arbitrary $v \in K(t^j)$ *and grid function vectors* v_h^i *fulfilling* $v_h^i \in K_h^i$ *for* $i = 0, \ldots, N$.

Proof. First let us prove the second a priori estimate.

Considering the evolutionary inequality formulation (3.1), (3.2) together with its finite element discretization (at the moment in the form (5.5), (5.6)) at the time instant $t = t^j$, we obtain the relation

$$
\begin{aligned}
& a(t^j; P^G(u_h^j - v_h^j), P^G(u_h^j - v_h^j)) \\
& \leq \Big[a(t^j; u(t^j), v - u(t^j)) - (f(t^j), v - u(t^j)) \\
& \quad - (g_N(t^j), v - u(t^j))_{\Gamma_N} - \sum_{i=0}^{j-1} \int_{t^i}^{t^{i+1}} b(t'; u(t'), v - u(t^j))\, dt' \Big] \\
& + \Big[- a(t^j; P^G v_h^j, P^G(u_h^j - v_h^j)) + (f(t^j), P^G(u_h^j - v_h^j)) \\
& \quad + (g_N(t^j), P^G(u_h^j - v_h^j))_{\Gamma_N} + \sum_{i=0}^{j-1} b_G^i(P^G u_h, P^G(u_h^j - v_h^j)) \Big]
\end{aligned}
$$

for arbitrary $v \in K(t^j)$ and $v_h^j \in K_h^j$. In this inequality we insert the following representation with respect to the elliptic bilinear form a,

$$
\begin{aligned}
& a(t^j; u(t^j), v - u(t^j)) - a(t^j; P^G v_h^j, P^G(u_h^j - v_h^j)) \\
& = a(t^j; u(t^j), v - P^G u_h^j) + a(t^j; u(t^j), P^G v_h^j - u(t^j)) \\
& \quad + a(t^j; u(t^j) - P^G v_h^j, P^G(u_h^j - v_h^j))
\end{aligned}
$$

and the following decomposition with respect to the memory term,

$$
\begin{aligned}
& b_G^i(P^G u_h,\ P^G(u_h^j - v_h^j)) - \int_{t^i}^{t^{i+1}} b(t'; u(t'), v - u(t^j))\, dt' \\
& = b_G^i(P^G(u_h - v_h),\ P^G(u_h^j - v_h^j)) \\
& + \Big[b_G^i(P^G v_h, P^G(u_h^j - v_h^j)) - \int_{t^i}^{t^{i+1}} b(t'; u(t'), P^G(u_h^j - v_h^j))\, dt' \Big] \\
& - \int_{t^i}^{t^{i+1}} \big[b(t'; u(t'), v - P^G u_h^j) + b(t'; u(t'), P^G v_h^j - u(t^j)) \big]\, dt'.
\end{aligned}
$$

At this point, we refer to definition (5.6) to make clear that the notation

$$
b_G^i(P^G u_h, P^G v_h)
$$

expresses that both $P^G u_h^i$ and $P^G u_h^{i+1}$ appear in the corresponding bilinear form. Substituting the last two decompositions into the above inequality we get the following relation.

$$
\begin{aligned}
& a(t^j; P^G(u_h^j - v_h^j),\ P^G(u_h^j - v_h^j)) \le a(t^j; u(t^j) - P^G v_h^j,\ P^G(u_h^j - v_h^j)) \\
& + \sum_{i=0}^{j-1} b_G^i(P^G(u_h - v_h), P^G(u_h^j - v_h^j)) \\
& + \Big[\ a(t^j; u(t^j), v - P^G u_h^j) - (f(t^j), v - P^G u_h^j) \\
& \quad - (g_N(t^j), v - P^G u_h^j)_{\Gamma_N} - \int_0^{t^j} b(t'; u(t'), v - P^G u_h^j)\, dt'\ \Big] \\
& + \Big[\ a(t^j; u(t^j), P^G v_h^j - u(t^j)) - (f(t^j), P^G v_h^j - u(t^j)) \\
& \quad - (g_N(t^j), P^G v_h^j - u(t^j))_{\Gamma_N} - \int_0^{t^j} b(t'; u(t'), P^G v_h^j - u(t^j))\, dt'\ \Big] \\
& + \sum_{i=0}^{j-1} \Big[\ b_G^i(P^G v_h, P^G(u_h^j - v_h^j)) - \int_{t^i}^{t^{i+1}} b(t'; u(t'), P^G(u_h^j - v_h^j))\, dt'\ \Big].
\end{aligned}
$$

Finally, we consider the bilinear forms

$$
a(t^j; P^G(u_h^j - v_h^j), P^G(u_h^j - v_h^j)), \qquad \sum_{i=0}^{j-1} b_G^i(P^G(u_h - v_h), P^G(u_h^j - v_h^j))
$$

in the foregoing relation at the time instant t^j. As already indicated in Remark 5.2, we collect the terms containing $P^G(u_h^j - v_h^j)$ as (trial !) function in the left-hand side. As a result we end up with the second a priori estimate stated in the lemma.

The first a priori estimate is obtained in a similar way, but now we take into account only the finite element approximation (5.5), (5.6) at the time instant t^j. Using the abbreviation

$$
w_h = u_h^j - v_h^j \quad \text{for arbitrary} \quad v_h^j \in K_h^j,
$$

we have

$$
\begin{aligned}
& a(t^j; P^G w_h, P^G w_h) \le (f(t^j), P^G w_h) + (g_N(t^j), P^G w_h)_{\Gamma_N} \\
& \quad - a(t^j; P^G v_h^j, P^G w_h) + \sum_{i=0}^{j-1} b_G^i(P^G u_h, P^G w_h) \quad \text{for} \quad v_h^j \in K_h^j.
\end{aligned}
$$

We insert the following decompositions in this inequality.

$$
-a(t^j; P^G v_h^j, P^G w_h) = a(t^j; u(t^j) - P^G v_h^j, P^G w_h) - a(t^j; u(t^j), P^G w_h)
$$

and with respect to the memory term,

$$
\begin{aligned}
& b_G^i(P^G u_h, P^G w_h) = b_G^i(P^G(u_h - v_h), P^G w_h) + \int_{t^i}^{t^{i+1}} b(t'; u(t'), P^G w_h)\, dt' \\
& + \Big[b_G^i(P^G v_h, P^G w_h) - \int_{t^i}^{t^{i+1}} b(t'; u(t'), P^G w_h)\, dt' \Big].
\end{aligned}
$$

The proof is now completed by the same procedure as for the second a priori estimate. More precisely, at the time instant t^j, we collect all the terms containing $P^G(u_h^j - v_h^j)$ as trial function in the left-hand side. ■

Let us now switch over to the derivation of similar a priori estimates for the finite volume approximation (5.9), (5.10).

Lemma 5.14 *Let the assumptions (5.1) and (5.2) be satisfied. Let u and $P^G y_h$ be the solutions of problem (3.1), (3.2) and its finite volume/box approximation (5.9), (5.10). Furthermore, by E_B^j, the expression*

$$\begin{aligned}
E_B^j &= \Big\{ (f(t^j), P^B(y_h^j - v_h^j) - P^G(y_h^j - v_h^j)) \\
&+ (g_N(t^j), P^B(y_h^j - v_h^j) - P^G(y_h^j - v_h^j))_{\Gamma_N} \\
&+ a(t^j; P^G v_h^j,\ P^G(y_h^j - v_h^j)) - a_B(t^j; P^G v_h^j,\ P^B(y_h^j - v_h^j)) \\
&+ \sum_{i=0}^{j-1} \Big[\ b_B^i(P^G v_h,\ P^B(y_h^j - v_h^j)) - b_G^i(P^G v_h,\ P^G(y_h^j - v_h^j))\ \Big] \Big\}
\end{aligned}$$

is denoted for $j = 0, \dots, N$.

Then, for all $j = 0, \dots, N$, the following a priori estimate holds.

$$\begin{aligned}
&\langle \Lambda_B^j\ (y_h^j - v_h^j),\ y_h^j - v_h^j \rangle \le a(t^j; u(t^j) - P^G v_h^j,\ P^G(y_h^j - v_h^j)) \\
&+ \sum_{i=0}^{j-1} \langle (\Lambda_B^{i+1} - \Lambda_B^i)\ (y_h^i - v_h^i),\ y_h^j - v_h^j \rangle \\
&+ \Big[\ a(t^j; u(t^j), P^G(v_h^j - y_h^j)) - (f(t^j), P^G(v_h^j - y_h^j)) \\
&\quad - (g_N(t^j), P^G(v_h^j - y_h^j))_{\Gamma_N} - \int_0^{t^j} b(t'; u(t'), P^G(v_h^j - y_h^j))\, dt'\ \Big] \\
&+ \sum_{i=0}^{j-1} \Big[\ b_G^i(P^G v_h, P^G(y_h^j - v_h^j)) - \int_{t^i}^{t^{i+1}} b(t'; u(t'), P^G(y_h^j - v_h^j))\, dt'\ \Big] \\
&+ E_B^j
\end{aligned}$$

with arbitrary grid function vectors v_h^i satisfying

$$v_h^i \in K_h^i \quad \textit{for} \quad i = 0, \dots, N.$$

A second a priori estimate holds in the form that, for all $j = 0, \dots, N$, *one has the relation*

$$
\begin{aligned}
&\langle \Lambda_B^j\, (y_h^j - v_h^j),\ y_h^j - v_h^j\rangle \le a(t^j; u(t^j) - P^G v_h^j,\ P^G(y_h^j - v_h^j)) \\
&+ \sum_{i=0}^{j-1} \langle (\Lambda_B^{i+1} - \Lambda_B^i)\, (y_h^i - v_h^i),\ y_h^j - v_h^j\rangle \\
&+ \Big[\, a(t^j; u(t^j), v - P^G y_h^j) - (f(t^j), v - P^G y_h^j) \\
&\quad - (g_N(t^j), v - P^G y_h^j)_{\Gamma_N} - \int_0^{t^j} b(t'; u(t'), v - P^G y_h^j)\, dt' \,\Big] \\
&+ \Big[\, a(t^j; u(t^j), P^G v_h^j - u(t^j)) - (f(t^j), P^G v_h^j - u(t^j)) \\
&\quad - (g_N(t^j), P^G v_h^j - u(t^j))_{\Gamma_N} - \int_0^{t^j} b(t'; u(t'), P^G v_h^j - u(t^j))\, dt' \,\Big] \\
&+ \sum_{i=0}^{j-1} \Big[\, b_G^i(P^G v_h, P^G(y_h^j - v_h^j)) - \int_{t^i}^{t^{i+1}} b(t'; u(t'), P^G(y_h^j - v_h^j))\, dt' \,\Big] \\
&+ E_B^j
\end{aligned}
$$

with arbitrary $v \in K(t^j)$ *and grid function vectors* v_h^i *fulfilling* $v_h^i \in K_h^i$ *for all* $i = 0, \dots, N$.

Proof. To prove the second estimate, let us consider the evolutionary inequality formulation (3.1), (3.2) together with its box discretization in the form (5.7), (5.8) at the time instant $t = t^j$.
Similar to the proof of Lemma 5.13, we obtain the following inequality.

$$
\begin{aligned}
&a(t^j; P^G(y_h^j - v_h^j), P^B(y_h^j - v_h^j)) \le \Big[\, a(t^j; u(t^j), v - u(t^j)) - (f(t^j), v - u(t^j)) \\
&- (g_N(t^j), v - u(t^j)) - \sum_{i=0}^{j-1} \int_{t^i}^{t^{i+1}} b(t'; u(t'), v - u(t^j))\, dt' \,\Big] \\
&+ \Big[-\, a(t^j; P^G v_h^j, P^B(y_h^j - v_h^j)) + (f(t^j), P^B(y_h^j - v_h^j)) \\
&\quad + (g_N(t^j), P^B(y_h^j - v_h^j))_{\Gamma_N} + \sum_{i=0}^{j-1} b_B^i(P^G y_h, P^B(y_h^j - v_h^j)) \,\Big]
\end{aligned}
$$

for arbitrary $v \in K(t^j)$ and $v_h^j \in K_h^j$.
In view of interpolation properties which will be applied later for the evaluation of the a priori estimates, we insert the following decompositions, i.e.,

$$
\begin{aligned}
&(f(t^j),\ P^B(y_h^j - v_h^j) - v + u(t^j)) = (f(t^j),\ P^G y_h^j - v) \\
&+ (f(t^j),\ u(t^j) - P^G v_h^j) + (f(t^j),\ P^B(y_h^j - v_h^j) - P^G(y_h^j - v_h^j)),
\end{aligned}
$$

an analogous representation for the right-hand side g_N in the Neumann/Newton boundary condition

$$(g_N(t^j), P^B(y_h^j - v_h^j) - v + u(t^j))_{\Gamma_N}$$

and

$$\begin{aligned}
&a(t^j; u(t^j),\ v - u(t^j)) - a_B(t^j; P^G v_h^j,\ P^B(y_h^j - v_h^j)) \\
&= a(t^j; u(t^j),\ v - P^G y_h^j) \\
&\quad + a(t^j; u(t^j),\ P^G v_h^j - u(t^j)) + a(t^j; u(t^j) - P^G v_h^j,\ P^G(y_h^j - v_h^j)) \\
&\quad + a(t^j; P^G v_h^j,\ P^G(y_h^j - v_h^j)) - a_B(t^j; P^G v_h^j,\ P^B(y_h^j - v_h^j)).
\end{aligned}$$

Finally, the memory term will be decomposed as follows.

$$\begin{aligned}
&b_B^i(P^G y_h, P^B(y_h^j - v_h^j)) - \int_{t^i}^{t^{i+1}} b(tn'; u(t'), v - u(t^j))\, dt' \\
&= b_B^i(P^G(y_h - v_h), P^B(y_h^j - v_h^j)) \\
&\quad + \left[\ b_B^i(P^G v_h, P^B(y_h^j - v_h^j)) - b_G^i(P^G v_h, P^G(y_h^j - v_h^j))\ \right] \\
&\quad + \left[\ b_G^i(P^G v_h, P^G(y_h^j - v_h^j)) - \int_{t^i}^{t^{i+1}} b(t'; u(t'), P^G(y_h^j - v_h^j))\, dt'\ \right] \\
&\quad - \int_{t^i}^{t^{i+1}} \left[\ b(t'; u(t'), v - P^G y_h^j) + b(t'; u(t'), P^G v_h^j - u(t^j))\ \right] dt'.
\end{aligned}$$

Substituting these decompositions in the above inequality, we obtain

$$\begin{aligned}
&a_B(t^j; P^G(y_h^j - v_h^j),\ P^B(y_h^j - v_h^j)) \le a(t^j; u(t^j) - P^G v_h^j,\ P^G(y_h^j - v_h^j)) \\
&\quad + \sum_{i=0}^{j-1} b_B^i(P^G(y_h - v_h), P^B(y_h^j - v_h^j)) \\
&\quad + \Big[\ a(t^j; u(t^j), v - P^G y_h^j) - (f(t^j), v - P^G y_h^j) \\
&\quad - (g_N(t^j), v - P^G y_h^j)_{\Gamma_N} - \int_0^{t^j} b(t'; u(t'), v - P^G y_h^j)\, dt'\ \Big] \\
&\quad + \Big[\ a(t^j; u(t^j), P^G v_h^j - u(t^j)) - (f(t^j), P^G v_h^j - u(t^j)) \\
&\quad - (g_N(t^j), P^G v_h^j - u(t^j))_{\Gamma_N} - \int_0^{t^j} b(t'; u(t'), P^G v_h^j - u(t^j))\, dt'\Big] \\
&\quad + \sum_{i=0}^{j-1} \Big[b_G^i(P^G v_h, P^G(y_h^j - v_h^j)) - \int_{t^i}^{t^{i+1}} b(t'; u(t'), P^G(y_h^j - v_h^j))\, dt'\ \Big]\ +\ E_B^j.
\end{aligned}$$

Now, we consider, similar to the proof of Lemma 5.13, the bilinear forms

$$a_B(t^j; P^G(y_h^j - v_h^j), P^B(y_h^j - v_h^j)), \qquad \sum_{i=0}^{j-1} b_B^i(P^G(y_h - v_h), P^B(y_h^j - v_h^j))$$

at the time instant t^j. Collecting the terms containing $P^G(y_h^j - v_h^j)$ as trial function in the left-hand side, we arrive at the second a priori estimate stated in the present lemma.

To prove the first a priori estimate, let us denote

$$w_h = y_h^j - v_h^j$$

for arbitrary $v_h^j \in K_h^j$. Then, the finite volume approximation (5.7), (5.8) is rewritten as

$$\begin{aligned} a_B(t^j; P^G w_h, P^B w_h) &\le (f(t^j), P^B w_h) + (g_N(t^j), P^B w_h)_{\Gamma_N} \\ &- a_B(t^j; P^G v_h^j, P^B w_h) + \sum_{i=0}^{j-1} b_B^i(P^G y_h, P^B w_h) \quad \text{for} \quad v_h^j \in K_h^j. \end{aligned}$$

We insert the following decompositions in the above inequality, i.e.,

$$(f(t^j), P^B w_h) = (f(t^j), P^G w_h) + (f(t^j), P^B w_h - P^G w_h),$$

an analogous representation for the right-hand side in the Neumann/Newton boundary condition $(g_N(t^j), P^B(y_h^j - v_h^j))_{\Gamma_N}$ and

$$\begin{aligned} &-a_B(t^j; P^G v_h^j, P^B w_h) = -a(t^j; u(t^j), P^G w_h) + a(t^j; u(t^j) - P^G v_h^j, P^G w_h) \\ &+ \left[\, a(t^j; P^G v_h^j, P^G w_h) - a_B(t^j; P^G v_h^j, P^B w_h) \,\right]. \end{aligned}$$

Furthermore, with respect to the memory term, we take

$$\begin{aligned} &b_B^i(P^G y_h, P^B w_h) = b_B^i(P^G(y_h - v_h), P^B w_h) + \int_{t^i}^{t^{i+1}} b(t'; u(t'), P^G w_h)\, dt' \\ &+ \left[\, b_G^i(P^G v_h, P^G w_h) - \int_{t^i}^{t^{i+1}} b(t'; u(t'), P^G w_h)\, dt' \,\right] \\ &+ \left[\, b_B^i(P^G v_h, P^B w_h) - b_G^i(P^G v_h, P^G w_h) \,\right]. \end{aligned}$$

The proof is now completed in the same way as for the second a priori estimate, i.e., for the time instant t^j we collect all the terms containing $P^G(y_h^j - v_h^j)$ in the left-hand side. ∎

Remark 5.15 The a priori estimates established in Lemmas 5.13 and 5.14 for the finite element and box approximations (5.3) and (5.9), respectively can be interpreted as follows. The difference

$$a(t^j; u(t^j) - P^G v_h^j, P^G(u_h^j - v_h^j)) \quad (\text{resp. } a(t^j; u(t^j) - P^G v_h^j, P^G(y_h^j - v_h^j)) \,)$$

and the terms appearing in square brackets on the right-hand sides have to be estimated, whereas

$$\sum_{i=0}^{j-1} \langle (\Lambda_G^{i+1} - \Lambda_G^i)\,(u_h^i - v_h^i), u_h^j - v_h^j \rangle \quad (\text{resp. } \sum_{i=0}^{j-1} \langle (\Lambda_B^{i+1} - \Lambda_B^i)\,(y_h^i - v_h^i), y_h^j - v_h^j \rangle \,)$$

can be treated by means of Gronwall's Lemma in its discrete form.

Furthermore, it is not difficult to see that the introduced notation E_B^j for $j = 0, \ldots, N$ contains additional error terms caused by the usage of the box approximation. □

5.2.3 Convergence rate for the finite element and finite volume solutions

We will now study the accuracy of the finite element solution u_h of (5.3) and the finite volume solution y_h of (5.9). Assuming a reasonable regularity (cf. Chapter 3) of the solution u of the evolutionary variational inequality (3.1), (3.2) we will evaluate the a priori estimates derived in Lemmas 5.13 and 5.14.

As main result we will show that both the finite element and the finite volume solution have a first order accuracy of the form

$$\|u - I_\tau^0(P^G u_h)\|_{L_\infty(0,T;H^1(\Omega))} \leq C_1\, h + C_2\, \tau \quad \text{and}$$
$$\|u - I_\tau^0(P^G y_h)\|_{L_\infty(0,T;H^1(\Omega))} \leq C_3\, h + C_4\, \tau,$$

where $I_\tau^0(P^G u_h)$ (resp. $I_\tau^0(P^G y_h)$) denote the piecewise constant in time function (cf. Remark 5.7[ii]) of the Galerkin prolongations $P^G u_h$ and $P^G y_h$, respectively.

We refer to Sections 3.3, 3.4 and, in particular, to the summary and the additional remarks contained in Section 3.5 concerning detailed regularity investigations for the solution u of the evolutionary variational inequality (3.1), (3.2). In agreement with the statements derived there, our error analysis will be based on the following assumptions.

$$\left. \begin{array}{l} u \in \big(W_\infty^1(0,T;H^1(\Omega)) \,\cap\, L_\infty(0,T;H^2(\Omega))\big), \\ u(t) \in H^2(\Omega) \quad \forall\, t \in [0,T] \quad \text{with} \\ \|u(t)\|_{H^2(\Omega)} \text{ is bounded independent of } t \end{array} \right| \qquad (5.18)$$

(cf. Section 3.5). Furthermore, owing to the conditions

$$l(0)\ \varphi \le g_N(0) \le g_N(t) \quad \text{on the Neumann/Newton boundary} \quad \Gamma_N \subset \partial\Omega$$

(cf. Theorems 3.9, 3.13 and Lemmas 3.8, 3.11), the solution u of the evolutionary variational inequality (3.1), (3.2) fulfills a complementarity problem of the following form (cf. Section 3.5).

$$\left.\begin{array}{l}
(Au)(t) \ge f(t) + \displaystyle\int_0^t (Bu)(t')\ dt', \quad u(t) \ge \varphi, \\
\left((Au)(t) - f(t) - \displaystyle\int_0^t (Bu)(t')\ dt' \right) (u(t) - \varphi) = 0 \quad \text{a.e in} \quad \Omega; \\
(k\ \nabla u \cdot \vec{n})(t) + \kappa\ u(t) = g_N(t) + \displaystyle\int_0^t (\partial_t k\ \nabla u \cdot \vec{n})(t')\ dt' \quad \text{a.e. on} \quad \Gamma_N; \\
u(t) = g_D(t) \quad \text{on} \quad \Gamma_D; \qquad \forall\, t \in [0,T],
\end{array}\right| \tag{5.19}$$

where the differential operators A and B are defined in (3.10).

In view of the application of Green's formula and interpolation properties we will work with, besides our standard assumptions (5.1), (5.2), the following conditions on the data.

$$k \in W^1_\infty(0,T; C^1(\overline{\Omega})), \quad \varphi \in H^2(\Omega). \tag{5.20}$$

We are now in position to formulate the main convergence results. At first, this will be done for the finite element solution u_h of (5.3) in Theorem 5.16 and afterwards in Theorem 5.17 for the box solution y_h of (5.9). Due to the imposed non-homogeneous, time-dependent Dirichlet boundary conditions $g_D = g_D(x,t)$ on Γ_D, we prefer to evaluate the first a priori estimates derived in Lemmas 5.13 and 5.14.

Theorem 5.16 *Let the assumptions (5.1), (5.2), (5.18), (5.19) and (5.20) be fulfilled. Then, the estimate*

$$\left\| u - I^0_\tau(P^G u_h) \right\|_{L_\infty(0,T;H^1(\Omega))} \le C_1\ h + C_2\ \tau$$

holds for the error between the finite element solution u_h of (5.3), (5.4) and the solution u of the evolutionary variational inequality (3.1), (3.2).

Proof. As already indicated in Remark 5.15, the evaluation of the first a priori estimate of Lemma 5.13 comprises the estimation of the following three terms.

$$\begin{aligned}
(i): \quad a(t^j; u(t^j) - P^G v^j_h, P^G(u^j_h - v^j_h)) \ &\le C\ \left\| u(t^j) - P^G v^j_h \right\|_{H^1(\Omega)} \\
&* \left\| P^G(u^j_h - v^j_h) \right\|_{H^1(\Omega)}
\end{aligned}$$

and

$$(ii): \quad \left| \begin{array}{l} a(t^j; u(t^j), P^G(v_h^j - u_h^j)) - (f(t^j), P^G(v_h^j - u_h^j)) \\ - (g_N(t^j), P^G(v_h^j - u_h^j))_{\Gamma_N} - \int_0^{t^j} b(t'; u(t'), P^G(v_h^j - u_h^j)) \, dt' \end{array} \right.$$

as well as the memory term

$$(iii): \quad \sum_{i=0}^{j-1} \left[b_G^i(P^G v_h, P^G(u_h^j - v_h^j)) - \int_{t^i}^{t^{i+1}} b(t'; u(t'), P^G(u_h^j - v_h^j)) \, dt' \right]$$

for all $j = 0, \dots, N$ and grid function vectors $v_h^i \in K_h^i, \; i = 0, \dots, N$.

We will discuss these terms in reverse order, i.e., we start with the treatment of the error between the memory term and its discretized form (see (iii)), then we consider (ii) and, finally, (i) will be treated.
We shall take $v_h^i = \big((\Pi_h u(t^i))(\xi)\big)_{\xi \in \overline{\omega}}$ with Π_h as the (piecewise linear) Lagrange interpolation operator. Since we have assumed $u(t) \in H^2(\Omega)$ for all $t \in [0, T]$, this choice is permissible and, moreover, we have $K_h^i \ni P^G v_h^i = \Pi_h u(t^i)$ for all $i = 0, \dots, N$.

Compared to the corresponding error analysis for elliptic obstacle problems (cf. Sections 4.1.2 and 4.3), now an essential additional problem consists in the estimation of the memory term (see (iii)) for which we use the splitting

$$\sum_{i=0}^{j-1} \Big[\int_{t^i}^{t^{i+1}} b(t'; u(t'), P^G(u_h^j - v_h^j)) \, dt' - b_G^i(P^G v_h, P^G(u_h^j - v_h^j)) \Big] = \sum_{i=0}^{j-1} \sum_{l=1}^{3} (K_l^i + Q_l^i).$$

We refer to (5.6) with respect to the definition of the just mentioned memory term in its discretized form $b_G^i(P^G v_h, P^G w_h), \; i = 0, \dots, N-1$.

For the sake of brevity we denote

$$z_h^G = P^G(u_h^j - v_h^j)$$

and suppress the dependence on x. Then, the error terms K_l^i and Q_l^i can be represented as follows.

$$\left. \begin{array}{rl} K_1^i &= \displaystyle\int_{t^i}^{t^{i+1}} \Big(\int_\Omega \partial_t k(t') \left[\alpha \, \nabla\big(u(t') - u(t^{i+1})\big) + (1 - \alpha) \, \nabla\big(u(t') - u(t^i)\big) \right] \\ & \quad \cdot \nabla z_h^G \, dx \Big) \, dt', \\ Q_1^i &= \displaystyle\int_{t^i}^{t^{i+1}} \Big(\int_\Omega \partial_t q(t') \left[\alpha \, \big(u(t') - u(t^{i+1})\big) + (1 - \alpha) \, \big(u(t') - u(t^i)\big) \right] \\ & \quad * \, z_h^G \, dx \Big) \, dt' \end{array} \right| \tag{5.21}$$

and

$$
\left.
\begin{aligned}
K_2^i &= \int_{t^i}^{t^{i+1}} \Big(\int_\Omega \partial_t k(t') \Big[\alpha \, \nabla u(t^{i+1}) + (1-\alpha) \, \nabla u(t^i) \Big] \, \nabla z_h^G \, dx \Big) \, dt' \\
&\quad - \int_\Omega \big(k(t^{i+1}) - k(t^i)\big) \Big[\alpha \, \nabla u(t^{i+1}) + (1-\alpha) \, \nabla u(t^i) \Big] \, \nabla z_h^G \, dx, \\
Q_2^i &= \int_{t^i}^{t^{i+1}} \Big(\int_\Omega \partial_t q(t') \Big[\alpha \, u(t^{i+1}) + (1-\alpha) \, u(t^i) \Big] \, z_h^G \, dx \Big) \, dt' \\
&\quad - \int_\Omega \big(q(t^{i+1}) - q(t^i)\big) \Big[\alpha \, u(t^{i+1}) + (1-\alpha) \, u(t^i) \Big] \, z_h^G \, dx
\end{aligned}
\right| \tag{5.22}
$$

as well as, finally,

$$
\left.
\begin{aligned}
K_3^i &= \int_\Omega \big(k(t^{i+1}) - k(t^i)\big) \Big[\alpha \, \nabla\big(u(t^{i+1}) - P^G v_h^{i+1}\big) \\
&\qquad + (1-\alpha) \, \nabla\big(u(t^i) - P^G v_h^i\big) \Big] \, \nabla z_h^G \, dx, \\
Q_3^i &= \int_\Omega \big(q(t^{i+1}) - q(t^i)\big) \Big[\alpha \, \big(u(t^{i+1}) - P^G v_h^{i+1}\big) \\
&\qquad + (1-\alpha) \, \big(u(t^i) - P^G v_h^i\big) \Big] \, z_h^G \, dx
\end{aligned}
\right| \tag{5.23}
$$

for $i = 0, \dots, N-1$.

We make use of the regularity $u \in W_\infty^1(0,T;H^1(\Omega))$ for the treatment of the error terms K_1^i and Q_1^i, $i = 0, \dots, N-1$. Hence, u is Lipschitz continuous as a mapping from the time interval $[0,T]$ to the Sobolev space $H^1(\Omega)$, such that we have

$$\|u(t) - u(\bar{t})\|_{H^1(\Omega)} \le L_u \, |t - \bar{t}| \quad \text{for all} \quad t, \ \bar{t} \in [0,T] \quad \text{and fixed} \quad L_u \ge 0;$$

cf. [ZEIDII90], Corollary 4.23. Applying this property to the error terms K_1^i and Q_1^i, we derive the estimate

$$
\left.
\begin{aligned}
|K_1^i| + |Q_1^i| \ &\le \Big(\|\partial_t k\|_{L_\infty(0,T;L_\infty(\Omega))} + \|\partial_t q\|_{L_\infty(0,T;L_\infty(\Omega))} \Big) \, L_u \\
&\quad * \ \|z_h^G\|_{H^1(\Omega)} \, (t^{i+1} - t^i)^2 \quad \text{for} \quad i = 0, \dots, N-1.
\end{aligned}
\right| \tag{5.24}
$$

Owing to the assumption $k, \ q \in W_\infty^1(0,T;L_\infty(\Omega))$, a close look at K_2^i and Q_2^i defined in (5.22) shows that we have $K_2^i = Q_2^i = 0$ for $i = 0, \dots, N-1$.

As indicated above we take $v_h^i = \big((\Pi_h u(t^i))(\xi)\big)_{\xi \in \varpi}$ (analogously for the time instant t^{i+1}) for the evaluation of the terms K_3^i and Q_3^i. Furthermore, we exploit the Lipschitz continuity of the coefficient functions k and q as mappings from $[0,T]$ to $L_\infty(\Omega)$. The corresponding Lipschitz constants are denoted by L_k and L_q (cf. Lemma 5.4[ii]). Now, by means of standard interpolation properties of the form

$$\|w - \Pi_h w\|_{H^1(\Omega)} \le C_1 \, h \, |w|_{H^2(\Omega)} \quad \text{and} \quad \|w - \Pi_h w\|_{L_2(\Omega)} \le C_2 \, h^2 \, |w|_{H^2(\Omega)}$$

for $w \in H^2(\Omega)$ (cf. [CIAR78], Chapter 3 and [BRSC96], Chapter 4.4), we get

$$\left.\begin{aligned} |K_3^i| &\le C_1\ L_k\ (t^{i+1}-t^i)\ h\ \left|z_h^G\right|_{H^1(\Omega)} \\ &\quad * \left(\alpha\ \left|u(t^i)\right|_{H^2(\Omega)} + (1-\alpha)\ \left|u(t^{i+1})\right|_{H^2(\Omega)}\right), \\ |Q_3^i| &\le C_2\ L_q\ (t^{i+1}-t^i)\ h^2\ \left\|z_h^G\right\|_{L_2(\Omega)} \\ &\quad * \left(\alpha\ \left|u(t^i)\right|_{H^2(\Omega)} + (1-\alpha)\ \left|u(t^{i+1})\right|_{H^2(\Omega)}\right) \end{aligned}\right| \tag{5.25}$$

for $i = 0, \ldots, N-1$.

Summarizing the results derived in (5.24) and (5.25), we arrive at

$$\sum_{i=0}^{j-1}\sum_{l=1}^{3} |K_l^i| + |Q_l^i| \le C\ \left\|z_h^G\right\|_{H^1(\Omega)}\ \left[L_u\ t^j\ \tau\ +\ h\ \tau \sum_{i=0}^{j} \left|u(t^i)\right|_{H^2(\Omega)}\ \right].$$

Consequently, the discrepancy between the memory term and its discretized form (cf. (iii)) can be written as

$$\left.\begin{aligned} &\left|\sum_{i=0}^{j-1}\left[b_G^i(P^G v_h, P^G(u_h^j - v_h^j)) - \int_{t^i}^{t^{i+1}} b(t'; u(t'), P^G(u_h^j - v_h^j))\ dt'\ \right]\right| \\ &\le C\ \left\|P^G u_h^j - \Pi_h u(t^j)\right\|_{H^1(\Omega)}\ T\ \left(L_u\ \tau + h\ \max_{i=0,\ldots,j}\{\left|u(t^i)\right|_{H^2(\Omega)}\}\right) \end{aligned}\right| \tag{5.26}$$

for $j = 0, \ldots, N$.

Let us now switch over to the consideration of (ii). Due to our assumptions we may apply Green's formula. Making use of the equality on the Neumann/Newton part Γ_N of the boundary $\partial\Omega$ (cf. the complementarity problem (5.19)) and the discussion in Section 3.5), we obtain

$$\begin{aligned} &a(t^j; u(t^j), w) - (f(t^j), w) - (g_N(t^j), w)_{\Gamma_N} - \int_0^{t^j} b(t'; u(t'), w)\ dt' \\ &= \ \left((Au)(t^j) - f(t^j) - \int_0^{t^j} (Bu)(t')\ dt'\ ,\ w\right) \end{aligned}$$

for $w \in V$ and $t^j \in [0,T]$, $j = 0, \ldots, N$. Now, we take

$$w = \left(P^G(v_h^j - u_h^j)\right) \in V_h^G \subset V \quad \text{and} \quad v_h^j = \left((\Pi_h u(t^j))(\xi)\right)_{\xi \in \overline{\omega}}.$$

Taking into account the complementarity problem (5.19), the representation

$$\Pi_h u(t^j) - P^G u_h^j = \Pi_h u(t^j) - u(t^j) + u(t^j) - \varphi + \varphi - \Pi_h\varphi + \Pi_h\varphi - P^G u_h^j$$

and $\Pi_h\varphi \leq P^G u_h^j$ as well as using again interpolation properties, we find

$$
\begin{aligned}
&\Big((Au-f)(t^j) - \int_0^{t^j} (Bu)(t')\,dt' \;,\; \Pi_h u(t^j) - P^G u_h^j\Big) \\
&\leq \Big((Au-f)(t^j) - \int_0^{t^j} (Bu)(t')\,dt' \;,\; \Pi_h u(t^j) - u(t^j)\Big) \\
&\quad + \Big((Au-f)(t^j) - \int_0^{t^j} (Bu)(t')\,dt' \;,\; \varphi - \Pi_h\varphi\Big) \\
&\leq \Big\|(Au-f)(t^j) - \int_0^{t^j} (Bu)(t')\,dt'\Big\|_{L_2(\Omega)} C\,h^2\,\big(|u(t^j)|_{H^2(\Omega)} + |\varphi|_{H^2(\Omega)}\big)
\end{aligned}
\tag{5.27}
$$

as the final result for (ii).

Finally, we return to the beginning of the proof and consider (i). We obtain

$$
\begin{aligned}
a(t^j; u(t^j) - P^G v_h^j,\; P^G(u_h^j - v_h^j)) \;&\leq C\,h\,|u(t^j)|_{H^2(\Omega)} \\
&\quad * \Big\|P^G u_h^j - \Pi_h u(t^j)\Big\|_{H^1(\Omega)}
\end{aligned}
\tag{5.28}
$$

by means of $v_h^j = \big((\Pi_h u(t^j))(\xi)\big)_{\xi\in\overline{\omega}}$ and the above mentioned interpolation properties.

Now, we connect the first a priori estimate of Lemma 5.13 with the results (5.28), (5.27) and (5.26) derived for the error terms (i), (ii) and (iii).
Taking

$$
v_h^j = \big((\Pi_h u(t^j))(\xi)\big)_{\xi\in\overline{\omega}}\,, \quad j = 0,\ldots,N,
$$

we get the following intermediate result.

$$
\begin{aligned}
&\langle \Lambda_G^j\,(u_h^j - v_h^j)\;,\; u_h^j - v_h^j\rangle \leq \sum_{i=0}^{j-1} \langle (\Lambda_G^{i+1} - \Lambda_G^i)\,(u_h^i - v_h^i)\;,\; u_h^j - v_h^j\rangle \\
&+ C_1\,h\,|u(t^j)|_{H^2(\Omega)}\,\Big\|P^G u_h^j - \Pi_h u(t^j)\Big\|_{H^1(\Omega)} \\
&+ C_2\,h^2\,\Big\|(Au-f)(t^j) - \int_0^{t^j} (Bu)(t')\,dt'\Big\|_{L_2(\Omega)}\,\big(|u(t^j)|_{H^2(\Omega)} + |\varphi|_{H^2(\Omega)}\big) \\
&+ C_3\,\Big\|P^G u_h^j - \Pi_h u(t^j)\Big\|_{H^1(\Omega)}\;T\,\big(L_u\,\tau + h \max_{i=0,\ldots,j}\{|u(t^i)|_{H^2(\Omega)}\}\big)
\end{aligned}
\tag{5.29}
$$

for all $j = 0,\ldots,N$.

The application of Lemma 5.4[i], [ii] and Young's inequality leads to

$$
\begin{aligned}
C_4 \left\| P^G u_h^j - \Pi_h u(t^j) \right\|_{H^1(\Omega)}^2 &\leq C_5\, \tau \sum_{i=0}^{j-1} \left\| P^G u_h^i - \Pi_h u(t^i) \right\|_{H^1(\Omega)}^2 \\
&+ \tilde{C}_1\, h^2\, \left| u(t^j) \right|_{H^2(\Omega)}^2 \\
&+ C_2\, h^2 \left\| (Au - f)(t^j) - \int_0^{t^j} (Bu)(t')\, dt' \right\|_{L_2(\Omega)} \left(\left| u(t^j) \right|_{H^2(\Omega)} + |\varphi|_{H^2(\Omega)} \right) \\
&+ \tilde{C}_3 \left((L_u\, \tau)^2 + h^2 \left(\max_{i=0,\dots,j} \{ \left| u(t^i) \right|_{H^2(\Omega)} \} \right)^2 \right) \quad \text{for all } j = 0, \dots, N.
\end{aligned}
$$

In the next step of the proof we make use of Gronwall's inequality in its discrete form (see Lemma 3.1). Thus, we arrive at the estimate

$$
\left\| P^G u_h^j - \Pi_h u(t^j) \right\|_{H^1(\Omega)}^2 \leq \exp(c_1\, t^j) \left(c_2\, \tau^2 + c_3\, h^2 \right) \quad \text{for all } j = 0, \dots, N. \tag{5.30}
$$

Recalling the definition of the piecewise constant in time function $I_\tau^0(P^G u_h)$ from Remark 5.7[ii], the following estimate for the difference

$$
z := u - I_\tau^0(P^G u_h)
$$

holds.

$$
\|z(t)\|_{H^1(\Omega)} \leq \begin{cases} \left\| u(t) - u(t^0) \right\|_{H^1(\Omega)} + \left\| u(t^0) - P^G u_h^0 \right\|_{H^1(\Omega)} & \text{for I}, \\ \left\| u(t) - u(t^i) \right\|_{H^1(\Omega)} + \left\| u(t^i) - P^G u_h^i \right\|_{H^1(\Omega)} & \text{for II}, \\ \left\| u(t) - u(t^N) \right\|_{H^1(\Omega)} + \left\| u(t^N) - P^G u_h^N \right\|_{H^1(\Omega)} & \text{for III}, \end{cases}
$$

where I corresponds to $0 \leq t < t^0 + \tau/2$, II stands for $t^i - \tau/2 \leq t < t^i + \tau/2$, $i = 1, \dots, N-1$ and III is related to $t^N - \tau/2 \leq t \leq T$.

From that, we get

$$
\begin{aligned}
&\left\| u - I_\tau^0(P^G u_h) \right\|_{L_\infty(0,T;H^1(\Omega))} \leq (L_u\, \tau)/2 \\
&+ \max_{j=0,\dots,N} \left(\left\| u(t^j) - P^G v_h^j \right\|_{H^1(\Omega)} + \left\| P^G v_h^j - P^G u_h^j \right\|_{H^1(\Omega)} \right)
\end{aligned}
$$

using again $u \in W_\infty^1(0, T; H^1(\Omega))$ (Lipschitz continuity).

Taking $v_h^j = \left((\Pi_h u(t^j))(\xi) \right)_{\xi \in \overline{\omega}}$ for all $j = 0, \dots, N$, using interpolation properties and inserting (5.30), we end up with the claimed first order estimate of the form $O(h) + O(\tau)$. ■

Now we study the accuracy of the solution of the box approximation (5.9) using a similar line of reasoning. But, instead of the (first) a priori estimate from Lemma 5.13 for the just performed finite element error analysis, we concentrate now on the evaluation of the (first) a priori estimate from Lemma 5.14. An essential additional point consists in the treatment of the error terms E_B^j, $j = 0, \ldots, N$ caused by the box approximation.

Theorem 5.17 *Let the conditions (5.1), (5.2), (5.18), (5.19), (5.20) and*

$$\kappa \in W^1_\infty(\Gamma_N), \quad g_N \in C([0,T]; H^1(\Gamma_N))$$

be fulfilled. Furthermore, let the boxes be constructed according to $\mathcal{D}_h(1)$. *Then, the estimate*

$$\left\| u - I^0_\tau(P^G y_h) \right\|_{L_\infty(0,T;H^1(\Omega))} \leq C_1 \, h + C_2 \, \tau$$

holds for the error between the box solution y_h *of (5.9), (5.10) and the solution* u *of the evolutionary variational inequality (3.1), (3.2).*

Proof. Comparison of the a priori estimates for the finite element approximation (5.3) (cf. Lemma 5.13) and the box approximation (5.9) (cf. Lemma 5.14) shows that the essential difference consists in the terms E_B^j for $j = 0, \ldots, N$, which have to be estimated additionally for the box approximation (5.9).

Denoting for the sake of brevity $z_h = y_h^j - v_h^j$, these terms read as follows (cf. Lemma 5.14).

$$\begin{aligned} E_B^j \; &= \Big\{ (f(t^j), P^B z_h - P^G z_h) + (g_N(t^j), P^B z_h - P^G z_h)_{\Gamma_N} \\ &\quad + a(t^j; P^G v_h^j, P^G z_h) - a_B(t^j; P^G v_h^j, P^B z_h) \\ &\quad + \sum_{i=0}^{j-1} \big(\, b_B^i(P^G v_h, P^B z_h) - b_G^i(P^G v_h, P^G z_h) \, \big) \Big\}. \end{aligned}$$

We refer to the estimates (4.27) concerning the treatment of the first three terms. Hence, there remains to consider the difference

$$b_B^i(P^G v_h, P^B z_h) - b_G^i(P^G v_h, P^G z_h).$$

Recalling the definitions (5.6) and (5.8) for these bilinear forms and using the equality of the finite element and box stiffness matrices, we obtain

$$\begin{aligned} & b_B^i(P^G v_h, P^B z_h) - b_G^i(P^G v_h, P^G z_h) \\ &= \int_\Omega \left(q(t^{i+1}) - q(t^i) \right) \Big[\, \alpha \, \left(P^B v_h^{i+1} \; P^B z_h - P^G v_h^{i+1} \; P^G z_h \right) \\ &\quad + (1-\alpha) \, \left(P^B v_h^i \; P^B z_h - P^G v_h^i \; P^G z_h \right) \Big] \; dx. \end{aligned}$$

Consequently, by means of the Lipschitz continuity of q as a mapping from $[0,T]$ to $L_\infty(\Omega)$ and Corollary 4.7, Lemma 4.10[i], we derive

$$| b^i_B(P^G v_h, P^B z_h) - b^i_G(P^G v_h, P^G z_h) | \le C\, L_q\, h\, \tau\, \|P^G z_h\|_{H^1(\Omega)}$$
$$* \left(\alpha\, \|P^G v_h^{i+1}\|_{H^1(\Omega)} + (1-\alpha)\, \|P^G v_h^i\|_{H^1(\Omega)}\right).$$

As indicated above, we connect this estimate with the results derived in (4.27). Hence, we get the following estimate for the terms E^j_B caused by the box approximation.

$$\begin{aligned}
|E^j_B| \le h\, \|P^G z_h\|_{H^1(\Omega)} &\Big\{ c_1\, \|f(t^j)\|_{L_2(\Omega)} \\
&+ c_2\, h^{1/2} \Big(\sum_{\sigma\in\sigma_{h,2}} |g_N(t^j)|^2_{H^1(\sigma)} \Big)^{1/2} \\
&+ \Big[c_3\, \|q(t^j)\|_{L_\infty(\Omega)} + c_4\, \big(\|\kappa\|_{L_\infty(\Gamma_N)} + \max_{\sigma\in\sigma_{h,2}} \{|\kappa|_{W^1_\infty(\sigma)}\}\big)\Big]\, \|P^G v_h^j\|_{H^1(\Omega)} \\
&+ c_5\, L_q\, \tau \sum_{i=0}^{j} \|P^G v_h^i\|_{H^1(\Omega)} \Big\} \quad \text{for } j=0,\dots,N \quad \text{and with} \quad z_h = y_h^j - v_h^j.
\end{aligned} \tag{5.31}$$

Returning to the first a priori estimate of Lemma 5.14, it can be easily seen that the terms, apart from the just discussed additional error E^j_B, can be treated analogously to the error analysis for the finite element approximation, cf. the proof of Theorem 5.16. Consequently, instead of (5.29), we deduce now the following intermediate estimate for the box approximation (5.9).

$$\begin{aligned}
\langle \Lambda^j_B\, (y_h^j - v_h^j)\,,\, y_h^j - v_h^j\rangle &\le \sum_{i=0}^{j-1} \langle (\Lambda^{i+1}_B - \Lambda^i_B)\, (y_h^i - v_h^i)\,,\, y_h^j - v_h^j\rangle + E^j_B \\
&+ C_1\, h\, |u(t^j)|_{H^2(\Omega)}\, \|P^G y_h^j - \Pi_h u(t^j)\|_{H^1(\Omega)} \\
&+ C_2\, h^2\, \Big\|(Au-f)(t^j) - \int_0^{t^j} (Bu)(t')\, dt'\Big\|_{L_2(\Omega)} \big(|u(t^j)|_{H^2(\Omega)} + |\varphi|_{H^2(\Omega)} \big) \\
&+ C_3\, \|P^G y_h^j - \Pi_h u(t^j)\|_{H^1(\Omega)}\; T\, \big(L_u\, \tau + h \max_{i=0,\dots,j} \{|u(t^i)|_{H^2(\Omega)}\}\big)
\end{aligned} \tag{5.32}$$

with $v_h^j = \big((\Pi_h u(t^j))(\xi)\big)_{\xi\in\bar\omega}$ and E^j_B (estimated in (5.31)) for all $j=0,\dots,N$.

On the basis of this estimate the remainder of the proof is performed similar to the proof of Theorem 5.16 (finite element approximation). Instead of the properties of the matrices Λ^i_G in the finite element case, we use the analogous properties of the box matrices Λ^i_B, $i=0,\dots,N$; cf. Lemma 5.4[i], [ii]. Consequently, we also end up with the first order accuracy of the form $O(h)+O(\tau)$ for the finite volume approximation (5.9), (5.10). ■

Remark 5.18 [i] The second a priori estimates derived in Lemmas 5.13 and 5.14 can be evaluated in the absence of Dirichlet boundary conditions or when these boundary conditions have a special form (e.g., g_D = const. on Γ_D, cf. Remark 4.17). The crucial point is to guarantee the property $v \in K(t^j)$ for $j = 0, \dots, N$, if one takes $v = \max\{P^G u_h^j, \varphi\}$ (resp. $v = \max\{P^G y_h^j, \varphi\}$) in the corresponding a priori estimates.

[ii] Since we work in the proofs of Theorems 5.16 and 5.17 'only with the first time-derivative' of u, the weighting factor $\alpha \in [0, 1]$ introduced in (5.4), (5.6), (5.8) and (5.10) has no direct influence on the convergence rate. However, from the practical point of view, we recommend taking $\alpha = 0.5$. □

5.3 Two penalization methods for the solution of the evolutionary finite volume inequalities

The main aim of this section consists in substitution of the discrete evolutionary inequality problems by a family of nonlinear equations and the demonstration that their solutions converge to the solutions of the discrete inequalities.

In other words, similar to Section 4.4, we will approach the finite volume inequalities (5.9) by different penalization (regularization) methods. In combination with Theorem 5.17 we will establish a first order convergence rate for the total error between the solutions of the evolutionary inequality (3.1) and the penalization problems (cf. Remark 5.23). Furthermore, iteration methods will be analysed for the solution of the (nonlinear) penalization problems.

Again, we will exploit the discrete maximum principle introduced in Section 4.4.1. Therefore, we restrict ourselves to considering the finite volume approximation (5.9), (5.10) for a weakly acute triangulation.

5.3.1 Discussion of penalization techniques

On the one hand, we consider a penalization problem of the form

$$\left|\begin{array}{l} \text{Find } \ y_{h\varepsilon}^j \in L_2(\overline{\omega}) : \quad (y_{h\varepsilon}^j)_\xi = g_D(\xi, t^j) \quad \forall \xi \in \gamma_D \\ \langle \Lambda_B^j \, y_{h\varepsilon}^j, v_h \rangle + \langle \varepsilon^{-1} \, \beta_{1h}(y_{h\varepsilon}^j - \varphi_h), v_h \rangle = \langle f_{Bh}^j + g_{Bh}^j, v_h \rangle \\ + \displaystyle\sum_{i=0}^{j-1} \langle (\Lambda_B^{i+1} - \Lambda_B^i) \, y_{h\varepsilon}^i, v_h \rangle \quad \forall v_h \in L_2(\tilde{\omega}), \quad j = 0, \dots, N \\ \text{with } \ \varepsilon > 0, \quad \big(\beta_{1h}(w_h)\big)_{\xi \in \tilde{\omega}} = \big(\text{meas}_n(D_\xi) \, (w_{h,\xi})^-\big)_{\xi \in \tilde{\omega}}. \end{array}\right. \tag{5.33}$$

As already pointed out for the elliptic problems (cf. Remark 4.26 and the consequence from Theorem 4.27), we will discuss the usage of penalization parameters $\varepsilon = \varepsilon_\xi$ depending on the nodes $\xi \in \tilde{\omega}$.

On the other hand, in agreement with our analytical investigations made in Chapter 3 (see, in particular, problem (3.8)), we will analyse a Lewy–Stampacchia type regularization in the discrete case.

$$\left|\begin{array}{l} \text{Find} \quad z^j_{h\varepsilon} \in L_2(\overline{\omega}): \quad (z^j_{h\varepsilon})_\xi = g_D(\xi, t^j) \quad \forall \xi \in \gamma_D \\ \langle \Lambda^j_B \, z^j_{h\varepsilon}, v_h \rangle + \langle \mu_h \, \beta_{2h}(\frac{z^j_{h\varepsilon} - \varphi_h}{\varepsilon}), v_h \rangle = \langle \mu_h + f^j_{Bh} + g^j_{Bh}, v_h \rangle \\ \quad + \sum_{i=0}^{j-1} \langle (\Lambda^{i+1}_B - \Lambda^i_B) \, z^i_{h\varepsilon}, v_h \rangle \quad \forall v_h \in L_2(\tilde{\omega}), \quad j = 0, \ldots, N; \quad \varepsilon > 0. \end{array}\right. \tag{5.34}$$

Similar to the elliptic problem (4.32) we define

$$\big(\beta_{2h}(w_h)\big)_{\xi \in \tilde{\omega}} = \big(w^+_{h,\xi}/(1 + w_{h,\xi})\big)_{\xi \in \tilde{\omega}} \quad \text{and} \quad 0 \le \mu_h = \big(\Lambda^0_B \, \varphi_h - f^0_{Bh} - g^0_{Bh}\big)^+.$$

We note that one can take other auxiliary vectors $\mu_h \ge 0$. For instance, one can put

$$\mu^j_h = \big(\Lambda^0_B \, \varphi_h - f^j_{Bh} - g^j_{Bh}\big)^+, \quad j = 0, \ldots, N.$$

The proofs of the following results have to be only slightly modified for this time-dependent choice.

Owing to the monotonicity of the penalization terms $\beta_{1h}(w_h)$ and $\beta_{2h}(w_h)$ (cf. Section 4.4) we deduce the unique solvability of the problems (5.33) and (5.34) by an *inductive* argument, because the problems (5.33) and (5.34) represent a (nonlinear) elliptic equation at each time instant t^j, $j = 0, \ldots, N$. Moreover, the boundedness and stability statements established in Theorems 5.6 and 5.8 for the solutions of the (discrete) evolutionary obstacle problems can be transferred to the penalization problems (5.33) and (5.34).

Let us now switch over to consider the time evolution of the solutions $y_{h\varepsilon}$ and $z_{h\varepsilon}$ of the problems (5.33) and (5.34). In Theorem 5.10 we have stated that the solution of the finite volume inequality formulation (5.9) is nondecreasing with respect to time. Now we show that the solutions of the penalization problems (5.33) and (5.34) also have this behaviour concerning their time evolution.

Lemma 5.19 *Let the assumptions of Theorem 5.10 be fulfilled. Then, the solutions of the problems (5.33) and (5.34) are such that*

$$(y^{j+1}_{h\varepsilon})_\xi \ge (y^j_{h\varepsilon})_\xi \quad \textit{and} \quad (z^{j+1}_{h\varepsilon})_\xi \ge (z^j_{h\varepsilon})_\xi \quad \forall \xi \in \overline{\omega}, \quad j = 0, \ldots, N-1$$

hold.

Proof. Subtracting the penalization equations in (5.33) for two successive time instants t^j and t^{j+1}, we obtain

$$\begin{aligned} &\Lambda^{j+1}_B \, (y^{j+1}_{h\varepsilon} - y^j_{h\varepsilon}) + \varepsilon^{-1} \, \big(\beta_{1h}(y^{j+1}_{h\varepsilon} - \varphi_h) - \beta_{1h}(y^j_{h\varepsilon} - \varphi_h)\big) \\ &= \; f^{j+1}_{Bh} + g^{j+1}_{Bh} - f^j_{Bh} - g^j_{Bh} \ge 0. \end{aligned}$$

Owing to the Dirichlet boundary conditions nondecreasing with respect to time we have $(y_{h\varepsilon}^{j+1} - y_{h\varepsilon}^{j})_\xi \geq 0$ for Dirichlet nodes $\xi \in \gamma_D$. Therefore, the above relation can be multiplied by

$$0 \geq ((y_{h\varepsilon}^{j+1} - y_{h\varepsilon}^{j})^-) \in L_2(\tilde{\omega}).$$

By virtue of

$$\left\langle \operatorname{meas}_n(D_\xi)\ \left((y_{h\varepsilon}^{j+1} - \varphi_h)^- - (y_{h\varepsilon}^{j} - \varphi_h)^-\right)\ ,\ (y_{h\varepsilon}^{j+1} - y_{h\varepsilon}^{j})^- \right\rangle \geq 0$$

and the maximum principle (4.30), we get $(y_{h\varepsilon}^{j+1} - y_{h\varepsilon}^{j})^- = 0$. Hence, the statement for the problem (5.33) is proved.

The statement for the regularization method (5.34) can be obtained in a similar way by subtracting the equations for two successsive time instants, which gives

$$\begin{aligned}
&\Lambda_B^{j+1}\ (z_{h\varepsilon}^{j+1} - z_{h\varepsilon}^{j}) + \mu_h\ \left(\beta_{2h}((z_{h\varepsilon}^{j+1} - \varphi_h)/\varepsilon) - \beta_{2h}((z_{h\varepsilon}^{j} - \varphi_h)/\varepsilon)\right)\\
&= \ f_{Bh}^{j+1} + g_{Bh}^{j+1} - f_{Bh}^{j} - g_{Bh}^{j} \geq 0.
\end{aligned}$$

Multiplying this relation by $0 \geq ((z_{h\varepsilon}^{j+1} - z_{h\varepsilon}^{j})^-) \in L_2(\tilde{\omega})$ as well as taking into account $\mu_h \geq 0$ and

$$\left\langle \mu_h\ \left(\beta_{2h}(v_h/\varepsilon) - \beta_{2h}(w_h/\varepsilon)\right)\ ,\ (v_h - w_h)^- \right\rangle \geq 0,$$

we obtain $(z_{h\varepsilon}^{j+1} - z_{h\varepsilon}^{j})^- = 0$. Hence, the inequality $z_{h\varepsilon}^{j+1} \geq z_{h\varepsilon}^{j}$ is proved. ■

Let us now study the convergence rate of the penalization method (5.33) applied to the finite volume inequalities (5.9).

Lemma 5.20 *Let the assumptions of Theorem 5.10 be fulfilled.*
[i] *Then, using penalization parameters $\varepsilon = \varepsilon_\xi$ depending on the nodes $\xi \in \tilde{\omega}$ of the triangulation in problem (5.33), one has the estimate*

$$\left\|(y_{h\varepsilon}^{j} - \varphi_h)^-\right\|_{0,\overline{\omega}} \leq \left\|(y_{h\varepsilon}^{0} - \varphi_h)^-\right\|_{0,\overline{\omega}} \leq M_{\varphi_h}\ \max_{\xi \in \tilde{\omega}} \varepsilon_\xi \quad \textit{for all} \quad j = 0, \ldots, N$$

with

$$\begin{aligned}
M_{\varphi_h} &= \left\|(\Lambda_B^0\ \varphi_h - f_{Bh}^0 - g_{Bh}^0)/\operatorname{meas}_n(D)\right\|_{0,\overline{\omega}}\\
&= \left\|P^B\left((\Lambda_B^0\ \varphi_h - f_{Bh}^0 - g_{Bh}^0)/\operatorname{meas}_n(D)\right)\right\|_{L_2(\Omega)}.
\end{aligned}$$

[ii] *Let the estimates*

$$\left\|(y_{h\varepsilon}^{j} - \varphi_h)^-\right\|_{1,\overline{\omega}} \leq M_j$$

be fulfilled with constants M_j, which may depend on the time instants t^j for $j = 0,\ldots,N$. Then, there exist constants $C_i > 0$, $i = 1,2$, such that the penalization error in problem (5.33) can be estimated by

$$\left\| y_h^j - y_{h\varepsilon}^j \right\|_{1,\varpi} \le C_1 \exp\left(C_2\, t^j\right) \max_{i=0,\ldots,j} M_i \quad \textit{for all} \quad j = 0,\ldots,N.$$

Proof. *[i].* We take $v_h = (y_{h\varepsilon}^0 - \varphi_h)^-$ in problem (5.33) for $j = 0$, which leads to

$$\left.\begin{aligned} &\langle \Lambda_B^0\, (y_{h\varepsilon}^0 - \varphi_h)\,,\, (y_{h\varepsilon}^0 - \varphi_h)^- \rangle + \sum_{\xi\in\tilde{\omega}} \frac{\mathrm{meas}_n(D_\xi)}{\varepsilon_\xi} \left((y_{h\varepsilon}^0 - \varphi_h)_\xi^-\right)^2 \\ &= \langle f_{Bh}^0 + g_{Bh}^0 - \Lambda_B^0\, \varphi_h\,,\, (y_{h\varepsilon}^0 - \varphi_h)^- \rangle. \end{aligned}\right| \tag{5.35}$$

The maximum principle (4.30) implies the non-negativity of the first term on the left-hand side. Consequently, we get

$$\left\| (y_{h\varepsilon}^0 - \varphi_h)^- \right\|_{0,\varpi} \le M_{\varphi_h} \max_{\xi\in\tilde{\omega}} \varepsilon_\xi.$$

Furthermore, due to Lemma 5.19, we immediately deduce

$$|(y_{h\varepsilon}^{j+1} - \varphi_h)_\xi^-| \le |(y_{h\varepsilon}^j - \varphi_h)_\xi^-| \le |(y_{h\varepsilon}^0 - \varphi_h)_\xi^-| \quad \forall \xi \in \tilde{\omega}, \quad j = 0,\ldots,N-1,$$

which gives the result for all time instants t^j, $j = 1,\ldots,N$.

[ii]. We represent the difference of the solutions of the inequality (5.9) and the penalization problem (5.33) in the form

$$y_h^j - y_{h\varepsilon}^j = r_{h\varepsilon}^j - (y_{h\varepsilon}^j - \varphi_h)^- \quad \text{for all} \quad j = 0,\ldots,N.$$

Therefore, we are allowed to take

$$v_h = r_{h\varepsilon}^j \in L_2(\tilde{\omega}) \quad \text{in (5.33)} \quad \text{and} \quad v_h = (y_h^j - r_{h\varepsilon}^j) \in K_h^j \quad \text{in (5.9)}$$

for $j = 0,\ldots,N$. The addition of equation (5.33) and inequality (5.9) yields

$$\begin{aligned} \langle \Lambda_B^j\, (y_h^j - y_{h\varepsilon}^j),\, r_{h\varepsilon}^j \rangle \;&\le \langle \varepsilon^{-1}\, \beta_{1h}(y_{h\varepsilon}^j - \varphi_h),\, r_{h\varepsilon}^j \rangle \\ &+ \sum_{i=0}^{j-1} \langle (\Lambda_B^{i+1} - \Lambda_B^i)\, (y_h^i - y_{h\varepsilon}^i),\, r_{h\varepsilon}^j \rangle \quad \text{for } j = 0,\ldots,N. \end{aligned}$$

Due to $\langle \beta_{1h}(y_{h\varepsilon}^j - \varphi_h), r_{h\varepsilon}^j \rangle \le 0$ and the ellipticity of Λ_B^j (cf. Lemma 5.4[i]), we derive

$$C_3\, \left\| r_{h\varepsilon}^j \right\|_{1,\varpi}^2 \le \langle \Lambda_B^j\, (y_{h\varepsilon}^j - \varphi_h)^-,\, r_{h\varepsilon}^j \rangle + \sum_{i=0}^{j-1} \langle (\Lambda_B^{i+1} - \Lambda_B^i)\, (y_h^i - y_{h\varepsilon}^i),\, r_{h\varepsilon}^j \rangle.$$

Now, we apply again the results of Lemma 5.4[i], [ii]. Hence, we arrive at

$$\left\|y_h^j - y_{h\varepsilon}^j\right\|_{1,\overline{\omega}} \leq C_4 \left\|(y_{h\varepsilon}^j - \varphi_h)^-\right\|_{1,\overline{\omega}} + C_5\, \tau \sum_{i=0}^{j-1} \left\|y_h^i - y_{h\varepsilon}^i\right\|_{1,\overline{\omega}} .$$

Consequently, the proof is completed by the application of Gronwall's inequality in its discrete form (Lemma 3.1). ∎

By means of Lemma 5.20 we are now able to derive two results for the error caused by application of the penalization approach (5.33) to the finite volume inequality (5.9).

We assume the inverse property of the triangulation $\mathcal{S}_h$ (only !) in part [i] of the following theorem, i.e., there should hold $h/h_S \leq \nu$ for all $S \in \mathcal{S}_h$ (cf. Remark 4.26 and [CIAR78], p. 140).

Theorem 5.21 *Let the conditions of Theorem 5.10 be satisfied.*
[i] *Under the assumption of the inverse property on the triangulation $\mathcal{S}_h$,*

$$\left\|(y_{h\varepsilon}^j - \varphi_h)^-\right\|_{1,\overline{\omega}} \leq C_1\, M_{\varphi_h}\, \varepsilon\, h^{-1}$$

and

$$\left\|y_h^j - y_{h\varepsilon}^j\right\|_{1,\overline{\omega}} \leq C_2\, \exp\left(C_3\, t^j\right)\, M_{\varphi_h}\, \varepsilon\, h^{-1}$$

hold for all $j = 0, \ldots, N$, where M_{φ_h} is given in Lemma 5.20.

[ii] *Taking the penalization parameters $\varepsilon = \varepsilon_\xi$ depending on the nodes $\xi \in \tilde{\omega}$ according to*

$$\varepsilon_\xi = \breve{\varepsilon} \min_{S\in\mathcal{S}_h : S\subset\Omega_\xi} h_S^2,$$

the penalization error in problem (5.33) can be estimated by

$$\left\|(y_{h\varepsilon}^j - \varphi_h)^-\right\|_{1,\overline{\omega}} \leq C_4\, M_{\varphi_h}\, \breve{\varepsilon}\, h$$

and

$$\left\|y_h^j - y_{h\varepsilon}^j\right\|_{1,\overline{\omega}} \leq C_5\, \exp\left(C_6\, t^j\right)\, M_{\varphi_h}\, \breve{\varepsilon}\, h$$

for all $j = 0, \ldots, N$ (M_{φ_h} is given in Lemma 5.20).

Proof. [i]. We exploit the following (global) inverse estimate

$$\left|P^G v_h\right|^2_{H^1(\Omega)} \leq C\, h^{-2} \left\|P^G v_h\right\|^2_{L_2(\Omega)} \quad \text{for arbitrary} \quad v_h \in L_2(\overline{\omega})$$

(cf. [CIAR78], Section 3.2). Consequently, using Lemma 4.1[i], we have

$$\begin{aligned}\|v_h\|^2_{1,\overline{\omega}} &= \left|P^G v_h\right|^2_{H^1(\Omega)} + \left\|P^B v_h\right\|^2_{L_2(\Omega)}\\ &\leq (C_1/h^2+1)\, \left\|P^B v_h\right\|^2_{L_2(\Omega)} = (C_1/h^2+1)\, \|v_h\|^2_{0,\overline{\omega}} .\end{aligned}$$

Combining this property with the result derived in Lemma 5.20[i] (with $\varepsilon_\xi = \varepsilon$ independent of $\xi \in \tilde{\omega}$), we obtain

$$\left\|(y^j_{h\varepsilon} - \varphi_h)^-\right\|_{1,\overline{\omega}} \leq \frac{C_2}{h} \left\|(y^j_{h\varepsilon} - \varphi_h)^-\right\|_{0,\overline{\omega}} \leq \frac{C_2}{h} \|(y^0_{h\varepsilon} - \varphi_h)^-\|_{0,\overline{\omega}} \leq C_2\, M_{\varphi_h}\, \frac{\varepsilon}{h}.$$

The application of Lemma 5.20[ii] with $M_j = C_2\, M_{\varphi_h}\, \varepsilon\, h^{-1}$ (independent of $j = 0, \ldots, N$) yields the statement for the penalization error $y^j_h - y^j_{h\varepsilon}$ for all $j = 0, \ldots, N$ measured in the energy norm.

[ii]. As the first step we will show

$$\left|P^G((y^j_{h\varepsilon} - \varphi_h)^-)\right|^2_{H^1(\Omega)} = \left|(y^j_{h\varepsilon} - \varphi_h)^-\right|^2_{1,\overline{\omega}} \leq C_1\, (\breve{\varepsilon}\, h\, M_{\varphi_h})^2. \tag{5.36}$$

For the sake of brevity let us introduce the notation

$$w^j_h = (y^j_{h\varepsilon} - \varphi_h)^- \quad \text{for} \quad j = 0, \ldots, N.$$

Using the relation $|w^j_{h,\xi}| \leq |w^0_{h,\xi}|$ for all nodes $\xi \in \overline{\omega}$, $j = 0, \ldots, N$ (cf. the proof of Lemma 5.20[i]) and the box condition

$$\text{meas}_n(D_\xi) / \text{meas}_n(\Omega_\xi) \geq \delta > 0, \qquad \text{(cf. (4.12))},$$

we derive

$$\begin{aligned}
&\sum_{\xi \in \tilde{\omega}} \text{meas}_n(D_\xi)\, \left|w^0_{h,\xi}\right|^2 \varepsilon_\xi^{-1} \geq \sum_{\xi \in \tilde{\omega}} \text{meas}_n(D_\xi)\, \left|w^j_{h,\xi}\right|^2 \varepsilon_\xi^{-1} \\
&\geq \delta \sum_{\xi \in \tilde{\omega}} \text{meas}_n(\Omega_\xi)\, \left|w^j_{h,\xi}\right|^2 \varepsilon_\xi^{-1} \geq \delta \sum_{S \in \mathcal{S}_h} \text{meas}_n(S) \sum_{\eta \in (\partial S \cap \tilde{\omega})} \left|w^j_{h,\eta}\right|^2 \varepsilon_\eta^{-1} \\
&\geq (n+1)\, \delta \sum_{S \in \mathcal{S}_h} \breve{\varepsilon}^{-1}\, h_S^{-2} \left\|P^G w^j_h\right\|^2_{L_2(S)}.
\end{aligned}$$

We refer to Lemma 4.1 for an explicit expression of $\left\|P^G w^j_h\right\|_{L_2(S)}$ and recall that the notation $\eta \in (\partial S \cap \tilde{\omega})$ comprises the vertices η of a simplex $S \in \mathcal{S}_h$. Now, we use the inverse relation connecting the H^1-seminorm and the L_2-norm *locally* (i.e., for each simplex $S \in \mathcal{S}_h$). Therefore, proceeding with our last estimate, we obtain

$$\left|P^G w^j_h\right|^2_{H^1(\Omega)} \leq C\, \breve{\varepsilon} \sum_{\xi \in \tilde{\omega}} \text{meas}_n(D_\xi)\, \left|w^0_{h,\xi}\right|^2 \varepsilon_\xi^{-1}.$$

Recalling (5.35) and Lemma 5.20[i], we deduce

$$\left|P^G w^j_h\right|^2_{H^1(\Omega)} \leq C\, \breve{\varepsilon}\, M_{\varphi_h}\, \|w^0_h\|_{0,\overline{\omega}} \leq C\, \breve{\varepsilon}\, M^2_{\varphi_h} \max_{\xi \in \tilde{\omega}} \varepsilon_\xi,$$

such that the estimate (5.36) is shown.

Combining (5.36) with the result of Lemma 5.20[i] for the L_2-norm of w_h^j, we end up with

$$\left\|w_h^j\right\|_{1,\varpi} = \left\|(y_{h\varepsilon}^j - \varphi_h)^-\right\|_{1,\varpi}^2 \le c\,(\breve{\varepsilon}\; h\; M_{\varphi_h})^2 \quad \text{for all} \quad j = 0,\dots,N.$$

Finally, the statement with respect to the penalization error

$$y_h^j - y_{h\varepsilon}^j, \quad j = 0,\dots,N$$

is concluded by the application of Lemma 5.20[ii]. ■

Let us note that we have discussed such a choice of the penalization parameters depending on the nodes of the triangulation already for the elliptic case in Remark 4.26.

Recalling the discussion of the regularization method (4.32) applied to the elliptic obstacle problem in its finite volume approximation (4.16) (cf. Theorem 4.24[ii]), we will now consider the passage to the limit $\varepsilon \longrightarrow 0$ in the evolutionary situation of the problems (5.9) and (5.34). We will show that the solution $z_{h\varepsilon}$ of problem (5.34) preserves the constraints and that the convergence rate is $O(\sqrt{\varepsilon})$.

Theorem 5.22 *Under the assumptions of Theorem 5.10 the solution* $z_{h\varepsilon}$ *of problem (5.34) fulfills*

$$z_{h\varepsilon}^j \in K_h^j \quad \textit{for all} \quad j = 0,\dots,N.$$

There exist constants $C_i > 0$, $i = 1, 2$, *such that*

$$\left\|y_h^j - z_{h\varepsilon}^j\right\|_{1,\varpi} \le C_1\left[\exp\left(C_2\, t^j\right)\langle \mu_h, 1\rangle\right]^{1/2}\sqrt{\varepsilon} \quad \textit{for all} \quad j = 0,\dots,N$$

holds for the difference in the energy norm of $z_{h\varepsilon}$ *and* y_h *as solutions of (5.34) and (5.9), respectively.*

Proof. It can be easily checked that for $j = 0$ the (discrete) regularization problem (5.34) coincides with the (discrete) elliptic problem (4.32). Hence, the inclusion $z_{h\varepsilon}^0 \in K_h^0$ is an immediate consequence of Theorem 4.24[ii]. Owing to $z_{h\varepsilon}^j \ge z_{h\varepsilon}^0$ stated in Lemma 5.19, we have $z_{h\varepsilon}^j \ge \varphi_h$ and consequently $z_{h\varepsilon}^j \in K_h^j$ for all $j = 0,\dots,N$ is shown.

Proceeding with the evaluation of the penalization error for problem (5.34), we put

$$v_h = (y_h^j - z_{h\varepsilon}^j) \in L_2(\tilde{\omega})$$

in the finite volume inequality formulation (5.9). Furthermore, owing to $z_{h\varepsilon}^j \in K_h^j$, we are allowed to take $v_h = z_{h\varepsilon}^j$ in problem (5.34). With this choice we add equation (5.34) and inequality (5.9) for the time instant t^j. As a result we get

$$\begin{aligned}\langle \Lambda_B^j\,(y_h^j - z_{h\varepsilon}^j),\, y_h^j - z_{h\varepsilon}^j\rangle \;&\le \langle \mu_h\,(\beta_{2h}((z_{h\varepsilon}^j - \varphi_h)/\varepsilon) - 1),\, y_h^j - z_{h\varepsilon}^j\rangle \\ &+ \sum_{i=0}^{j-1} \langle (\Lambda_B^{i+1} - \Lambda_B^i)\,(y_h^i - z_{h\varepsilon}^i),\, y_h^j - z_{h\varepsilon}^j\rangle .\end{aligned}$$

From $z_{h\varepsilon}^j \ge \varphi_h$ and $-y_h^j \le -\varphi_h$, it follows that

$$\begin{aligned}\langle \mu_h\,(\beta_{2h}((z_{h\varepsilon}^j - \varphi_h)/\varepsilon) - 1),\, y_h^j - z_{h\varepsilon}^j\rangle \;&= \langle \frac{z_{h\varepsilon}^j - y_h^j}{\varepsilon + z_{h\varepsilon}^j - \varphi_h}, \varepsilon\,\mu_h\rangle \\ &\le \langle \frac{z_{h\varepsilon}^j - \varphi_h}{\varepsilon + z_{h\varepsilon}^j - \varphi_h}, \varepsilon\,\mu_h\rangle .\end{aligned}$$

Combining the last two inequalities and applying the results of Lemma 5.4[i], [ii], we derive

$$C_3\,\left\| y_h^j - z_{h\varepsilon}^j \right\|_{1,\varpi}^2 \le \varepsilon\,\langle \mu_h, 1\rangle + C_4\,\tau\,\left\| y_h^j - z_{h\varepsilon}^j \right\|_{1,\varpi} \sum_{i=0}^{j-1} \| y_h^i - z_{h\varepsilon}^i \|_{1,\varpi} .$$

Exploiting Young's inequality we obtain

$$\left\| y_h^j - z_{h\varepsilon}^j \right\|_{1,\varpi}^2 \le C_5\;\varepsilon\,\langle \mu_h, 1\rangle + C_6\,\tau \sum_{i=0}^{j-1} \| y_h^i - z_{h\varepsilon}^i \|_{1,\varpi}^2 .$$

Therefore, the application of Gronwall's inequality in its discrete form (Lemma 3.1) completes the proof. ■

Remark 5.23 [i] Recalling from Remark 5.7[ii] the definition of $I_\tau^0(P^G v_h)$ as the piecewise constant in time function of $P^G v_h$, the estimates stated in Theorems 5.21[ii] and 5.22 for the penalization error can be rewritten as

$$\| I_\tau^0(P^G(y_h - y_{h\varepsilon})) \|_{L_\infty(0,T;H^1(\Omega))} \le\; C_5\;\exp(C_6\;T)\;M_{\varphi_h}\;\breve{\varepsilon}\;h$$

(analogously in the situation of Theorem 5.21[i]) and

$$\| I_\tau^0(P^G(y_h - z_{h\varepsilon})) \|_{L_\infty(0,T;H^1(\Omega))} \le C_1\,\left[\exp(C_2\;T)\,\langle \mu_h, 1\rangle\right]^{1/2}\,\sqrt{\varepsilon}.$$

[ii] From the main statements of Theorems 5.21 and 5.22 it can be easily seen that the terms M_{φ_h} and $\langle \mu_h, 1\rangle$ coincide with the corresponding terms occuring in the penalization error estimates for the elliptic problems (4.31) and (4.32) (cf. Theorem 4.24). Therefore, the discussion of these terms in Remark 4.28 for the elliptic problems (4.31) and (4.32) can be transferred to the evolutionary situation considered in the present section.

In other words, the estimates (4.36) and (4.37) hold also for the penalization errors arising in the application of the problems (5.33) and (5.34) to the evolutionary (box) obstacle problem (5.9). □

Remark 5.24 Finally, an important consequence consists in the connection of the convergence result $O(h) + O(\tau)$ (see Theorem 5.17) for the discrepancy between the solution u of (3.1), (3.2) and the box solution y_h of (5.9), (5.10) with the results for the penalization error discussed in the present section.

Coupling the discretization parameters with the penalization parameters we obtain a *first order convergence* $O(h)$ of the form

$$\left\|u - I_\tau^0(P^G(y_{h\varepsilon}))\right\|_{L_\infty(0,T;H^1(\Omega))} \leq C_1\ h \quad \text{and}$$
$$\left\|u - I_\tau^0(P^G(z_{h\varepsilon}))\right\|_{L_\infty(0,T;H^1(\Omega))} \leq C_2\ h$$

for the overall error between the solution u and the penalization solutions

$$I_\tau^0(P^G(y_{h\varepsilon})) \quad \text{resp.} \quad I_\tau^0(P^G(z_{h\varepsilon}))$$

measured in the norm of the space $L_\infty(0,T;H^1(\Omega))$. □

5.3.2 Iterative solution of the penalization problems

Extending the considerations of Section 4.4.3 to the evolutionary problems, we will analyse two iteration processes in order to compute the penalization solutions of (5.33) and (5.34) effectively. Note, that in the present section a superscript (j or i) denotes the time step and a subscript m (in contrast to Section 4.4.3) refers to the iteration step.

We propose the following iteration scheme for the solution of the penalization equations (5.33).

$$\left.\begin{aligned}
&\Lambda_B^j\, y_{h,m+1}^j + \varepsilon^{-1}\, r_{1h}(y_{h,m}^j)\,(y_{h,m+1}^j - \varphi_h) = f_{Bh}^j + g_{Bh}^j \\
&+ \sum_{i=0}^{j-1}(\Lambda_B^{i+1} - \Lambda_B^i)\, y_{h\varepsilon}^i \quad \text{for} \quad \xi \in \tilde{\omega}, \\
&(y_{h,m+1}^j)_\xi = g_D(\xi, t^j) \quad \text{for} \quad \xi \in \gamma_D; \quad m \geq 0, \quad j = 0, \dots, N.
\end{aligned}\right| \tag{5.37}$$

We should select an arbitrary initial vector $y_{h,0}^j \in L_2(\overline{\omega})$ fulfilling the Dirichlet conditions $(y_{h,0}^j)_\xi = g_D(\xi, t^j)$ for $\xi \in \gamma_D$ for each time instant t^j, $j = 0, \dots, N$.

From Section 4.4.3 (elliptic case) we recall the definition of the grid function vector $r_{1h}(v_h) \in L_2(\tilde{\omega})$, i.e.,

$$(r_{1h}(v_h))_\xi = \begin{cases} 0 & \text{if} \quad v_{h,\xi} > \varphi_{h,\xi}, \\ \text{meas}_n(D_\xi) & \text{if} \quad v_{h,\xi} \leq \varphi_{h,\xi}. \end{cases}$$

It is easy to see that the iteration process (5.37) can be written equivalently in the form

$$(\Lambda_B^j + \varepsilon^{-1}\, r_{1h}(y_{h,m}^j))\,(y_{h,m+1}^j - y_{h,m}^j) = f_{Bh}^j + g_{Bh}^j$$
$$- \Lambda_B^j\, y_{h,m}^j - \varepsilon^{-1}\, \beta_{1h}(y_{h,m}^j - \varphi_h) + \sum_{i=0}^{j-1}(\Lambda_B^{i+1} - \Lambda_B^i)\, y_{h\varepsilon}^i.$$

In extension of the method (4.40) the iteration process for the solution of the evolutionary penalization problem (5.34) reads as follows.

$$\left.\begin{aligned}
&(\Lambda_B^j + \mu_h\, r_{2h}(z_{h,m}^j))\,(z_{h,m+1}^j - z_{h,m}^j) = f_{Bh}^j + g_{Bh}^j - \Lambda_B^j\, z_{h,m}^j \\
&+ \sum_{i=0}^{j-1}(\Lambda_B^{i+1} - \Lambda_B^i)\, z_{h\varepsilon}^i + \mu_h\Big[1 - \beta_{2h}\big(\frac{z_{h,m}^j - \varphi_h}{\varepsilon}\big)\Big] \quad \text{for} \quad \xi \in \tilde{\omega}, \\
&(z_{h,m+1}^j)_\xi = g_D(\xi, t^j) \quad \text{for} \quad \xi \in \gamma_D; \quad m \geq 0, \quad j = 0, \ldots, N.
\end{aligned}\right| \qquad (5.38)$$

The initial vectors for this iteration method should be taken in the following way.

$$(z_{h,0}^0)_\xi = \varphi_{h,\xi} \quad \text{for} \quad \xi \in \tilde{\omega}, \qquad (z_{h,0}^0)_\xi = g_D(\xi, 0) \quad \text{for} \quad \xi \in \gamma_D;$$
$$(z_{h,0}^j)_\xi = (z_{h\varepsilon}^{j-1})_\xi \quad \text{for} \quad \xi \in \tilde{\omega}, \qquad (z_{h,0}^j)_\xi = g_D(\xi, t^j) \quad \text{for} \quad \xi \in \gamma_D$$

for $j = 1, \ldots, N$.

Analogously to Section 4.4.3, the grid function vector $r_{2h}(v_h) \in L_2(\tilde{\omega})$ is defined by

$$(r_{2h}(v_h))_\xi = \begin{cases} \dfrac{\varepsilon}{((\varepsilon + v_{h,\xi} - \varphi_{h,\xi})^2)} & \text{if} \quad v_{h,\xi} \geq \varphi_{h,\xi}, \\ 0 & \text{if} \quad v_{h,\xi} < \varphi_{h,\xi}. \end{cases}$$

For the sake of completeness we mention that the iteration (5.38) can be expressed equivalently by

$$\Lambda_B^j\, z_{h,m+1}^j + \mu_h\, r_{2h}(z_{h,m}^j)\,(z_{h,m+1}^j - \varphi_h) = f_{Bh}^j + g_{Bh}^j + \sum_{i=0}^{j-1}(\Lambda_B^{i+1} - \Lambda_B^i)\, z_{h\varepsilon}^i$$
$$+ \mu_h\, \Big[1 + r_{2h}(z_{h,m}^j)\,(z_{h,m}^j - \varphi_h) - \beta_{2h}\big(\frac{z_{h,m}^j - \varphi_h}{\varepsilon}\big)\Big].$$

Now, we study the convergence of the iteration methods (5.37) and (5.38). We derive global convergence results both for method (5.37) and (5.38).

Theorem 5.25 *Let the assumptions of Theorem 5.10 be fulfilled.*
[i] *The solutions* $y_{h\varepsilon}^j$ *of (5.33) and* $y_{h,m}^j$ *of (5.37) fulfill*

$$y_{h,m}^j \leq y_{h,m+1}^j \leq y_{h\varepsilon}^j \quad \textit{for} \quad m \geq 1$$

for each time instant t^j, $j = 0, \ldots, N$. *Moreover,* $y^j_{h,M} = y^j_{h\varepsilon}$ *holds after a finite number* M *of iterations (for fixed triangulation) for each time instant* t^j, $j = 0, \ldots, N$.

[ii] *The solutions* $z^j_{h\varepsilon}$ *of the penalization problem (5.34) and* $z^j_{h,m}$ *of the iteration method (5.38) satisfy*

$$\varphi_h \le z^0_{h,0} \le z^0_{h,m} \le z^0_{h,m+1} \le z^0_{h\varepsilon} \quad \text{and}$$

$$\varphi_h \le z^{j-1}_{h\varepsilon} \le z^j_{h,0} \le z^j_{h,m} \le z^j_{h,m+1} \le z^j_{h\varepsilon}, \quad j = 1, \ldots, N, \quad m \ge 0.$$

Furthermore, r_2 *is simplified to* $r_{2h}(z^j_{h,m}) = \varepsilon/((\varepsilon + z^j_{h,m} - \varphi_h)^2)$, $m \ge 0$ *and the sequence* $\{z^j_{h,m}\}_{m\in\mathbb{N}}$ *converges towards* $z^j_{h\varepsilon}$ *for each* $j = 0, \ldots, N$.

Proof. [i]. We repeat the proof of Theorem 4.31[i] with a modified right-hand side

$$f^j_{Bh} + g^j_{Bh} + \sum_{i=0}^{j-1} (\Lambda^{i+1}_B - \Lambda^i_B)\, y^i_{h\varepsilon}$$

for each time instant t^j, $j = 0, \ldots, N$.

[ii]. The relation $z^j_{h\varepsilon} \ge \varphi_h$ for $j = 0, \ldots, N$ has already been proved in Theorem 5.22. At the initial time $t^0 = 0$ the iteration method (5.38) coincides with the method (4.40) considered in Section 4.4.3. Therefore, the statements for $j = 0$ have been established in Theorem 4.31[ii].

Let us now consider an arbitrary time instant $j = 1, \ldots, N$. As the first step we show

$$z^j_{h,1} \ge z^{j-1}_{h\varepsilon}.$$

In this context we recall the initial vector $z^j_{h,0} = z^{j-1}_{h\varepsilon}$ (apart from Dirichlet nodes) in the iteration (5.38) at the time t^j. We subtract the equation for $z^{j-1}_{h\varepsilon}$ in (5.34) from the equation for $z^j_{h,1}$ in (5.38) (note the above mentioned equivalent form of the iteration (5.38)). This leads to

$$\Lambda^j_B\, (z^j_{h,1} - z^{j-1}_{h\varepsilon}) + \mu_h\, r_{2h}(z^{j-1}_{h\varepsilon})\, (z^j_{h,1} - z^{j-1}_{h\varepsilon}) = f^j_{Bh} - f^{j-1}_{Bh} + g^j_{Bh} - g^{j-1}_{Bh} \ge 0.$$

Hence, we verify

$$\varphi_h \le z^{j-1}_{h\varepsilon} \le z^j_{h,0} \le z^j_{h,1}$$

by means of the discrete maximum principle (4.30).

As the second step we prove $z^j_{h,m+1} \ge z^j_{h,m}$, performing an induction with respect to $m > 0$. Just now we have shown $z^j_{h,0} \le z^j_{h,1}$. Assuming $z^j_{h,m} \ge z^j_{h,m-1} \ge \varphi_h$, the iteration process (5.38) implies

$$(\Lambda^j_B + \mu_h\, r_{2h}(z^j_{h,m}))\, (z^j_{h,m+1} - z^j_{h,m}) = \mu_h \Big[\, r_{2h}(z^j_{h,m-1})\, (z^j_{h,m} - z^j_{h,m-1})$$
$$+ \beta_{2h}\Big(\frac{z^j_{h,m-1} - \varphi_h}{\varepsilon}\Big) - \beta_{2h}\Big(\frac{z^j_{h,m} - \varphi_h}{\varepsilon}\Big) \Big] \ge 0.$$

The non-negativity is a consequence of $\beta_{2h}(v_h)$ being a concave function. Consequently, the application of the maximum principle (4.30) gives $z^j_{h,m+1} \geq z^j_{h,m}$.

Finally we show the relation $z^j_{h,m} \leq z^j_{h\varepsilon}$ for $m \geq 0$. Recalling Lemma 5.19 we have $z^{j-1}_{h\varepsilon} = z^j_{h,0} \leq z^j_{h\varepsilon}$ for nodes $\xi \in \tilde{\omega}$. By subtraction of the equations (5.38) from the penalization equations in (5.34) we obtain

$$\Lambda^j_B\,(z^j_{h\varepsilon} - z^j_{h,m+1}) + \mu_h\,\Big[\,\beta_{2h}(\frac{z^j_{h\varepsilon} - \varphi_h}{\varepsilon}) - \beta_{2h}(\frac{z^j_{h,m+1} - \varphi_h}{\varepsilon})\,\Big]$$
$$= \mu_h\Big[\,\beta_{2h}(\frac{z^j_{h,m} - \varphi_h}{\varepsilon}) - \beta_{2h}(\frac{z^j_{h,m+1} - \varphi_h}{\varepsilon}) + r_{2h}(z^j_{h,m})\,(z^j_{h,m+1} - z^j_{h,m})\,\Big] \geq 0.$$

Now, the maximum principle (4.30) yields the claimed relation $z^j_{h,m+1} \leq z^j_{h\varepsilon}$. Since $\{z^j_{h,m}\}_{m\in\mathbb{N}}$ is a monotone and bounded sequence, we obtain the convergence of $\{z^j_{h,m}\}$. Moreover, comparing (5.38) and (5.34), we find that $\{z^j_{h,m}\}$ converges towards $z^j_{h\varepsilon}$. ■

Remark 5.26 As already emphasized in Remark 4.32[ii], the number of iterations which are necessary to arrive at $y^j_{h,M} = y^j_{h\varepsilon}$ in the penalty-iteration method (5.33), (5.37) depends essentially on the quality of the agreement of $\omega^-(y^j_{h,0})$ and $\omega^-(y^j_{h\varepsilon})$, i.e.,

$$\omega^-(y^j_{h,0}) \approx \omega^-(y^j_{h\varepsilon})$$

is desirable. We recall the definition (4.41) for the (discrete) coincidence set $\omega^-(v_h)$. If no additional information about $\omega^-(y^j_{h\varepsilon})$ is available, we recommend the choice

$$\omega^-(y^j_{h,0}) = \omega^-(y^{j-1}_{h\varepsilon}).$$

On the other hand, in our application problems (injection and compression moulding, see Applications 2.2, 2.3) the difference of the right-hand sides of two successive time instants is essentially determined by the volume of plastic melt which will be injected in the time interval (resp. by the closing speed).

By means of this information the set $\omega^-(y^j_{h,0})$ is extrapolated on the basis of the obtained set

$$\omega^-(y^{j-1}_{h,M}) = \omega^-(y^{j-1}_{h\varepsilon}).$$

From numerical experiments (cf. Sections 5.4 and, in particular, Chapter 6) it has been observed that such a simple extrapolation leads to 'good results', i.e., in general only 2–5 iterations at each time instant are necessary. □

5.4 Numerical experiments for evolutionary variational inequalities

In Section 4.6 we have discussed numerical results for three elliptic obstacle problems. Similar to this, but now for two evolutionary inequality problems of the form (3.1), (3.2) (with known exact solution u), we report on the results of numerical experiments. Both evolutionary inequality problems can be interpreted as the result of the application of the Baiocchi transformation to the general free boundary problem (2.1) (cf. Chapter 2).

We have performed a number of numerical experiments including a systematic variation of the discretization and penalization parameters in order to observe the errors between exact, finite element and finite volume solution as well as the error behaviour of the penalty-iteration methods (5.33), (5.37) and (5.34), (5.38).

The first example is an evolutionary obstacle problem on the unit square with homogeneous Dirichlet boundary conditions. Consequently, we have a time-independent convex set K for this example. In contrast to this, our second example contains mixed (time-dependent) Dirichlet and Newton (Robin) boundary conditions. The domain Ω for this example is the (unit) circle.

The boxes for the finite volume method are chosen exclusively as barycentric domains (concerning the advantages see Section 4.6). Furthermore, the weighting parameter α is taken as $\alpha = 0.5$ for assembling the finite volume and finite element matrices (cf. (5.4) and (5.10)).

We have chosen a sequence of mesh-size parameters $h = h_i$ lying in the range $[1/200, 1/20]$ for the first example and in $[1/160, 1/20]$ for the second one. According to the convergence statements established in Section 5.2.3, the space and time discretization parameters are coupled by $h = \tau$.

On the one hand, we have solved the two evolutionary inequality problems by means of the penalty-iteration schemes (5.33), (5.37) and (5.34), (5.38). On the other hand, notwithstanding the enormous cpu-time necessary for the SOR method with projection (successive over-relaxation with projection, see Section 4.6 and [GLT81], [GLOW84]), this standard relaxation procedure has been applied to observe the error behaviour between the exact, finite element and finite volume solutions.

Similar to Section 4.6 we are also in the present section mainly interested in the (numerical) examination of the predicted convergence rates between exact, finite element, finite volume and (finite volume) penalization solutions. We refer to Remarks 5.26 and 4.32 for computational aspects of the penalty-iterations (number of necessary penalty iterations).

The computations were carried out by means of the C++ finite element toolbox *FeliCs* (see [BAUE99]) and cf. Section 4.6).

5.4.1 Two evolutionary variational inequalities and the related free boundary problems

Example 5.27 Taking Ω as the unit square and the time interval $[0,T] = [0,1]$, we consider the evolutionary inequality (3.1), (3.2) with the coefficients

$$k(x,t) = 1 - \rho(x)\, d(t), \quad q(x,t) = d(t) + \rho(x), \quad x = (x_1, x_2),$$
$$\rho(x) = (x_1 - 0.5)^2 + (x_2 - 0.5)^2, \quad d(t) = t^2/4.$$

The right-hand side $f = f(x,t)$ in the inequality formulation (3.1), (3.2) is chosen according to

$$f(x,t) = \begin{cases} 10^3 \left[\, 4\, d\, (1 - \rho\, d) + d^3/3 - d\, \rho^2 - 8\, \rho\, (1 - \rho^2) + 2\, \rho^3/3 \,\right], & \text{(i)}, \\ 10^3 \left[\, 4\, \rho\, (d\, \rho - 1) \,\right], & \text{(ii)}. \end{cases}$$

where (i) corresponds to $\rho(x) \le d(t)$ and (ii) to $\rho(x) \ge d(t)$. Note, that due to our change in the notation from problem (2.11) to the more general formulation (3.1) (right-hand side from F to f), the just given right-hand side $f(x,t)$ for problem (3.1) corresponds to (capital) $F = F(x,t)$ in (2.11).
We take homogeneous Dirichlet boundary conditions $g_D(x,t) = 0$ on the whole boundary $\Gamma = \Gamma_D = \partial\Omega$ of the unit square Ω. The obstacle φ is also identical with zero.
Then, the exact solution $u = u(x,t)$ of problem (3.1), (3.2) turns out to be

$$u(x,t) = 500\, \left[\left(d(t) - \rho(x)\right)^+ \right]^2 = \begin{cases} 500\, \left[d(t) - \rho(x)\right]^2 & \text{for } \rho(x) \le d(t), \\ 0 & \text{for } \rho(x) \ge d(t). \end{cases}$$

The free boundary $\Gamma_f(t)$ is described by the circumference of the circle with centre $(0.5, 0.5)$ and (time-dependent) radius $\sqrt{d(t)}$ for $t \in [0,1]$.

Recalling the introduction of the general free boundary value problem (2.1) in Section 2.1, this evolutionary variational inequality corresponds to problem (2.1) with

$$\theta(x,t) = \partial_t u = 10^3\, (d - \rho)\, \partial_t d \quad \text{in } \Omega(t) = \{x \in \Omega : \rho(x) \le d(t)\}, \quad \eta = \gamma_D = 0.$$

The right-hand side f and the 'latent heat' λ in problem (2.1) are now given by

$$f(x,t) = 10^3 \left[\, 4\, \partial_t d - 8\, \rho\, d\, \partial_t d + d^2\, \partial_t d - \rho^2 \,\right] \quad \text{and} \quad \lambda(x,t) = 10^3 \left[\, 4\, \rho\, (1 - d\, \rho) \,\right],$$

respectively. □

Example 5.28 Our second example is characterized by mixed boundary conditions and time-dependent convex sets $K = K_{0,g_D}(t)$. The domain Ω is given as the circle with centre in $(0.5, 0.5)$ and radius $r = 0.5$. The time interval $[0, T]$ is again $[0, 1]$. We take an obstacle φ which is zero. The coefficients k and q are defined by

$$k(x,t) = 4 + \rho(x)\ (4 - d(t)), \quad q(x,t) = d(t) + \rho(x), \quad x = (x_1, x_2),$$
$$\rho(x) = x_1 + x_2, \quad d(t) = \sqrt{2}\, t + 1 - \sqrt{2}/2$$

for the evolutionary inequality (3.1), (3.2) (resp. (2.11)). The right-hand side $f = f(x,t)$ in (3.1) (resp. $F(x,t)$ in (2.11)) is chosen according to

$$f(x,t) = \begin{cases} -8 + 8\,d - d^2 - 16\,\rho + 3\,\rho^2 + d^3/3 + 2\,\rho^3/3 - \rho^2 d, & \rho(x) \le d(t), \\ -8 - 8\,\rho + 2\,d\,\rho, & \rho(x) \ge d(t). \end{cases}$$

The boundary $\partial\Omega$ of the circle Ω is partitioned into a Dirichlet and a Newton part, i.e., $\partial\Omega = \Gamma_D \cup \Gamma_N$. To be more precise, Γ_N is given as that part of $\partial\Omega$ for which $x_1 + x_2 \le 0.5$ holds ('a quarter' of the circumference $\partial\Omega$). We take Newton boundary conditions (cf. e.g., (5.19)) on Γ_N with $\kappa(x) = \rho(x) = x_1 + x_2$ and

$$g_N(x,t) = \begin{cases} \rho\ (d-\rho)^2/2 + \left[(1-\rho)\ (d-\rho)\ (4 + 4\rho - \rho\, d/2 - \rho^2/2)\right]/\ \delta, & \text{(i)} \\ 0, & \text{(ii)}. \end{cases}$$

where (i) corresponds to $\rho(x) \le d(t)$ and (ii) to $\rho(x) \ge d(t)$. Furthermore, δ is

$$\delta = \delta(x) = \left((x_1 - 0.5)^2 + (x_2 - 0.5)^2\right)^{1/2}.$$

Time-dependent Dirichlet conditions of the form

$$g_D(x,t) = 0.5\ \left[\left(d(t) - \rho(x)\right)^+\right]^2$$

are imposed on the remaining part Γ_D of the boundary $\partial\Omega$.

Then, the exact solution $u = u(x,t)$ of problem (3.1), (3.2) is given by

$$u(x,t) = 0.5\ \left[\left(d(t) - \rho(x)\right)^+\right]^2 = \begin{cases} 0.5\ \left[d(t) - \rho(x)\right]^2 & \text{for} \quad \rho(x) \le d(t), \\ 0 & \text{for} \quad \rho(x) \ge d(t). \end{cases}$$

The free boundary $\Gamma_f(t)$ is described by (the line segment) $\rho(x) = x_1 + x_2 = d(t)$ for $t \in [0, 1]$.

Returning to the free boundary value problem (2.1), we have

$$\theta(x,t) = \partial_t u = (d-\rho)\,\partial_t d \quad \text{in} \quad \Omega(t) = \{x \in \Omega:\ \rho(x) \le d(t)\}, \quad \eta = 0,$$
$$f(x,t) = 2\,\partial_t d\,(4-d) + \partial_t d\,(d^2-\rho^2).$$

The boundary conditions are given by Newton (Robin) conditions on Γ_N with $\kappa(x) = \rho(x)$ and

$$\gamma_N(x,t) = \begin{cases} \rho\,\theta + k\,(1-\rho)\,\partial_t d\,/\delta & \text{for} \quad \rho(x) \le d(t), \\ 0 & \text{for} \quad \rho(x) \ge d(t). \end{cases}$$

Dirichlet conditions $\gamma_D(x,t) = \theta(x,t)$ are imposed on the remaining part of the boundary characterized by $\rho(x) \ge 0.5$. Finally, the function λ is defined by $\lambda(x,t) = 8 + 8\,\rho - 2\,d\,\rho$. □

5.4.2 Numerical results for the errors between exact, finite element and finite volume solution

To compare the exact, the finite element and the finite volume solutions of Examples 5.27 and 5.28 we have solved these evolutionary variational inequalities by means of the standard SOR-method with projection.

In case of Example 5.27 the computations have been performed for mesh-size parameters $h = h_i$ given by

$$h_1 = 1/20, \quad h_2 = 1/40, \quad h_3 = 1/80, \quad h_4 = 1/120, \quad h_5 = 1/160, \quad h_6 = 1/200,$$

where structured (spatial) meshes consisting of right-angled triangles were used for the unit square.

For Example 5.28 (circle) we have used unstructured meshes characterized by

$$h_1 = 1/20, \quad h_2 = 1/40, \quad h_3 = 1/80, \quad h_4 = 1/100 \quad \text{and} \quad h_5 = 1/160.$$

According to the predicted convergence rate $O(h)+O(\tau)$ for the finite element and finite volume approximation (cf. Section 5.2.3) the spatial and time discretization parameters have been coupled by $h_i = \tau_i$.

We have determined the errors

$$\max_{j=0,\dots,N} \left\| P^G(u_h^j - y_h^j) \right\|_{H^1(\Omega)}, \quad \max_{j=0,\dots,N} \left\| \Pi_h u(t^j) - P^G u_h^j \right\|_{H^1(\Omega)} \quad \text{and}$$
$$\max_{j=0,\dots,N} \left\| \Pi_h u(t^j) - P^G y_h^j \right\|_{H^1(\Omega)},$$

where u is the exact solution of (3.1), (3.2), u_h^j denotes the finite element solution of (5.3) and y_h^j is the finite volume (box) solution of (5.9). Furthermore, j refers

	fem - box		exact - fem		exact - box	
$h = \tau$	error	error/h^2	error	er./$h^{1.513}$	error	er./$h^{1.512}$
1/20	1.025E0	4.099E+2	1.058E0	9.843E+1	1.028E0	9.533E+1
1/40	2.533E-1	4.054E+2	3.733E-1	9.908E+1	3.730E-1	9.863E+1
1/80	6.376E-2	4.080E+2	1.229E-1	9.308E+1	1.246E-1	9.396E+1
1/120	2.834E-2	4.088E+2	7.053E-2	9.867E+1	6.788E-2	9.451E+1
1/160	1.594E-2	4.081E+2	4.479E-2	9.684E+1	4.423E-2	9.514E+1
1/200	1.024E-2	4.098E+2	3.203E-2	9.706E+1	3.173E-2	9.562E+1

Table 5.1: *Error behaviour between exact, finite element (fem) and finite volume (box) solution. Evolutionary inequality – Example 5.27.*

	fem - box		exact - fem		exact - box	
$h = \tau$	error	er./$h^{1.963}$	error	er./$h^{1.147}$	error	er./$h^{1.148}$
1/20	1.257E-3	4.501E-1	3.392E-3	1.054E-1	3.372E-3	1.051E-1
1/40	3.310E-4	4.620E-1	1.409E-3	9.693E-2	1.423E-3	9.824E-2
1/80	8.571E-5	4.664E-1	6.129E-4	9.338E-2	6.023E-4	9.216E-2
1/100	5.529E-5	4.663E-1	4.819E-4	9.484E-2	4.793E-4	9.477E-2
1/160	2.136E-5	4.533E-1	3.009E-4	1.015E-1	2.974E-4	1.008E-1

Table 5.2: *Error behaviour between exact, finite element (fem) and finite volume (box) solution. Evolutionary inequality – Example 5.28.*

to the time instant t^j $(N\tau = 1)$ and $\Pi_h u(t^j)$ is the piecewise linear (Lagrange) interpolant of the solution u at the time $t = t^j$.

The results for Example 5.27 are presented in Table 5.1, whereas Table 5.2 shows the obtained errors for the second Example 5.28. For a more illustrative impression, the numerically determined convergence behaviour is depicted in Figure 5.2 (error between finite element and finite volume solution), in Figure 5.1 (discrepancy between exact and finite element solution) and, finally, in Figure 5.3 (discrepancy between exact and finite volume solution).

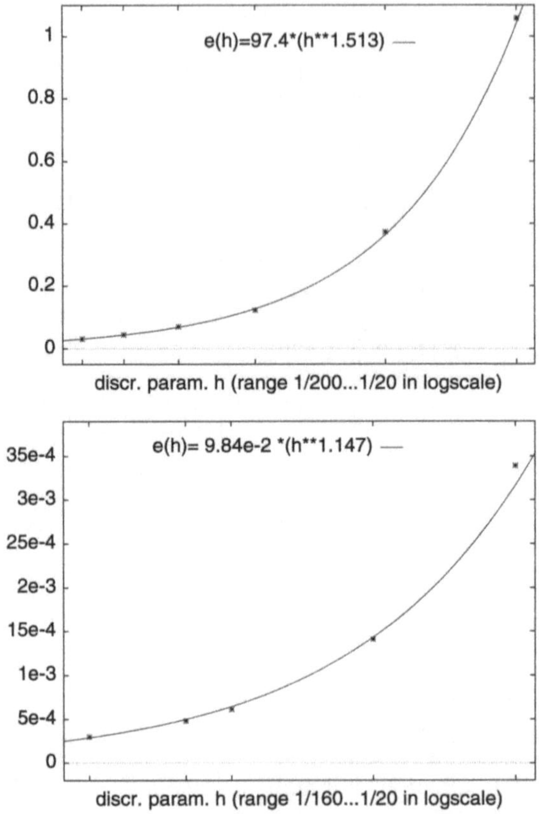

Figure 5.1: *Error behaviour between exact* (u) *and finite element* (u_h) *solution. Solid lines: comparison curves* $e(h)$*. Stars* $(*)$*: error* $\max_{j=0,\ldots,N} \|\Pi_h u(t^j) - P^G u_h^j\|_{H^1(\Omega)}$ *for different discretizations* $h_i = \tau_i$*. Top: Example 5.27. Bottom: Example 5.28.*

The exponent δ in

$$\max_{j=0,\ldots,N} \|.\|_{H^1(\Omega)} \approx C\, h^\delta$$

(see the comparison curves in the corresponding figures) has been determined by averaging the single exponents

$$\delta_{il} = (\log e_i - \log e_l) \;/\; (\log h_i - \log h_l),$$

where e_i denotes the error obtained for the discretization parameters $h_i = \tau_i$.

Summarizing, our numerical results show the second order convergence rate $O(h^2)$ for the error between finite element and finite volume solution (cf. Theorem 5.11 and Remark 5.12). We mention that the errors between the finite volume and

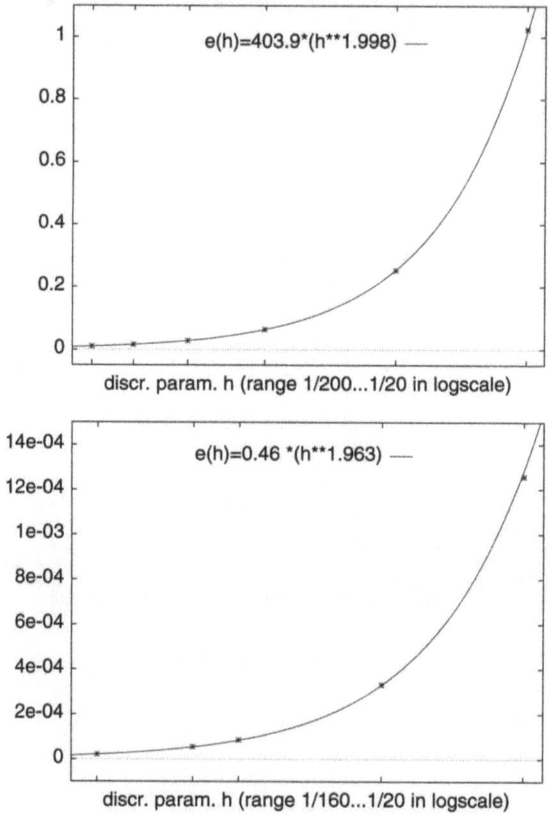

Figure 5.2: *Error behaviour between finite element (u_h) and finite volume (y_h) solution. Solid lines: comparison curves e(h). Stars (*): error* $\max_{j=0,\dots,N} \|P^G u_h^j - P^G y_h^j\|_{H^1(\Omega)}$ *for different discretizations* $h_i = \tau_i$. *Top: Example 5.27. Bottom: Example 5.28.*

the corresponding finite element penalization solutions (cf. Section 4.6.2 for the computation of finite element penalization solutions) exhibit the same behaviour.

The error behaviour between exact and finite element solution as well as between exact and finite volume solution is nearly identical. The results presented in Tables 5.1, 5.2 and in Figures 5.1, 5.3 point out a convergence order of about $O(h^{1.5})$ for Example 5.27 and $O(h^{1.1})$ for Example 5.28. In this context we also refer to the discussion in Section 4.6.2 of the numerically obtained convergence rate between exact and finite volume solution for elliptic obstacle problems (see Examples 4.37, 4.38, 4.39).

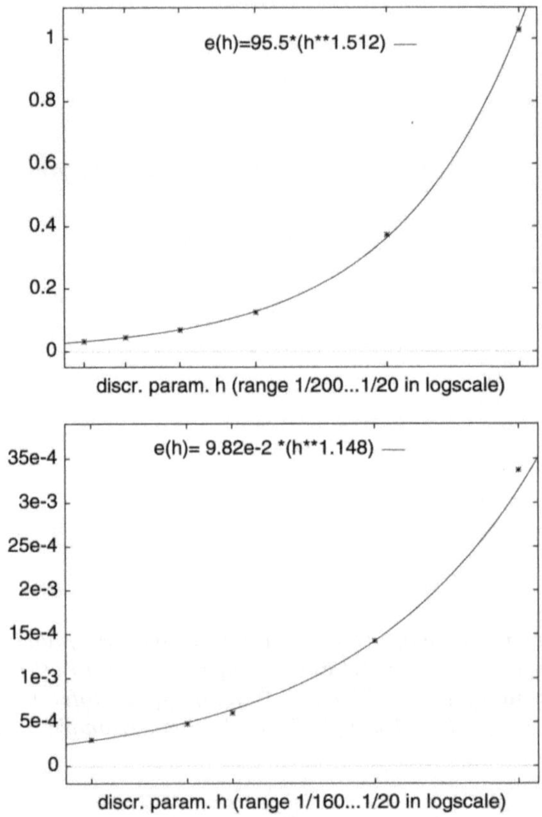

Figure 5.3: *Error behaviour between exact (u) and finite volume (y_h) solution. Solid lines: comparison curves e(h).*
Stars (∗): error $\max_{j=0,\dots,N} \|\Pi_h u(t^j) - P^G u_h^j\|_{H^1(\Omega)}$ *for different discretizations $h_i = \tau_i$. Top: Example 5.27. Bottom: Example 5.28.*

5.4.3 Error behaviour of the penalization solutions

To check the error behaviour of the solutions of the penalty-iteration methods (5.33), (5.37) and (5.34), (5.38) for the Examples 5.27 and 5.28, we have chosen the same (space and time) discretization parameters $h_i = \tau_i$ as in Section 5.4.2.

According to our investigations made in Section 5.3.1 (cf., in particular, Remark 5.23), these parameters are coupled with the penalization parameters ε by

$$\varepsilon = c\,h^2.$$

To choose suitable constants c for both examples, we have taken into account the absolute size of the terms M_{φ_h} and $\langle \mu_h, 1 \rangle$ (cf. Remark 5.23). Note, that in Example 5.27 (due to the scaling factor 1000) these terms are greater than in Example 5.28.
Some of the obtained numerical results are shown in Table 5.3 and depicted in Figures 5.4, 5.5. To be more precise, Table 5.3 contains the exponents δ_{il} which we have computed by

$$\delta_{il} = (\log e_i - \log e_l)/(\log h_i - \log h_l).$$

Example 5.27			Example 5.28		
	exponent δ_{il}			exponent δ_{il}	
$h_i : h_l$	p-i (5.33)	p-i (5.34)	$h_i : h_l$	p-i (5.33)	p-i (5.34)
1/20 : 1/40	1.509	1.527	1/20 : 1/40	1.301	1.192
1/40 : 1/80	1.551	1.476	1/40 : 1/80	1.237	1.176
1/80 : 1/160	1.485	1.171	1/80 : 1/160	1.066	0.988
1/40 : 1/120	1.527	1.317	1/40 : 1/100	1.217	1.136
1/120 : 1/200	1.606	1.252	1/100 : 1/160	1.024	0.985
average	≈ 1.54	≈ 1.34	average	≈ 1.17	≈ 1.10

Table 5.3: *Error behaviour* $\max_{j=0,\dots,N} \left\| \Pi_h u(t^j) - P^G y^j_{h\varepsilon} \right\|_{H^1(\Omega)} \approx C\ h^\delta$ *between exact solution* u *of (3.1) and finite volume solution* $y_{h\varepsilon}$ *of the penalty-iteration (p-i) method (5.33), (5.37). Error* $\max_{j=0,\dots,N} \left\| \Pi_h u(t^j) - P^G z^j_{h\varepsilon} \right\|_{H^1(\Omega)} \approx C\ h^\delta$ *with* $z_{h\varepsilon}$ *as solution of the penalty-iteration (p-i) method (5.34), (5.38).*

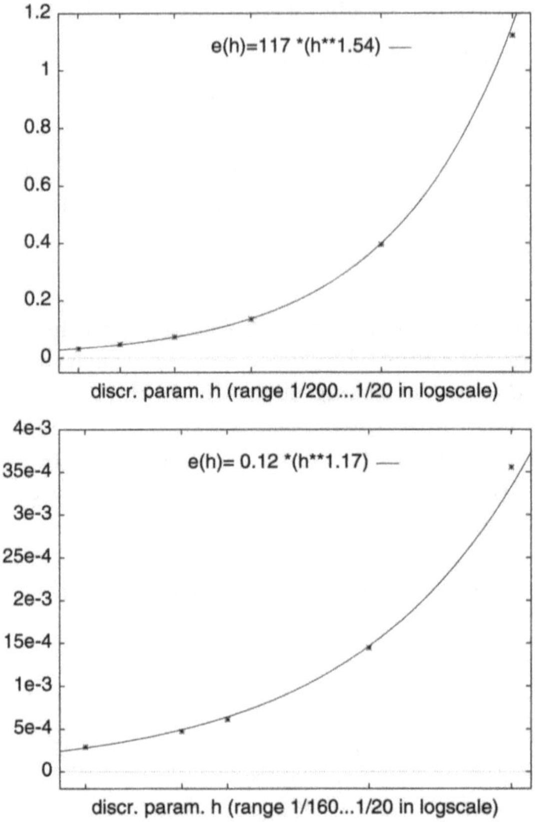

Figure 5.4: *Error behaviour between exact (u) and finite volume (penalty, $y_{h\varepsilon}$) solution (penalty-iteration (5.33), (5.37)). Solid lines: comparison curves $e(h)$. Stars ($*$): error* $\max_{j=0,\ldots,N} \|\Pi_h u(t^j) - P^G y^j_{h\varepsilon}\|_{H^1(\Omega)}$ *for different discretizations $h_i = \tau_i$. Top: Example 5.27. Bottom: Example 5.28.*

Here, e_i denotes the error

$$\max_{j=0,\ldots,N} \|\Pi_h u(t^j) - P^G y^j_{h\varepsilon}\|_{H^1(\Omega)} \quad \Big(\text{resp.} \quad \max_{j=0,\ldots,N} \|\Pi_h u(t^j) - P^G z^j_{h\varepsilon}\|_{H^1(\Omega)} \Big)$$

between the Lagrange interpolant $\Pi_h u(t^j)$ of the solution $u(t^j)$ of (3.1) and the Galerkin prolongation $P^G y^j_{h\varepsilon}$ of $y^j_{h\varepsilon}$ as the solution of the penalty-iteration method (5.33), (5.37) (resp. $z^j_{h\varepsilon}$ of (5.34), (5.38)) obtained for the discretization parameters $h_i = \tau_i$. Furthermore, in Figures 5.4 and 5.5 the error behaviour of the solution of the penalty-iterations (5.33), (5.37) and (5.34), (5.38) compared to the exact solution is illustrated.

Due to the different influences of the parameters (time and space discretization, penalization), one cannot expect to get numerically such a uniform convergence

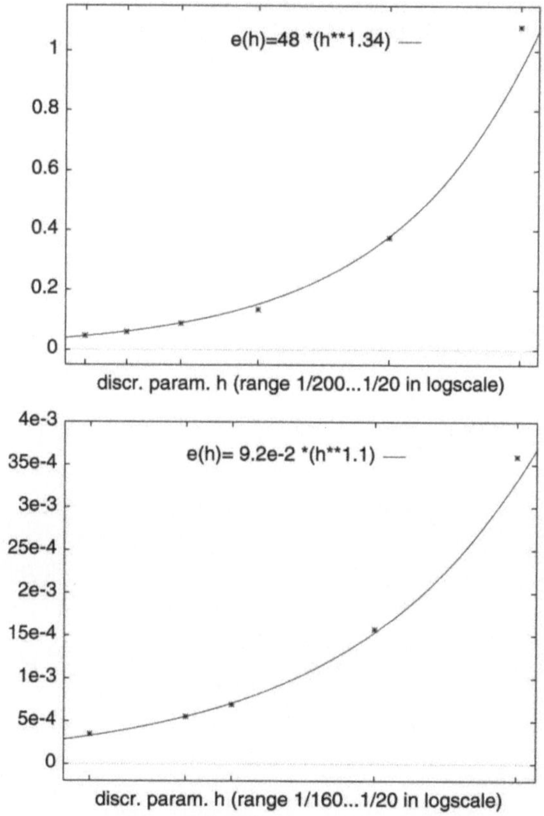

Figure 5.5: *Error behaviour between exact (u) and finite volume (penalty, $y_{h\varepsilon}$) solution (penalty-iteration (5.34), (5.38)). Solid lines: comparison curves e(h). Stars (∗): error* $\max_{j=0,\dots,N} \|\Pi_h u(t^j) - P^G y^j_{h\varepsilon}\|_{H^1(\Omega)}$ *for different discretizations* $h_i = \tau_i$. *Top: example 5.27. Bottom: example 5.28.*

rate as in the previous section for the discrepancy between finite element and finite volume solution.

Nevertheless, for both examples, the numerically determined convergence between exact and (finite volume) penalization solutions is (somewhat) better than the theoretically predicted first order convergence (cf. Remark 5.23).

International Series of Numerical Mathematics
Vol. 136,

Chapter 6

Injection and Compression Moulding as Application Problems

In addition to Section 2.2 we extend now the mathematical modelling and numerical simulation of the injection and compression moulding process (cf. Applications 2.2, 2.3) using a generalized Hele–Shaw flow which includes non-isothermal and possible non-Newtonian effects. Beside this, we will mention some recent developments concerning the solution of the full three-dimensional non-Newtonian Navier–Stokes equations without the Hele–Shaw simplifications.
We will discuss the numerical core problems occuring in simulation of the just mentioned application problems in polymer processing. Furthermore, we shall present some results of numerical simulations.

In Section 6.2 we will describe the following three mathematical models, including the main aspects of their numerical implementation, to simulate the injection moulding process.

- Finite element/volume modelling using a generalized Hele–Shaw flow and the variational inequality approach,
- geometrical modelling (distance concept) using a generalized Hele–Shaw flow,
- three-dimensional non-Newtonian Navier–Stokes equations without the Hele–Shaw simplifications.

In Section 6.3 we will study the influence of some basic geometrical effects and operating conditions for several real plastic products, e.g., variation in gate location (gate design), variation in thickness, thermal effects and the influence of a narrow flow region.
Most of the simulation results presented in this section are based on software developed by the author of this work for numerical simulation of the injection and compression moulding process using the variational inequality approach.

Furthermore, we shall compare some of our results with different mathematical models, namely the distance model based on a geometrical approach and the three-dimensional Navier–Stokes equations.

Before we come to the generalized Hele–Shaw flow in Section 6.2, we will start with a brief (by far not complete) overview of the mathematical study of the classical Hele–Shaw flow.

6.1 Some general remarks on classical Hele–Shaw flows and related moving boundary problems

Let us briefly recall the classical Hele–Shaw flow as a moving free boundary problem. The problem arises when an incompressible, viscous, isothermal liquid is moving within a (thin) region between two closely-spaced parallel plates. The mathematical problem in its simplest form is given by Laplace's or Poisson's equation for the pressure $p = p(x,t)$, $x = (x_1, x_2)$ in the liquid region $\Omega(t)$ together with conditions on the (unknown) moving boundary, i.e.,

$$\left.\begin{aligned} -\operatorname{div}(k\ \nabla p) = f \quad &\text{in } \Omega(t); \\ p = 0 \quad \text{and} \quad k\frac{\partial p}{\partial n} = -V_n \quad &\text{on } \partial\Omega(t); \quad k \equiv 1, \end{aligned}\right| \tag{6.1}$$

where V_n is the velocity of $\partial\Omega(t)$ in the (outward) normal direction $\vec{n}$.

The liquid movement is driven by sources or sinks introduced in the field equation by a right-hand side (sometimes point sources are used) or by using boundary conditions along a domain surrounded by the liquid. While the injection problem characterized by an expanding liquid region is well-posed, the problem with prescribed suction mechanism is ill-posed (see [ELOC82] for numerical experiments and [DIFR84] for analytical investigations).

We remark that this Hele–Shaw problem is formally identical with the (classical) one-phase Stefan problem when sensible heat is negligible (zero-specific heat). Besides front tracking/fixing and fixed domain methods known for the Stefan problem (cf. [CRAN84]), the complex variable theory also provides a starting point for the solution of the two-dimensional problem (for a survey see [HOWI92]).

Richardson (see [RICH72], [RICH81]) used a conformal mapping method (i.e., the domain $\Omega(t)$ is mapped onto a convenient region (unit disc)) to study the problem (6.1) with injection from several sources into an infinite cell (i.e., the expanding blob is not restricted by any container). Identifying $\mathbb{R}^2$ with $\mathbb{C}$ and assuming that the regions $\Omega(t)$ are simply connected, the problem can be formulated as a differential equation for the mapping function. Generalizations of this method can be found in [RICH94]. In this context we also refer to [REIS94] and [MAY94] for study of the application of the nonlinear abstract Cauchy–Kovalevskaja Theorem (cf. [NISH77]) to problems of the form (6.1) and consequences concerning the existence and uniqueness of analytic solutions.

For mathematical study and numerical development of front tracking techniques applied to free boundary problems of the form (6.1), we call attention to the monograph [CRAN84] and papers by Meyer (see [MEYE90] and [MEYE93]).

Results concerning the extension of the enthalpy method to degenerate one-phase Stefan type problems with 'zero-specific heat' can be found in [CROW79], [NIEZ82] and [ALLU83] (see also Remark 2.4[ii]).

The application of an integral (Baiocchi type) transformation with respect to time to problem (6.1) with $k = \text{const.}$ leads to elliptic variational inequalities, where the time variable occurs only as a parameter. Together with a tangential flow on the container walls (closed mould), i.e., a Neumann condition at the fixed boundary, this fixed domain formulation is analysed in [CIJA81], [ELJA81] and [ELJA83]. The inequality formulation with non-local injection conditions and an open mould (i.e., $p = 0$ on $\partial\Omega$) is considered in [PRRO93] (for more details see Remark 2.7).

Gustafsson (see [GUST85]) investigated the connections between different (classical and weak) formulations for problem (6.1) in an open mould. In the papers [BEGI86] and [BEGI87], Begehr and Gilbert considered Hele–Shaw flows in $\mathbb{R}^n, n \geq 2$. In their second paper the case of an anisotropic tensor $k_{ij} = k_{ij}(x)$ instead of $k = \text{const.}$ is investigated (see also [GISH89] for an anisotropic flow in a closed mould).

The close connection between Stefan and Hele–Shaw type problems is also expressed in phase field models. Caginalp (see e.g., [CAGI91], [CAGI89]) pointed out how one has to choose the scaling limit of the microscopic parameters contained in the phase field equation to obtain various Stefan and Hele–Shaw type problems as formal limits.

Moreover, different aspects of singularities of the free boundary, especially in the suction case, are studied in [HHHO94], [KLV96] and other papers by these authors (persistence of corners, cusp-development or cracks).

At the end of this survey, which is by far not complete, we should mention an essential reason for the extensive research and comprehensive literature concerning Hele–Shaw type models, namely their mathematical analogy to flows in porous media (see e.g., [SATA58], [HOWI92]). Similar problems appear also in the modelling of the electro-chemical machining process (see Application 2.4).

Finally, we mention that the (ill-posed) suction situation corresponds to the Stefan problem for super-cooled water (see [DIFR84]) and to the electro-forming process (cf. [LMGS87] and [KLV96]).

A more detailed discussion of theoretical aspects of solutions of classical Hele–Shaw flows and related moving boundary problems can be found in the lectures notes [HOHL96] based on a seminar course given by Hohlov at the Freiberg University of Mining and Technology in 1996.

6.2 Mathematical modelling of injection and compression moulding

Additionally to Section 2.2 we give now a summary of the mathematical modelling of the injection and compression moulding process (see Applications 2.2, 2.3). Besides the mathematical models which lead to the free boundary problems introduced in Section 2.2, we will deal with other approaches (geometrical approach and some recent developments for the full three-dimensional numerical solution).

We start with a short discussion of the technical background and the main objectives of the numerical simulation processes. In Sections 6.2.2 and 6.2.3 we summarize the general conservation laws (balance equations for mass, momentum and energy) and the constitutive equations describing the (rheological) behaviour of the materials. These equations are applicable to model fluid flow in general.

Most plastic parts are thin. In other words, most cavities for injection and compression moulding are narrow gaps whose thickness is much smaller than their lateral dimensions. Based on this geometric assumption, in Section 6.2.4 we derive a spatially 2.5-D mathematical model for the filling process in injection and compression moulding. We make use of lubrication theory and adapt the underlying simplifications to non-isothermal and non-Newtonian fluid flow. This approach, known as generalized Hele–Shaw flow, is the basis for most mould filling simulations in polymer processing today.

In Section 6.2.5 we point out the basic ideas leading to a geometrical approach for the injection moulding process.

Finally, in Section 6.2.6 we will give an overview of recent trends to solve the spatially three-dimensional flow problem (i.e., to renounce the Hele–Shaw simplifications).

It is beyond the scope of this work to give a detailed overview of the extensive engineering literature on injection and compression moulding. Here, we cite only the monographs [TAGO79], [BERN83], [PERI83], [WHCG85], [MANZ87], [ISAY87], [MACO89] [TUCK89], [AASC91], [ADVA94], [BACO95], [KENN95] and [PAGU97]. Among hundreds of available journal articles we mention only the basic papers [HISH80] (injection) and [LETU87], [LFTU84] (compression). The cited, comprehensive studies deal with different aspects of the injection and compression moulding process. As an example (which will not considered in this work), let us mention fibre-reinforced injection and compression moulding. The books [ADVA94] and [PAGU97] mainly treat the flow of fluids containing short, slender fibre suspensions (short-fibre composites). Advani and Tucker have introduced in [ADTU87] a model of fibre suspensions in which average fibre orientation is represented by even-order orientation tensors (as moments of a probability density function, see also [ALTA90] for a review and a lot of references). In a sense, this approach was the starting point for studies over the last decade concerning improvements of the mathematical model and also the numerical simulation (see e.g., [VERW98], [VTFG99] and [STKA00]) of fibre-reinforced injection and compression moulding.

6.2.1 Injection and compression moulding – Technical background and requirements on simulation

6.2.1.1 Technical background

Two of the most commonly used manufacturing techniques for plastic products are injection and compression moulding. These processes are used to manufacture an assortment of products, ranging in size and geometrical complexity from a variety of polymeric materials. In injection moulding the polymer melt is injected from gates into a cavity (metal mould) which is mounted in an injection moulding press, cf. Figure 6.1. In compression moulding a pre-measured initial charge is placed between the mould halves and then these mould halves are brought together to fill the cavity.

Both amorphous and semi-crystalline thermoplastics as well as thermosets and elastomers (rubber) can be the materials for injection and compression moulding. It is beyond the scope of this work to delineate the different processes in detail. More information on variants in injection and compression moulding can be found e.g., in [TAGO79] and [WHCG85]. Here, we will discuss only the main properties of thermoplastics, thermosets and elastomers (rubber).

Plastics are polymers which are built from monomers through a polymerization process involving a phase change process (thermoplastics) or a chemical (cross-linking, curing) reaction (thermosets and also elastomers). In other words, depending on the type of the polymerization process, plastics are classified as either thermoplastics or thermosets. Furthermore, depending on the polymer chain conformation (microstructure), thermoplastics can be subdivided into:

- amorphous (having disordered molecular orientation regardless of the state),
- semi-crystalline (having well-ordered regular lattice below the melting temperature) and
- liquid crystalline (having ordered, rod-like molecular orientation regardless of the state).

Thermoplastics can be made fluent by heating and they solidify by cooling. In contrast to thermosets, thermoplastic polymers are capable of being resoftened and reshaped repeatedly, such that they can be recycled. Important examples for thermoplastics are polyethylene, polypropylene, polystyrene and polyvinyl chloride.

Thermosets also become fluent by heating, but at some temperature a reactive chemical process (so-called cross-linking or curing) begins. As a consequence, the liquid irreversibly transforms from a viscous material to a rigid lattice (solid) structure. After the cure reaction the material cannot be resoftened and reprocessed by heating (at least without degrading some linkages). At this point we decline to describe the cure kinetics for this exothermic, crosslinking reaction (see [TUCK89], Section 2.8). We mention that elastomers (rubber) also exhibit such a material behaviour, where the cure reaction is only weakly exothermic. Usually, thermosets have greater mechanical strength and higher upper temperature limits than thermoplastics.

Consequently, the mould temperature is a main variable in the process conditions for thermosets, elastomers ('cold mould') and thermoplastics ('hot mould').

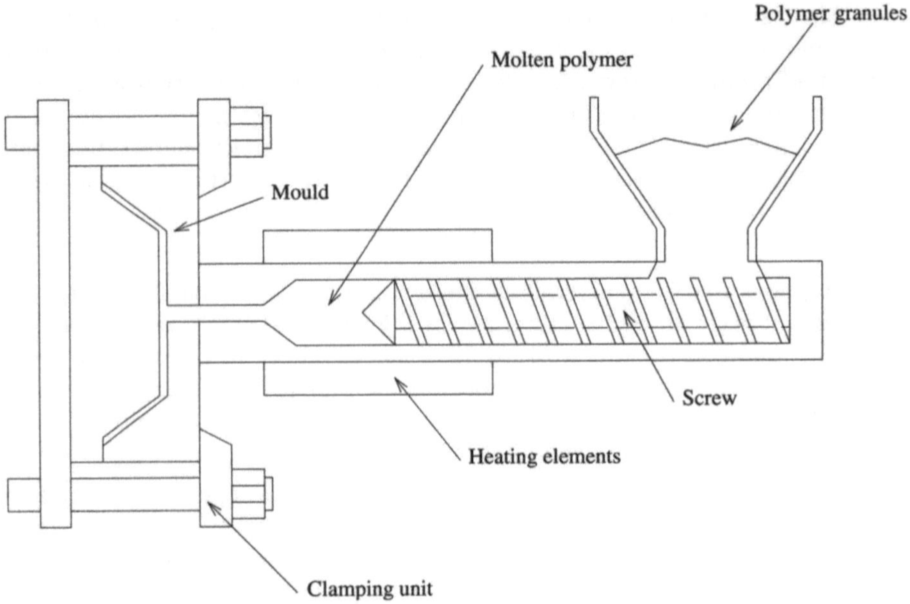

Figure 6.1: *Schematic representation of an injection-compression moulding machine*

We are now going to sketch the main features of the production methods for polymer parts, i.e., we will restrict ourselves to describing the peculiar aspects of a typical injection moulding cycle. An injection (compression) moulding machine schematically shown in Figure 6.1[1] consists of two main parts: the injection unit (screw rotating inside a heated barrel) and the clamping unit (holding the mould).

The production cycle in injection moulding can be subdivided into five major stages:

- Preplasticization,
- Filling,

[1]This diagram is taken from P. Johansson's thesis, [JOHA97]. Thanks to him.

- Packing,
- Cooling or Curing and
- Ejection of the product.

First, the barrel is filled with polymer granules which are melted by means of heating elements. The screw retreats until sufficient molten polymer (for a 'shot') is in front of it. Now, the screw acts as a ram (hydraulic action) and forces the molten polymer through nozzles (gates) into the mould (cavity) until the cavity is filled. This process is called the *filling phase* or 'shot'.

At the end of the filling stage, the pressure is maintained for a certain period of time: the *packing stage* within the cavity. During this packing phase a small amount of additional molten polymer enters the cavity to compensate the subsequent shrinkage of the material due to cooling (in case of thermoplastics). The final stage, the *cooling (or curing) phase*, starts when the melt is frozen (or cured) at the gates (nozzles). During this period the cooling (or curing) process of the polymer melt is intensified by means of cooling (or heating) channels in the mould. After the solidification (or cross-linking) of the material in the cavity the mould is opened and the formed product is *ejected*. At the same time of the cooling stage, new polymer granules are fed into the barrel, i.e., the next 'shot' is prepared. Then the mould is closed and the process cycle is repeated.

The entire process is highly automated and usually has cycle times in the range of seconds in case of thermoplastics. For rubber and thermosetting parts the cycle times can be longer. Important advantages of injection and compression moulding are the abilities to fill both simple, mass-produced parts in a rapid manner, but also complicated geometries. We mention, that depending on the geometry of the part, a mould can contain either a single cavity or multiple cavities.

In this book we restrict ourselves to the modelling and simulation of the *filling process* in injection and compression moulding.

During the filling process of geometrically complex industrial parts, the molten polymer flows through converging and diverging sections. Moreover, there can exist areas having changes in thickness and flow directions. Flow problems can be caused by flow in ribs, sudden contractions/expansions, flow in corners and around inserts.

6.2.1.2 Short comparison of injection/compression moulding and metal casting

Metal casting is another important application for the analysis of mould filling, solidification processes and stress evolution/behaviour including moving free surfaces. From the mould filling point of view, a main difference between standard metal casting (e.g., iron or steel) and polymer processing consists in the (usually) Newtonian behaviour of molten alloys (over the liquidus temperature). Furthermore, metal parts have typically volumetric geometries which do not allow application the Hele–Shaw (lubrication) approximation. Concerning several aspects of metal casting we refer to the proceedings of a series of conferences on modelling casting and welding processes, see [PIVK93], [CRCA95], [THBE98], [SAHC00].

Other metal casting processes such as the semi-solid processing of aluminium and magnesium alloys have much in common with polymer processing. This semi-solid processing of castings (thixocasting and thixomoulding) is rapidly gaining interest in the casting industry. The processing conditions of these manufacturing technologies lie between solid and liquid temperatures (lower thermal load of the mould). Therefore, the viscosity of the alloy shows in the semi-solid state a behaviour similar to that of polymer melts. Concerning semi-solid processing see e.g., [BMYM98] and [LISH00].

6.2.1.3 Some aims of the numerical simulation in injection and compression moulding

Let us mention some of the aims of the numerical simulation of the (injection or compression) filling process of a cavity. Matters of interest are:

- the advancement of the flow front (filling pattern),
- pressure analysis (reducing the necessary injection pressure required to fill the cavity),
- the placement of gates and air escape holes (vents) to avoid air traps,
- the location of weld lines formed by merging melt fronts (optical properties, surface appearance, possible structural weakness are influenced),
- the (possible) formation of frozen boundary layers during the filling process.

The filling pattern has the most important role in determining the quality of the part. An optimal filling pattern is characterized by a nearly constant melt front velocity (interface velocity) and by a simultaneous reaching of the extremities of the cavities. As usual in computer simulations our plots presented in Section 6.3 show the filling pattern (filling zones) as contours at equal time intervals. Therefore, the space between adjacent contours characterizes the interface velocity (closely (resp. widely) contours – hesitation (resp. acceleration)).

The injection pressure required for filling the delivery system (runner, gate) and the cavity depends on several material parameters (e.g., melt viscosity) and processing parameters (flow length, volumetric flow rate, gap thickness). For instance, multiple gates reduce the flow length such that the necessary injection pressure is also reduced. On the other hand, the usage of multiple gates usually has visible weld lines as a consequence.

Air traps ('enclosed air') occur when air is surrounded by flow fronts and cannot escape. One of the features characterizing a suitable flow behaviour is that the air in front of the flow fronts can escape from the cavity. This is accomplished by air vents at places (between the two mould halves) that are to be filled last. Common reasons for air traps are the lack of appropriate vents, unsuitable gate locations and the tendency of the melt to flow preferentially in thicker regions (for an example see Figure 6.3).

Weld (or knit) lines appear when separate flow fronts traveling in opposite directions meet (i.e., several fluid streams may join each other). Such situations can be caused by holes and inserts in the cavity, multiple gate regions and a varying thickness (hesitation and acceleration of the melt front). If possible, weld lines should be positioned at low-visibility areas and regions, where the part strength is not very important. Especially, in connection with fibre-reinforced materials, weld lines are undesirable because the fibres are usually oriented parallel to them.

Due to the influence of a cooling regime at the lower and upper surfaces of the cavity, especially in the case of long flow paths, there can arise a melt freezing process already during the filling process. Another reason for frozen boundary layers can be the stagnation (hesitation) of the melt in thin sections of the cavity.

We will discuss really three-dimensional effects like the fountain flow region and jetting in Section 6.2.6.

Despite a continuing fall in the cost of computer power which brings into sight the feasibility of a full three-dimensional filling analysis, numerical simulation of the non-Newtonian Navier–Stokes equations, including the free boundary/surface in geometrically complex, spatially three-dimensional regions, represents a formidable problem. Due to the geometrical feature of many plastic parts that one characteristic dimension (the thickness) is much smaller than the others, application of the Hele–Shaw (lubrication) approximation can be used with an accuracy satisfactory for most parts. This reduces the complexity of the governing equations leading to a drastic reduction in the demand on computer power.

Let us mention that, for about two decades commercial software products (based mainly on Hele–Shaw flow analysis) for the simulation of the injection and compression process have been available, e.g., Moldflow (see the book [KENN95]), Cadmould from the Institute for Polymer Processing at the RWTH Aachen, Germany (see [IKV98] (Chapter 13) and [IKV00]) and C-MOLD (in April 2000 the Moldflow Corporation announced that it finalized its acquisition of C-MOLD). These software packages are capable of simulating not only the filling process, but also other effects of the different stages described above.

Nevertheless, in our opinion, there exists a necessity for further developments of both the spatially 2.5-D (generalized Hele–Shaw flow, see Sections 6.2.4, 6.2.5) and 3-D simulation (see Section 6.2.6). In particular, we have in mind the application of special finite element (finite volume) techniques for simulating the filling process. These techniques should allow us to handle numerical difficulties of this highly nonlinear problem like: resolution of sharp moving fronts, discretization of non-self-adjoint differential operators and the need for accurate fluid velocities (pressure gradient). In this context see also Section 6.2.4.3 for the numerical core problems.

6.2.2 Balance and state equations

The underlying mathematical equations describing the physical phenomena in many engineering applications (e.g., injection and compression moulding) are the result of applying continuum mechanics techniques. These consist of a combination of general conservation laws and equations of state characteristic for the particular medium.
As a result one obtains the governing partial differential equations associated with boundary and initial conditions as well as discontinuity relations (jump-type conditions) at the free boundaries. Details of this general concept in the mathematical modelling of different physical phenomena can be found e.g., in [DALI90], [RODR87], Chapter 2 and [RODR94], Section 1.2.
At this stage we only formulate the differential equations without, at the moment, descriptions of initial, boundary and jump conditions (at the free boundary/surface). The basic principles governing the motion of a *viscous, incompressible* fluid are conservation of mass and momentum (Navier–Stokes equations) together with a constitutive equation.

$$\left.\begin{array}{ll} \operatorname{div}\vec{v} = 0, & \rho\,\dfrac{D\vec{v}}{Dt} = \operatorname{div}\underline{\sigma} + \rho\,\vec{g}, \\ \underline{\sigma} = -\,p\,\underline{I} + \underline{\tau}, & \underline{\tau} = \eta\,\underline{\dot{\gamma}}, \quad \underline{\dot{\gamma}} = \nabla\vec{v} + (\nabla\vec{v})^T, \end{array}\right| \tag{6.2}$$

where

- $\rho > 0$ is the density (here constant due to incompressibility),
- $\vec{v} = (v_1, v_2, v_3)$ denotes the velocity vector with the components v_i in x_i-direction. Here, x_1, x_2, x_3 denote the Cartesian coordinates of the particle considered at the time instant t (sometimes $x = (x_1, x_2)$ and $z = x_3$ will be used),
- $\underline{\sigma}$, $\underline{\tau}$ and $\underline{\dot{\gamma}}$ are the stress and the viscous/deviatoric stress tensors as well as the rate of strain (or shear rate) tensor,
- $\rho\,\vec{g}$ denotes the density of the external forces acting on the fluid (gravitational forces),
- p is the pressure,
- $\eta = \eta(\dot{\gamma}, \vartheta)$ is the shear (dynamic) viscosity of the fluid and
- D/Dt denotes the particle derivative (derivative following the motion).

As mentioned above we restrict ourselves to considering the filling phase in injection and compression moulding. Thus we assume that the fluid is incompressible. It is known that compressibility plays a major role in the packing stage, but its analysis lies beyond the goals of the present work.

The quantity $\dot{\gamma}$ (shear rate) is a scalar measure for the rate of deformation, i.e., $\dot{\gamma}$ is defined as the square root of the second invariant of the shear rate tensor,

$$\dot{\gamma} = \left(\frac{1}{2}\,\underline{\dot{\gamma}} : \underline{\dot{\gamma}}\right)^{1/2}. \tag{6.3}$$

The double dot product of two tensors $\underline{\sigma}$ and $\underline{\tau}$ is defined as usual

$$\underline{\sigma} : \underline{\tau} = \sum_i \sum_j \sigma_{ij} \, \tau_{ji}.$$

In general, molten polymers show non-Newtonian behaviour. To be more precise, polymer melts are considered as *generalized Newtonian fluids.* As usual in literature concerning polymer processing, the item generalized Newtonian fluid will be used for a fluid for which the viscosity η depends on the temperature ϑ and on the shear rate $\dot{\gamma}$.

Let us mention that we do not consider any viscoelastic effects in this work. Roughly speaking, viscoelastic effects can be the reason for flow phenomena which cannot be explained on the basis of linear or nonlinear purely viscous behaviour. A survey of the modelling of viscoelastic fluids in connection with polymer processing is given in Chapter 9 of the monograph [TUCK89], see also [CRDW84]. Furthermore, the rheological models discussed in Section 6.2.3 lack any time dependence (memory or thixotropic effects). The item thixotropy means that the stress in the fluid at any time is not only dependent on the current state of deformation, but also on its recent history. Often the above mentioned semi-solid slurries show thixotropic effects. In the simulation of polymer processing one usually neglects thixotropic effects such that changes in shear rate and temperature are immediately followed by corresponding changes in viscosity.

For the energy balance, which is of interest for the temperature equation,

$$\rho \, (e + \frac{1}{2}|\vec{v}|^2)$$

is the quantity being conserved per unit volume, where e denotes the internal energy. The conservation of energy is expressed by

$$\rho \, \frac{De}{Dt} + \operatorname{div} \vec{q} = \underline{\sigma} : \{\nabla \vec{v}\} + r, \qquad \vec{q} = -\tilde{\lambda} \, \nabla \vartheta, \tag{6.4}$$

where the classical Fourier law connects the temperature gradient $\nabla \vartheta$ with the heat flux $\vec{q}$ by the thermal conductivity $\tilde{\lambda} > 0$. An additional source term r can be present representing the heat released per unit volume and time (e.g., the curing heat evolved during the cross-linking (cure) reaction in thermosetting systems).

6.2.3 Rheological behaviour of polymer melts

In what follows we will discuss some aspects of the rheology of polymer melts and elastomers. We refer to [TUCK89], Chapter 2, [KENN95], Chapter 2, [PGLA95] and [TANN00] for a more detailed analysis of the rheological state equations, i.e., the viscosity relating the stress and shear rate tensor. The flow equations (6.2) and (6.4) are obtained using the principles of conservation of mass, momentum and energy. The constitutive law based on the material's properties gives the material's response to deformation. As mentioned above, the rheological behaviour of polymer melts can be characterized by non-isothermal and possibly non-Newtonian effects. Therefore, besides the Newtonian law, the constitutive laws used for flow analysis in injection and compression moulding are of generalized Newtonian type, i.e., the shear viscosity function η should have the form

$$\eta = \text{const}, \quad \eta = \eta(\vartheta) \quad \text{or} \quad \eta = \eta(\vartheta, \dot{\gamma}).$$

Let us repeat that the scalar magnitude $\dot{\gamma}$ of the strain rate tensor $\underline{\dot{\gamma}}$ is related to its second invariant by (6.3). We mention again that these models do not take into account any viscoelastic effects. Obviously, the viscosity increases when the temperature decreases. Virtually all polymer melts exhibit shear-thinning (pseudoplastic) behaviour which means that the viscosity diminishes as the shear rate increases. Let us give a rough idea of this behaviour. When the polymer melt is deformed, there take place some disentanglement of molecular chains and a molecular alignment in direction of the applied stress. As a consequence, resistance of the polymer melt to flow decreases with the deformation.

It is important to choose a model for a simulation process which is accurate over the processing range and for which data can be readily obtained (i.e., good fits can be obtained from experimental data).

A qualitative shear rate dependence of the viscosity for a typical polymer melt is illustrated in Figure 6.2 (note the double logarithmic scale). At low shear rate (region (a)) the melt behaves like a Newtonian fluid with the constant (so-called zero-shear rate) viscosity η_0. At higher shear rate (transition region (b)) the viscosity begins to decrease and goes into a straight line (power law region (c)). A further increase of the shear rate leads to another transition region (d), after which the melt behaves again like a Newtonian fluid with the constant (so-called infinite-shear rate) viscosity η_∞.

The temperature dependence of the viscosity is generally assumed to follow an Arrhenius-type relation of the form

$$\eta(\vartheta) = \eta_0(\vartheta) = m(\vartheta) = a_{\vartheta_R} \exp\left(\frac{\vartheta_A}{\vartheta} - \frac{\vartheta_A}{\vartheta_R}\right) \tag{6.5}$$

with the material constants: ϑ_R as reference temperature, ϑ_A as the Arrhenius parameter and the pre-exponential factor a_{ϑ_R} (with regard to ϑ_R).

In order to combine the non-isothermal with the non-Newtonian behaviour many models have been developed to represent experimentally observed viscosity-shear rate curves. Usually, one takes a basic law relating viscosity and shear rate (e.g., power-law, Carreau, Modified Cross, see below). Then, the temperature influence is taken into account using a temperature shift according to the Arrhenius relation (6.5) or a so-called WLF shift (Williams–Landel–Ferry equation, see [WILF55]),

$$\log b_{\vartheta_R,\vartheta_S}(\vartheta) = \frac{C_1\,(\vartheta_R - \vartheta_S)}{\vartheta_C + \vartheta_R - \vartheta_S} - \frac{C_1\,(\vartheta - \vartheta_S)}{\vartheta_C + \vartheta - \vartheta_S} \quad \text{with} \quad C_1 = 8.86,\ \vartheta_C = 101.6.$$

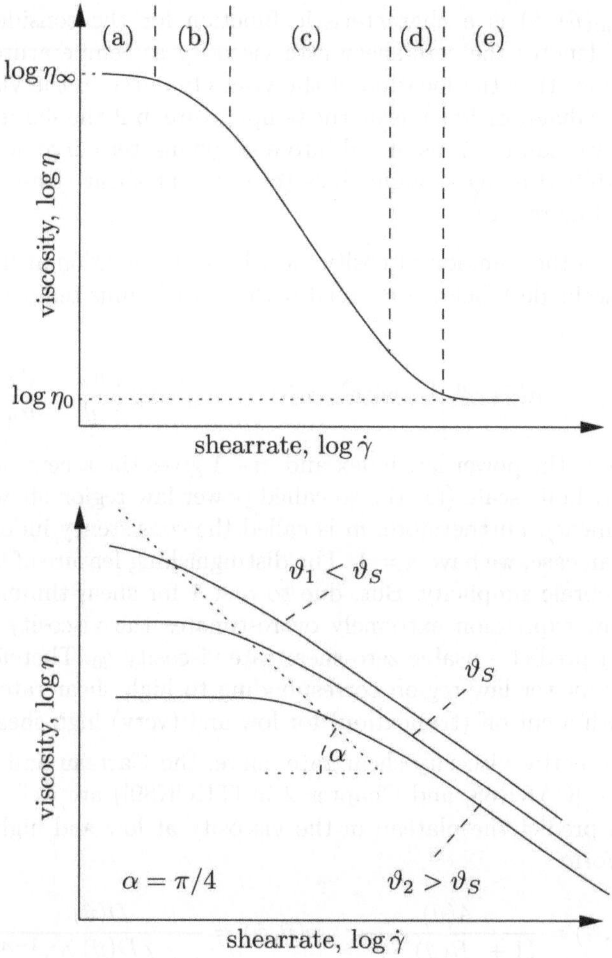

Figure 6.2: *Dynamic viscosity as a function of shear rate and temperature for a polymer melt.*

Here, ϑ_S is the (standard) temperature at which the main viscosity curve is defined and ϑ_R is a reference temperature at which a second viscosity curve is known. In other words, the first (resp. second) summand gives the shift between standard and reference temperature (resp. between standard and considered temperature). The first summand represents a scaling factor.

The fundamental idea for such a temperature shift is the assumption that the relation between viscosity, shear rate and temperature can be represented by

$$\frac{\eta(\dot\gamma, \vartheta)}{\eta_0(\vartheta)} = f(\eta_0(\vartheta)\,\dot\gamma),$$

where $f = f(\eta_0(\vartheta)\ \dot{\gamma})$ is a characteristic function for the considered polymer melt and $\eta_0(\vartheta)$ denotes the zero-shear rate viscosity at temperature ϑ. In other words, one assumes that the location of the viscosity curve in the viscosity/shear rate/temperature diagram depends on the temperature, but the shape is unaltered. All viscosity curves can be transmitted into a single master curve. More precisely, the curves are shifted along straight lines (of constant shear stress $\eta\ \dot{\gamma}$) with an ascent of -1 (cf. Figure 6.2).

Reviewing some of the common viscosity models, let us mention at first the power law model (Ostwald–de Waele) connected with an Arrhenius temperature shift of the form

$$\eta(\vartheta, \dot{\gamma}) = m(\vartheta)\ \dot{\gamma}^{n-1} \quad \text{with} \quad m(\vartheta) = a_{\vartheta_R}\ \exp\left(\frac{\vartheta_A}{\vartheta} - \frac{\vartheta_A}{\vartheta_R}\right). \tag{6.6}$$

The exponent n is the power law index and $n-1$ gives the ascent of the curve in the double logarithmic scale (for the so-called power law region showing the relationship to be linear). Furthermore, m is called the consistency index. Obviously, for the Newtonian case, we have $n = 1$. The distinguishing feature of the power law model is its algebraic simplicity. But, due to $n < 1$ for shear-thinning polymers, such a power law expression extremely overestimates the viscosity at low shear rates and cannot predict a usable zero-shear rate viscosity η_0. Therefore, it can be used only in the power law region corresponding to high shear rates and should be combined with a cut off (truncation) for low and (very) high shear rates.

To represent the entire viscosity-shear rate curve, the Carreau and the modified Cross model (see [CARR68] and Chapter 2 in [TUCK89]) are well suited. These two models can predict the plateau in the viscosity at low and high shear rates. They have the form

$$\eta(\vartheta, \dot{\gamma}) = \frac{A(\vartheta)}{(1 +\ B(\vartheta)\ \dot{\gamma})^{C}}, \quad \eta(\vartheta, \dot{\gamma}) = \frac{D(\vartheta)}{1 + \left(\frac{D(\vartheta)\ \dot{\gamma}}{\tau^*}\right)^{1-n}}. \tag{6.7}$$

One has $0 < n < 1$ and $0 < C < 1$ for shear-thinning behaviour, whereas $n = 1$ and $C = 0$ correspond to non-isothermal Newtonian behaviour.

Considering the well-known Carreau–WLF model (see [CARR68]) one takes

$$A(\vartheta) = p_1\ b_{\vartheta_R,\vartheta_S}(\vartheta), \quad B(\vartheta) = p_2\ b_{\vartheta_R,\vartheta_S}(\vartheta)$$

with the WLF temperature shift $b_{\vartheta_R,\vartheta_S}(\vartheta)$ such that then p_1 (resp. p_2) is the zero-shear rate (resp. reciprocal transition shear rate) at the reference temperature ϑ_R. The exponent $C = p_3$ corresponds to $1 - n$ in the power law model (6.6). Combining the modified Cross model with the Arrhenius temperature shift, one takes

$$D(\vartheta) = a_{\vartheta_R}\ \exp\left(\frac{\vartheta_A}{\vartheta} - \frac{\vartheta_A}{\vartheta_R}\right).$$

The term τ^* characterizes the position of the transition region (shear stress level related to the transition between Newtonian and non-Newtonian behaviour) at ϑ_R.

To use the discussed viscosity models in a simulation it is necessary to fit experimental data from measurements to the corresponding model and to determine the model parameters. Usually, one gets corresponding shear rate-viscosity pairs for several temperatures. It is advantageous to perform a logarithmic fitting, i.e., to use the logarithm in the equations (6.6), (6.7).

6.2.4 Temperature-dependent Hele–Shaw flow in the injection and compression moulding process

The filling phase of injection and compression moulding is characterized by a time-dependent flow with a free boundary. In other words, the basic equations (6.2) and (6.4) have to be solved in a time-dependent (three-dimensional) region. Its boundary is not known in advance and has to be determined as part of the solution of the free boundary problem.
As already indicated in Section 6.2.1.3 (see also Section 6.2.6) the simulation process of these *spatial three-dimensional* equations combined with the unknown domain makes high demands on computer hardware and software, especially in view of the geometrical complexity of common industrial components. Fortunately, many of the injection moulded parts are thin-walled, i.e., they satisfy the geometrical assumption that the gap thickness of the cavity is much less than its diameter (cf. Figure 6.3). Therefore, a number of hypotheses known as Hele–Shaw (or lubrication) approximation are made to reduce the computation time considerably.

Moreover, most of the effects suppressed by the following simplifications have minor consequences on the flow behaviour during the filling phase. For a more detailed discussion of three-dimensional effects see Section 6.2.6.

6.2.4.1 The generalized Hele–Shaw flow

Let us now sketch the essential ideas and assumptions which lead, based on the conservation of mass and momentum (6.2), to the free boundary problems for the pressure $p = p(x, t)$ in the injection (see problem (2.2)) and compression (see problem (2.3)) moulding process. Our derivation is based on the the author's papers [STEIa97], [STEIb96]) and, on top of this, we will make some further remarks.

As mentioned in Section 6.1 the classical Hele–Shaw flow concerns the isothermal, Newtonian flow between two plates with a narrow gap (see [LAMB93], Chapter VI, §330, §330a and Figure 2.2).
Let us now discuss this approximation for the case of cavities with a variable thickness and non-Newtonian flow. We emphasize that in the schematic representations of Figures 2.3 and 2.4 (see Section 2.2) the ratio of the gap thickness $\tilde{d}$ and the

diameter of Ω is not depicted realistically. We refer to Figure 6.3 for a more realistic situation. In the upper part of this figure the shape of a 'real-life' plastic part is illustrated (with inserts (two circled holes, grey highlighted) and a reduced thickness inside the dotted rectangular area). The essential geometrical assump-

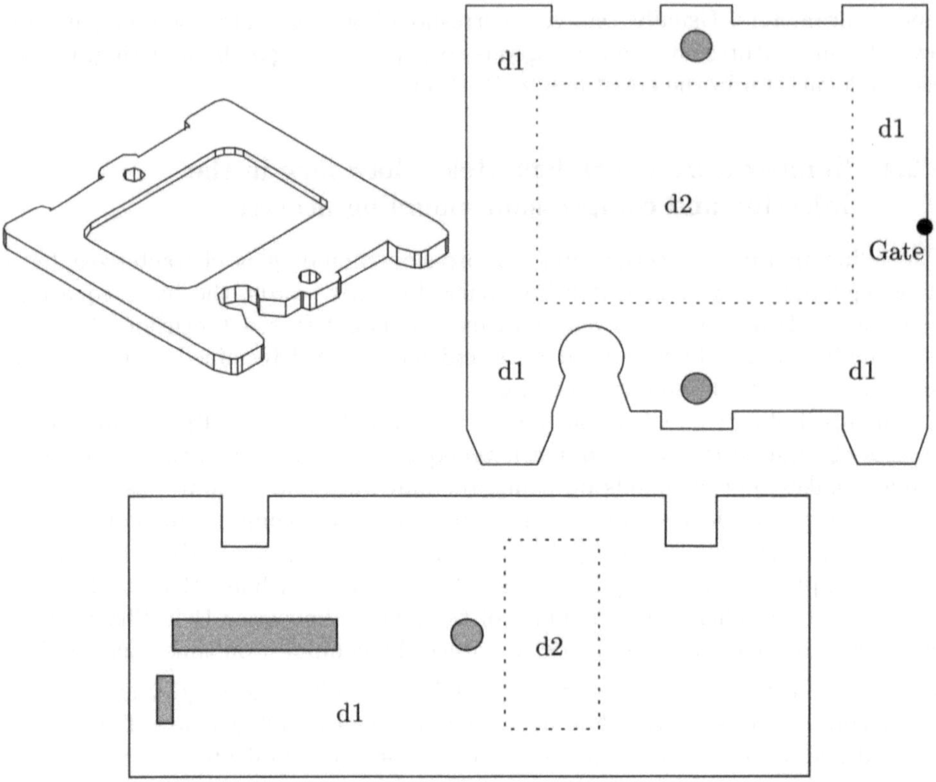

Figure 6.3: *Object and computational geometry (mid-plane) of plastic products. Top: object geometry and computational geometry (mid-plane) of a plastic part. Bottom: computational geometry of a second plastic part.*

tion allowing application of the lubrication theory (see e.g., [TUCK89], Chapter 3) which leads to a Hele–Shaw flow, is that the gap thickness of the cavity is much less than its diameter, i.e.,

$$\max_{x \in \overline{\Omega}} \tilde{d}(x) \ll \operatorname{diam} \Omega. \tag{6.8}$$

The following simplifications are based on Hele–Shaw's experiments made already in 1898 (see [LAMB93], Chapter VI, §330, §330a). These assumptions can also be verified by means of an order-of-magnitude analysis, i.e., non-dimensionalization of

(6.2) using (6.8) and introducing typical values for cavity length, cavity thickness, velocity and viscosity. Summarizing, it is assumed that:

- The flow is such that viscous forces and pressure differences dominate in the momentum equations (6.2). Therefore, external (body) and inertial forces are neglected (in comparison to friction forces, i.e., low Reynold's number as ratio of inertial force to viscous (friction) force). While the neglect of the inertial effects is justified for polymers and elastomers, it is not valid in metal casting.
- The pressure p is independent of z, i.e., $p = p(x,t)$.
- Due to the geometrical proportions and the no-slip conditions at the walls $z = \tilde{d}_i(x)$, $i = 1,2$, derivatives of v_i, $i = 1,2$ with respect to z are more relevant than derivatives with respect to x_i, $i = 1,2$. Therefore, one neglects the rates of variation of v_1, v_2 with respect to x compared to their rates of variation with respect to z in the shear rate tensor. Summarizing, the rate of the strain tensor takes the form,

$$\dot{\underline{\gamma}} = \begin{pmatrix} 0 & 0 & \partial v_1/\partial z \\ 0 & 0 & \partial v_2/\partial z \\ \partial v_1/\partial z & \partial v_2/\partial z & 0 \end{pmatrix}.$$

Consequently, the equation of momentum (6.2) is rewritten as

$$\frac{\partial p}{\partial x_i} = \frac{\partial}{\partial z}\left(\eta\,\frac{\partial v_i}{\partial z}\right), \quad i = 1,2 \quad \text{and} \quad \frac{\partial p}{\partial z} = 0. \tag{6.9}$$

We mention that the system (6.9) does not yet reduce the flow problem to a two-dimensional one, since v_i and η depend on z. In view of this fact, we now perform an averaging process (gapwise integration) of the equations (6.9) with respect to $(d_1(x,t), d_2(x,t))$ under consideration of no-slip conditions

$$\vec{v}(x, d_i(x,t), t) = 0 \quad \text{for} \quad i = 1,2$$

(cf. also Remark 6.2 below).

Analysing Green's function $G(\eta;.,.)$ for one-dimensional problems of the form

$$(\eta(x,t;z)\; w_z(z))_z = 1, \quad w(d_1) = w(d_2) = 0, \tag{6.10}$$

the velocity profile can be expressed by

$$\left.\begin{aligned} \vec{v}(x,z,t) &= \nabla p(x,t) \int_{d_1}^{d_2} G(\eta; z, \zeta)\, d\zeta \\ &= \nabla p(x,t) * \begin{cases} \displaystyle\int_{d_1}^{z} \frac{\zeta - \delta}{\eta(x,\zeta,t)} d\zeta & \text{for} \quad z \in (d_1, \delta), \\ \displaystyle\int_{z}^{d_2} \frac{\delta - \zeta}{\eta(x,\zeta,t)} d\zeta & \text{for} \quad z \in (\delta, d_2). \end{cases} \end{aligned}\right| \tag{6.11}$$

Here,

$$\delta = \Big(\int_{d_1}^{d_2} \frac{\zeta}{\eta(\zeta)}\, d\zeta\Big) \Big/ \Big(\int_{d_1}^{d_2} \frac{d\zeta}{\eta(\zeta)}\Big) \tag{6.12}$$

denotes the local value of z (depending on η and d_1, d_2) at which $\partial_z \vec{v}(x,\delta,t) = \vec{0}$ holds, or, in other words, the z value, at which $w = w(z)$ takes its minimum. The gapwise-averaged velocity $\bar{\vec{v}}$ can be obtained from

$$\left.\begin{aligned} \bar{\vec{v}}(x,t) &= \frac{1}{d(x,t)} \int_{d_1(x,t)}^{d_2(x,t)} \vec{v}(x,z,t)\; dz = \frac{-k(x,t)}{d(x,t)}\, \nabla p(x,t), \\ k(x,t) &= -\int_{d_1}^{d_2}\int_{d_1}^{d_2} G(\eta; z, \zeta)\; d\zeta\; dz = \int_{d_1}^{d_2} \frac{(z-\delta)^2}{\eta(x,z,t)}\; dz. \end{aligned}\right| \tag{6.13}$$

Remark 6.1 The coefficient function $k(x,t)$ is called flow conductance or fluidity. It is evident that k can also be expressed by

$$k(x,t) = -\int_{d_1}^{d_2} w(z)\; dz$$

with $w = w(z)$ as solution of (6.10). Thus, under the assumption $d_2 - d_1 > 0$ the maximum principle guarantees $k > 0$. Furthermore, in the symmetric situation characterized by

$$d_2(x,t) = -d_1(x,t) = d(x,t)/2 \quad \text{and} \quad \eta(x,z,t) = \eta(x,-z,t),$$

the relations (6.11), (6.13) with $\delta = 0$ can be written in a more convenient form

$$\left.\begin{aligned} \vec{v}(x,z,t) &= -\nabla p(x,t) \int_z^{d/2} \frac{\bar{z}\; d\bar{z}}{\eta(x,\bar{z},t)}, \qquad \bar{\vec{v}}(x,t) = -\frac{k(x,t)}{d(x,t)} \nabla p\,, \\ k(x,t) &= \int_0^{d/2} \frac{2\; z^2\; dz}{\eta(x,z,t)}. \end{aligned}\right| \tag{6.14}$$

Moreover, if $\tilde{d}(x) = d(x,t)$ (no frozen layers) holds, the equations (6.11), (6.13), (6.14) coincide with the usual Hele–Shaw relations (cf. e.g., [HISH80] and [TUCK89], Chapter 5 and 8). □

Taking into account (6.13) in the gapwise-integrated continuity equation (6.2) and using incompressibility, we obtain the governing free boundary problem (2.2) together with boundary conditions in the unknown domain $\Omega(t)$ for the pressure distribution $p = p(x,t)$ in the injection moulding process.

We remark that in the compression process the velocity component v_3 is small compared to v_i, $i = 1, 2$, but

$$v_3(x, d_i(x,t), t) = \partial_t\; d_i(x,t), \quad i = 1, 2 \tag{6.15}$$

(in contrast to injection). This influences the averaging of the continuity equation in compression moulding. Therefore, the free boundary problem for the pressure in compression moulding has the form (2.3).

Remark 6.2 In compression moulding the lower and upper walls are moved (or at least one of them; cf. Figure 2.4), such that the above mentioned averaging process is performed with respect to $(d_1(x,t), d_2(x,t))$. In contrast to this, in injection moulding the lower and upper walls are fixed. But, due to a possible formation of frozen boundary layers (cf. Figure 2.3 and Application 2.2), the effective thickness

$$d(x,t) = d_2(x,t) - d_1(x,t)$$

for the flow can be smaller than the gap thickness

$$\tilde{d}(x) = \tilde{d}_2(x) - \tilde{d}_1(x).$$

Consequently, it seems to be useful to perform the averaging process with respect to $(d_1(x,t), d_2(x,t))$ including the no-slip conditions at $d_1(x,t)$ and $d_2(x,t)$ also for injection moulding.

On the other hand, the shear viscosity η appears in the denominator of Green's function in (6.11), (6.13). Therefore, performing the averaging process in injection with respect to $(\tilde{d}_1(x), \tilde{d}_2(x))$, one has

$$k\big(\tilde{d}_1(x), \tilde{d}_2(x), \eta(x,z,t)\big) \approx k\big(d_1(x,t), d_2(x,t), \eta(x,z,t)\big)$$

in an approximate sense as result of an enormously increasing viscosity for temperatures $\vartheta = \vartheta(x,z,t)$ smaller than the no-flow temperature ϑ_{SO} of the melt in the corresponding viscosity models. □

Remark 6.3 At this point (see also Section 6.2.6) we need to note some restrictions caused by the lubrication (Hele–Shaw) approximation. Due to use of the gapwise-averaged velocity $\vec{v}$ for advancing the flow front in (2.2), (2.3), i.e., a plug flow as transverse velocity profile with respect to z (wetting angle is equal to $\frac{\pi}{2}$), the generalized Hele–Shaw model cannot predict the exact flow field near the flow front. However, this concerns only a thin layer region.

Furthermore, there exist geometrical situations in which the Hele–Shaw assumptions are not satisfied. For instance, in the case when very narrow sections (channels) occur in the cavity, cf. the example presented in Figure 6.9. □

6.2.4.2 Viscosity models and non-isothermal effects

The accuracy of an injection or compression moulding simulation depends on the viscosity model used, i.e., on the rheological constitutive equation relating the stress tensor to the deformation of the material. As already discussed in Section 6.2.3 the shear viscosity function should have the form

$$\eta = \text{const}, \quad \eta = \eta(\vartheta) \quad \text{or} \quad \eta = \eta(\vartheta, \dot{\gamma}).$$

The scalar magnitude $\dot{\gamma}$ of the strain rate tensor $\underline{\dot{\gamma}}$ is related to its second invariant by (6.3). Using again the Hele–Shaw simplifications, this quantity (shear rate) is given in a Hele–Shaw flow by

$$\dot{\gamma} = \sqrt{(\partial_z v_1)^2 + (\partial_z v_2)^2} = |\nabla p| \, \frac{|z - \delta|}{\eta} \tag{6.16}$$

(concerning δ see (6.12)).
Returning to the power law model (6.6), we have

$$\eta(\vartheta, \dot{\gamma}) = m(\vartheta)\, \dot{\gamma}^{n-1} \quad \Longrightarrow \quad \eta(\vartheta, \dot{\gamma}) = \eta(\vartheta, |\nabla p|) = (m(\vartheta))^{\frac{1}{n}} \, \big(|z - \delta| \, |\nabla p|\big)^{\frac{n-1}{n}},$$

where the second equation is obtained by substituting (6.16) for the shear rate $\dot{\gamma}$. Combining the Carreau and the modified Cross model (see (6.7)) with (6.16) and taking into consideration $0 < n < 1$ and $0 < C < 1$, one can apply Newton's method to solve the implicit equations in (6.7) for fixed temperature (for details see [STEIa97], Section 4.2).

Considering the non-isothermal Arrhenius law (6.5) or the non-Newtonian material models (6.6), (6.7), it remains to discuss the determination of the temperature ϑ. Here, we restrict ourselves to the basic concept of the temperature distribution.

While determination of the pressure distribution is, owing to the Hele–Shaw (lubrication) theory, a two-dimensional problem in space, the temperature has to be found in a moving, (in general) three-dimensional spatial region determined by the flow problem. Recalling the energy balance (6.4), the viscous heating term (dissipation) can be written as

$$\underline{\sigma} : \{\nabla \vec{v}\} = \eta \, \dot{\gamma}^2 = \frac{\eta}{2} \sum_{i,j=1}^{3} \left(\frac{\partial v_i}{\partial x_j} + \frac{\partial v_j}{\partial x_i} \right)^2 .$$

According to the simplifications in a Hele–Shaw flow with the shear rate $\dot{\gamma}$ given in (6.3), one obtains

$$\eta \, \dot{\gamma}^2 = |\nabla p|^2 \, \frac{(z - \delta)^2}{\eta} .$$

Neglecting at the moment a possible phase change process (due to the formation of frozen boundary layers) and the heat source term r ($r = 0$) as well as assuming a corresponding thermodynamic relation for the internal energy e and the temperature ϑ, the energy equation (6.4) in the (three-dimensional) melt region

$$\mathcal{L}^z := \Omega(t) \times (d_1(x,t), d_2(x,t)) \times (0, T)$$

becomes

$$\begin{aligned} &\rho\, c_p\, (\partial_t \vartheta + \vec{v}\cdot\nabla\vartheta) - \sum_{i=1}^{2} \frac{\partial}{\partial x_i}\Big(\tilde{\lambda}\,\frac{\partial\vartheta}{\partial x_i}\Big) - \frac{\partial}{\partial z}\Big(\tilde{\lambda}\,\frac{\partial\vartheta}{\partial z}\Big) \\ &= \tilde{r} := \eta\,\dot{\gamma}^2 = |\nabla p|^2\,\frac{(z-\delta)^2}{\eta}, \end{aligned} \qquad (6.17)$$

where c_p is the specific heat. For δ see (6.12).

On the left-hand side, besides the heat diffusion term, we observe the material derivative of the temperature including the convective term which plays an important role in our applications. This is because of a high Peclet number as ratio of heat convection to heat conduction

$$Pe = \frac{\rho\, c_p\, V\, L}{\tilde{\lambda}}$$

with V and L as characteristic velocity and dimension of the flow, respectively. On the right-hand side we recognize the viscous dissipation $\tilde{r} = \eta\,\dot{\gamma}^2$ which is a very sensitive term in many injection and compression moulding applications.

Besides $\mathcal{L}^z$ as the main coupling between the flow problem and the temperature equation, other non-linearities occur in the right-hand side

$$\tilde{r} = \tilde{r}(\eta, |\nabla p|^2)$$

of (6.17) and in the convective term requiring the viscosity (see (6.11) for $v_i, i = 1,2$). Furthermore, an order-of-magnitude analysis shows that in equation (6.17) the heat conduction is dominated by the z-direction, whilst the x_1x_2- flow has a determining influence on the convective term.

While in injection moulding $v_3 = \bar{v}_3 = 0$ approximately, in compression moulding the convection caused by the moving lower and upper surface walls should be taken into account by introducing an approximate velocity component v_3, e.g.,

$$v_3(x,z,t) = d(x,t)^{-1}\ \big[(z - d_1(x,t))\ \partial_t d_2(x,t) + (d_2(x,t) - z)\ \ \partial_t d_1(x,t)\big],$$

which satisfies (6.15) and implies

$$\bar{v}_3(x,t) = 0.5\ (\partial_t d_1(x,t) + \partial_t d_2(x,t)).$$

In [STEIa97] we have discussed different approaches for the determination of the temperature distribution (based on (6.17)).

Our first proposal uses an average temperature. To be more precise, performing the gapwise integration also in the temperature equation (6.17) and using the average velocity $\bar{\vec{v}}$ from (6.13), a two-dimensional nonlinear convection equation can be derived.

A second approach, three-dimensional in space, is also based on (6.17). A possible phase change process, due to the formation of frozen boundary layers (see Figure 2.3), is not explicitly considered. But, as already noted in Remark 6.2, for temperatures smaller than the no-flow temperature ϑ_{SO}, the viscosity increases enormously. Therefore, a freezing process is taken into account approximately in the flow problem by means of the gapwise integration in the derivation of the flow equation (see Section 6.2.4.1). Obviously, after the calculation the location of frozen regions can be observed by $\vartheta(x, z, t) < \vartheta_{SO}$.

Compared to the previous variants a third, more accurate, approach incorporates a possible formation of a solid frozen region in an explicit way by means of a two-phase Stefan problem for the temperature. Here, we do not go into further details. However, we emphasize that, in particular for the second and third approaches, a crucial point is the formulation of appropriate boundary conditions.

6.2.4.3 The numerical core problems

Similar to the close connection between classical Hele–Shaw flows and flows in porous media (cf. Section 6.1), the mathematical structure of the described complex model for injection and compression moulding has something in common with *nonlinear* flows in porous media (see e.g., the survey article [EWING85] for reservoir simulation in the petroleum industry). Therefore, similar numerical techniques can be used in both areas of application.

Let us sketch the numerical core problems in injection and compression moulding. In order to obtain the filling area $\Omega(t^n)$ for a given time instant t^n we recommend the solution of the corresponding evolutionary variational inequalities described in Chapter 5.

Since both the convective and viscous heating terms in the temperature problem (6.17) are governed by the velocity (pressure gradient), simulation of the injection and compression processes requires an accurate approximation of ∇p. This is a principal reason for using mixed methods for the pressure equation in the region $\Omega(t^n)$ (which can be determined, as just mentioned, by solving the inequality problem). The essential idea is to approximate the pressure p and the flux $k\ \nabla p$ simultaneously in a variational method, rather than to obtain the pressure gradient by differencing/differentiating. In this context we mention only the monograph [BRFO91] from the comprehensive literature on mixed finite element techniques.

The third core problem consists in treatment of the temperature equation as a convection-diffusion phenomenon. In Chapter 3 of the monograph [PIRO89] there is given a survey of several methods like upwinding, streamline diffusion and characteristics-oriented methods for handling the convective term.
We prefer the modified method of characteristics (MMOC) because the resulting matrices are symmetric, no additional upwinding or artificial diffusion is necessary, it works also in case of pure convection and, especially, the physical background of the moving domain for the temperature problem is well reflected. This method,

sometimes also called the Lagrange–Galerkin method, was proposed by Pironneau (see e.g., [PIRO89]) and Douglas, Russell in [DORU82].

Basically, in the MMOC one considers the hyperbolic part $\partial_t\vartheta + \vec{v}\cdot\nabla\vartheta$ in (6.17) as a directional derivative (direction is given by $(1,\vec{v})^T$). This derivative, in its discretized form, is combined with finite element/volume spatial discretizations for the remaining parts of (6.17). However, the method involves some practical problems. This concerns the influence of quadrature formulas, but mainly only for the pure convection equation (see [MPS88]). On the other hand, the efficiency of the method depends heavily on suitable retracing techniques for computation of the characteristics closely connected with the data structures used for the mesh. The diploma thesis [FRAN95] has reported on the development and implementation of an efficient retracing procedure (see also [PIRO89], p. 90). By means of the representation of the simplices of the mesh and the velocity field by barycentric coordinates, the characteristic is retraced stepwise from element (simplex) to element. Furthermore, when the characteristic intersects the boundary (edge or face) of a simplex, the velocity is updated by its local value.

6.2.5 The distance concept – a geometrical approach for injection moulding

A mathematical theory for the Hele–Shaw flow of isothermal, power law fluids (cf. (6.6)) was developed by Aronsson and Janfalk in [ARJA92] and [JANF96]. This theory was then applied to injection moulding of power law type thermoplastics in [ARON95] and [ARON96]. Furthermore, the numerical treatment and details of an efficient implementation of such geometrical models can be found in [BERN96] and [JOHA97].

The idea to exploit geometrical information exclusively for the determination of the filling pattern is also used in the engineering literature, see e.g., [TUCK89], Section 8.4.4 and [MENG80]. However, Aronsson and his co-workers have derived the mathematical basis and the connection between the constitutive law (cf. (6.6)) and the distance concept (pseudo-circle principle).

From the mathematical point of view one looks for an asymptotic solution (exponent $n \to 0^+$ in (6.6)) which is based on convex duality techniques applied to the p-Laplace equation. Performing the limit in the power law index as $n \to 0^+$, the basic result in this asymptotic approach is the pseudo-circle principle. Returning to the free boundary problem (2.2), we will give a brief description of the underlying ideas and assumptions in this section. We restrict ourselves to the application in the injection process. The transfer of the distance concept to compression moulding contains some difficulties, see [BERG98].

We assume an isothermal flow such that the energy equation is not needed. Consequently, the viscosity should not depend on the temperature. For the sake of simplicity let us consider the symmetric situation with respect to the plane $z = 0$ and $d(x,t) = \tilde{d}(x)$ (no frozen layers), cf. Remark 6.1. Then, the relations (6.11),

(6.12), (6.13) with $\delta = 0$ can be summarized according to (6.14). For ease of notation we will write G for $\Omega(t)$, when we are considering fixed time instants t. The domain G should be limited by Γ_1 (formerly the free boundary $\Gamma_f(t)$) and Γ_2 (consisting of parts of the external boundary $\Gamma = \partial\Omega$).

Now we remember the (isothermal) power law model (6.6) with a constant consistency index $m(\vartheta) = m_0$ and the shear rate expressed in a Hele–Shaw flow by (6.16), i.e.,

$$\eta(\dot\gamma) = m_0\ \dot\gamma^{n-1}, \quad \dot\gamma = \sqrt{(\partial_z v_1)^2 + (\partial_z v_2)^2} = |\nabla p|\ \frac{|z|}{\eta}.$$

By means of these two relations the (two-dimensional) velocity vector can be written as

$$\vec v = -e(\nabla p)\ |\nabla p|^{1/n} \frac{n}{n+1}\ \frac{1}{m_0^{1/n}}\ \left((\tilde d/2)^{1+1/n} - |z|^{1+1/n}\right) \quad \text{with}$$

$$e(\vec\omega) = \begin{cases} \dfrac{\vec\omega}{|\vec\omega|} & \text{if} \quad |\vec\omega| \neq 0, \\ 0 & \text{if} \quad |\vec\omega| = 0. \end{cases}$$

Furthermore, the velocity profile averaged with respect to the gap thickness can be represented by

$$\bar{\vec v} = -e(\nabla p)\ |\nabla p|^{1/n}\ \frac{n}{2n+1}\ \frac{1}{m_0^{1/n}}\ (\tilde d/2)^{1+1/n}. \tag{6.18}$$

Neglecting any source terms at the moment, the gapwise-integrated continuity equation now takes the form

$$\frac{2n}{2n+1}\ \frac{1}{m_0^{1/n}}\ \operatorname{div}\left((\tilde d(x)/2)^{2+1/n}\ |\nabla p(x,t)|^{1/n-1}\ \nabla p(x,t)\right) = 0. \tag{6.19}$$

We mention that this is a q-harmonic equation with $q = 1 + 1/n$ involving the weight function $(\tilde d/2)^{2+1/n}$. Because we use the traditional notation p for the pressure, the exponent will be denoted by $q = 1 + 1/n$ (instead of the more familiar notation p-harmonic or p-Laplace equation).

Let us now consider the injection at a point x_I (point source). In other words, we solve the equation (6.19) in the domain $G \backslash x_I$ with a prescribed singularity at x_I. This is equivalent to the solution of the problem

$$\operatorname{div}\left((\tilde d(x)/2)^{2+1/n}\ |\nabla p(x,t)|^{1/n-1}\ \nabla p(x,t)\right) = -c_I\ \delta_{x_I} \tag{6.20}$$

in the sense of distributions (with δ as the Dirac distribution). Comparing it with the free boundary problem (2.2) we have now to consider the following two problems.

- The *instantaneous problem*, which is the problem of determining the pressure and velocity field at a certain moment with a given position of the flow front (and the rate of injection or the pressure at the injection point x_I).
- The *evolution problem*, which is the problem of finding the filling pattern, i.e., the advancement of the flow front $\Gamma_1 = \Gamma_f(t)$ and the (increasing) filled region $G = \Omega(t)$.

The instantaneous problem is given by (6.20) completed by the boundary conditions

$$p = 0 \quad \text{on} \quad \Gamma_1 = \Gamma_f(t) \qquad \text{and} \qquad \nabla p \cdot \vec{n} = 0 \quad \text{on} \quad \Gamma_2$$

(compare problem (2.2) with $\kappa = 0$). The second (kinematic) boundary condition at the flow front $\Gamma_1 = \Gamma_f(t)$ describing mass conservation in the free boundary problem (2.2) is required for the evolution problem (advancing of the flow front).

Before we analyse the main ideas of the distance concept, let us briefly introduce the concept of interior distance. The *weighted interior distance* from a point x to Γ_1 is defined by

$$d(x, \Gamma_1) = \inf_{\gamma \in \Gamma(x, \Gamma_1)} \int_\gamma (\tilde{d}(x)/2)^{-1} \, ds,$$

where $\Gamma(x, \Gamma_1)$ is the set of all rectifiable curves γ in G having one endpoint on Γ_1 and x as the other end-point. The integration is done with respect to the arc-length.

For two arbitrary points x_a and x_b in G we define the weighted interior distance analogously by

$$d(x_a, x_b) = \inf_{\gamma \in \Gamma(x_a, x_b)} \int_\gamma (\tilde{d}(x)/2)^{-1} \, ds. \tag{6.21}$$

Now, we will look at the instantaneous problem (6.20) (q-harmonic equation with $q = 1 + 1/n$) and its associated extremum problem as well as the corresponding dual problem as obtained from convex analysis. Then, performing the formal limit $n \to 0^+$, we will give intuitive arguments for the basic fact in the distance model that the flow will follow shortest paths, i.e., at any given time the flow will go from the injection point to the nearest point on the flow front.

Returning to (6.20) we see that this is the Euler equation for the functional

$$J(p) = \frac{1}{1 + 1/n} \int_G (\tilde{d}(x)/2)^{2+1/n} \, |\nabla p(x,t)|^{1+1/n} \, dx - c_I \, p(x_I, t).$$

The primary extremum problem P_q ($q = 1 + 1/n$) is to minimize $J(p)$, i.e.,

$$P_q : \quad \text{Find} \quad \inf_{p \in U} J(p) = M_P, \quad U = \{u \in W_q^1(G) : u = 0 \text{ on } \Gamma_1\}. \tag{6.22}$$

Note that, due to the Sobolev embedding $W_q^1(G) \subset C(\bar{G})$ ($q = 1 + 1/n > 2$), this functional is well defined. Moreover, the functional J is strictly convex, coercive (i.e., $\lim_{\|v\|\to\infty} J(v) = \infty$) and (weakly) lower semi-continuous (see [CIAR78], Section 5.3). Therefore, using standard results concerning the minimization of convex functionals (see e.g., [EKTE76] and [GGZ74]), the above minimization problem (6.22) has one and only one solution.

Applying convex analysis (see [EKTE76], [ZEIDIII90], [BRFO91]), the dual problem to (6.22) can be derived (for details see [JANF96] and [ARON96]). This dual problem consists in maximizing the functional

$$J^*(\vec{r}) = -\frac{1}{1+n} \int_G (\tilde{d}(x)/2)^{-(1+2n)} \, |\vec{r}(x)|^{1+n} \, dx.$$

More precisely, the dual problem $D_{q'}$ with the conjugate exponent $q' = 1+n$ ($q^{-1} + q'^{-1} = 1$) takes the form

$$D_{q'}: \quad \text{Find} \quad \sup_{\vec{r}} J^*(\vec{r}) = M_D, \tag{6.23}$$

when $\vec{r}(x) = (r_1(x), r_2(x)) \in (L_{q'}(G))^2$ satisfies

$$\int_G \vec{r}(x) \cdot \nabla p(x) \, dx = -c_I \; p(x_I) \quad \forall p \in U \quad \text{and} \quad \vec{r} \cdot \vec{n} = 0 \quad \text{on} \quad \Gamma_2.$$

In other words, the last condition means that $\operatorname{div} \vec{r} = c_I \, \delta_{x_I}$ in the weak sense, such that the vector field $\vec{r}$ is divergence-free, except for a prescribed point source at x_I.

Now it follows from the theory of convex analysis that the dual problem $D_{q'}$ has the unique solution $\vec{r}^*$ satisfying

$$\vec{r}^* = -(\tilde{d}(x)/2)^{(2+1/n)} \, |\nabla p^*|^{1/n-1} \, \nabla p^*,$$

where p^* is the unique solution of the primary problem P_q. Furthermore, we have $M_P = M_D$ (for details see [ARON96] and [JANF96]). We see that the gapwise-integrated velocity $\vec{\bar{v}}$ (multiplied by the gap thickness, see (6.18)) is proportional to the solution $\vec{r}^*$ of the dual problem.

The main results of the distance concept are achieved by making a passage to the limit $n \to 0^+$ in the power law exponent n, i.e., $q \to \infty$ (using the assumption $\|(\tilde{d}/2)\nabla p\|_{L_\infty(G)} \le 1$) and $q' \to 1$ to obtain limit problems. Here, we restrict ourselves to describing some intuitive arguments. Again, we refer to [ARON96] and [JANF96] for a more detailed discussion and rigorous proofs.

We look at the family of dual problems (6.23), $D_{q'}$ with $1 < q' = 1+n < 2$ and the formal limit problem D_1. Roughly speaking, $D_{q'}$ and D_1 mean the minimization of the functionals

$$J^*_{q'} = \int_G (\tilde{d}(x)/2)^{1-2q'} \, |\vec{r}(x)|^{q'} \, dx, \quad J^*_1 = \int_G (\tilde{d}(x)/2)^{-1} \, |\vec{r}(x)| \, dx$$

under the same constraint

$$\operatorname{div} \vec{r} = c_I \; \delta(x_I).$$

The integrals $J^*_{q'}$ and J^*_1 can be interpreted as the 'cost' for the transportation of an incompressible fluid from the source point x_I to the 'exit part' Γ_1 of G. Some insight into the distance concept can be obtained from the following intuitive argument. For $q' > 1$ there is an extra cost for concentrating the flow, whereas at the optimal $\vec{r}^{\,*}_{q'}$ there is a balance between this extra cost and the gain obtained by letting the fluid flow along shorter paths to Γ_1 (with respect to the weighted metric caused by varying thickness $\tilde{d}(x)$). In the limit case $q' = 1$ there is no such extra cost. Hence, it is intuitively clear that the fluid will flow along shortest paths from x_I to Γ_1, i.e., along straight lines in G or straight line segments joined by parts of Γ_2.
This argument means that, when $q' \to 1$, the flow defined by $\vec{r}^{\,*}_{q'}$ will follow more and more straight lines from x_I to nearest points on Γ_1.

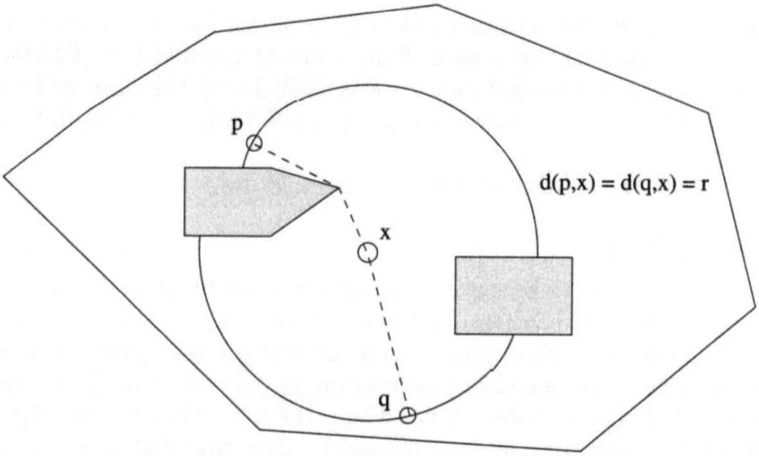

Figure 6.4: *Pseudo-circle shaped flow front*

Summarizing our line of reasoning and applying it to injection moulding we obtain the following basic pseudo-circle principle. Here, a pseudo-circle denotes a circle relative to the weighted interior distance, i.e., the set

$$C_r = \{x \in G : d(x_I, x) \leq r\}, \quad r > 0$$

is called a pseudo-circle with the center at x_I and the radius r (cf. Figure 6.4 concerning an illustration).

For small values of the power law exponent n, the fluid region in the cavity expands approximately like a family of pseudo-circles in $G = \Omega(t)$ with respect to the metric $(\tilde{d}/2)^{-1} ds$, each having its centre at the injection point x_I.

In other words, the level curves of the (weighted) interior distance function will represent the filling pattern for $n \to 0^+$.

As mentioned above the details of an efficient numerical implementation of the distance concept for both constant and varying thickness can be found in [BERN96] and [JOHA97]. Here, we only summarize the algorithms.

In computational geometry the problem of computing a Euclidean shortest path in a planar (polygonal) domain is well known with several applications, for example, in routing and robotics. Generalizations to the weighted situations (varying thickness) can be obtained by introducing some non-constant cost per unit travel in the domain.

The problem of finding a path with shortest weighted length in a planar domain is called the weighted region problem (see [MIPA91]. Its numerical solution is based on a computation of an appropriate discrete analogue d_h to the weighted interior distance defined in (6.21).

Let $\Omega \subset \mathbb{R}^2$ be a bounded polygonal Lipschitz domain (in our case the mid-plane of the cavity). Similar to the general finite element method (see [CIAR78], Section 2.1) a subdivision (triangulation) is established over the closure $\overline{\Omega}$ consisting of a finite number of closed convex subpolygons S_i with nonempty interior, such that

$$\overline{\Omega} = \bigcup S_i; \quad \operatorname{int} S_i \cap \operatorname{int} S_j = \emptyset \quad \text{for} \quad i \neq j,$$

(see also Section 4.1.2).

Under the assumption of a piecewise constant thickness function $\tilde{d}$ (with respect to the subpolygons S_i), the shortest path connecting two points in $\overline{\Omega}$ is a polygonal curve, which changes direction only at the edges of the triangulation. In order to introduce the above mentioned approximation d_h, each edge of the triangulation is divided equidistantly in node points. Then, the admissible curves $\Gamma_h(x_I, x)$ as subsets of $\Gamma(x_I, x)$ are given as the polygonal curves with end points at x and x_I (the injection point), which crosses the edges only at the prescribed node points.

Now, the discrete approximation $d_h(x_I, x)$ of (6.21) is defined as the infimum over the subset $\Gamma_h(x_I, x)$. An error estimate $O(h)$ with h as the distance between the node points at the edges can be found in [JOHA97].

To determine the flow fronts as level curves of the function $d_h(x_I, .)$, a two-step algorithm is performed. First, a graph is constructed, which represents the weighted interior distance from x_I to the given node points at the edges. If a pair of node points belongs to a subpolygon S, these points are connected (with the weighted distance). To find a shortest path tree for the graph, a variant of Dijkstra's algorithm is used (see [DIJK59]).

Secondly, an equidistant grid on $\overline{\Omega}$ is introduced. For each point x in this grid the approximation $d_h(x_I, x)$ is obtained by minimizing the sum of the distance from x to a node on the boundary of the surrounding subpolygon S and the distance from that node to the injection point.

We refer to [BERN96] and [JOHA97] for a detailed discussion of the accuracy and computational cost of this algorithm. Let us also mention that in a multiple-gate situation some additional difficulties occur in the computation of the flow fronts (see again [BERN96] and [JOHA97]).

Finally, we remember weld lines as the location where at least two flow fronts meet (merge). In terms of the distance concept, a weld line is represented by points in Ω characterized by a discontinuity of $\nabla d(x_I, .)$ (see [ARON95], [ARON96]). In other words, a point x is situated on a weld line, if there exist at least two shortest paths to x with different arrival angles. On the discrete grid over Ω the (approximating) weld lines are detected by setting a threshold for the difference in the arrival angle for neighbouring points (see [JOHA97], Section 8.2).

6.2.6 Recent three-dimensional simulation developments in injection moulding

As already indicated in this chapter, both commercial simulation programs and simulation tools developed at research institutes are usually dominated by the use of Hele–Shaw type flows. These simulations have been for about two decades important tools for providing moulding analysts with a deeper insight into the processes of injection and compression moulding. As a result of simulation, an essential qualitative increase in moulded parts has been achieved, connected with a drastic reduction of the cost in design and production of plastic parts.

Although this section gives an overview of recent developments in three-dimensional simulation tools, we want to emphasize at this point that Hele–Shaw based simulations give an accuracy sufficient for most industrially produced thin plastic parts (cf. geometric assumption (6.8)).

In our opinion the accompanying disadvantages of the Hele–Shaw simplifications can be divided into two main points:

- at first, the (automatic) creation of the two-dimensional mid-plane and,
- secondly, the neglect of some three-dimensional flow phenomena.

Let us recall that the application of the Hele–Shaw approximation requires a two-dimensional computational geometry (mid-plane) with the thickness as a parameter function ('layflat' of a single curved mid-surface). The mid-plane is the reference surface in which the melt is assumed to flow. This implies the identification of a dimension (thickness direction) in arbitrary geometries of cavities which is much smaller than the other two mutually perpendicular directions. On this mid-plane, the finite element mesh is generated. From the practical point of view one is interested in an automatic derivation of this mid-plane from the 3-D (solid) CAD geometry. But, certain geometrical features of the cavity (plastic part) can make the definition of a mid-plane difficult and ambiguous.

We recall that essential inherent factors of the generalized Hele–Shaw flow (lubrication approximation) consist in neglecting any inertial forces, simplifications in the

shear rate tensor (in-plane deformation rates, out-of-plane velocity components) and the use of a gapwise-averaged velocity as well as the geometrical assumption (6.8). Especially, in elastomer processing and gas-assisted injection moulding there are moulded parts which do not fulfill this requirement on the geometry (thin-walled). Furthermore, it is obvious that the use of the Hele–Shaw approximation leads to some loss of information regarding three-dimensional effects. In Remark 6.3 we referred to situations occuring in mould filling which cannot be accurately predicted by means of the generalized Hele–Shaw flow.
Let us mention: the fluid behaviour at the free surface (fountain flow region), near and at the walls and the flow kinematics in areas with sudden thickness changes and ribs.

At the flow front the fountain flow effect occurs. This is because the fluid near the centre is moving faster than the average velocity across the thickness. Therefore, a combination of shear and extensional flow leads to the phenomenon that hot fluid elements from the core area move toward the mould wall and subsequently solidify (but only a thin layer). Fountain flow effects have a direct influence on the fibre orientation.

Furthermore, jetting phenomena are three-dimensional effects. More precisely, jetting (as snake-like jetting stream) occurs when the melt is pushed with a high velocity through narrow regions (e.g., gates) into thicker areas, where the melt has no contact with the mould wall.

Before we mention some recent developments in three-dimensional simulation for mould filling of plastics and elastomers which are based on general conservation laws (cf. Section 6.2.2), let us summarize some main difficulties of this three-dimensional simulation.

- The computational domain has usually a complex shape (requirement of an automatic mesh generation using tetrahedra or hexahedra).
- The treatment of the free surface. Multiple interfaces may come in contact which each other. Usually front tracking algorithms have to be used (no fixed domain formulation is available).
- The treatment of phase changes.
- High demands on computer speed and memory (e.g., because of the non-Newtonian flow behaviour and the relatively high viscosity compared to metal casting).
- Implementation of a 3-D unstructured mesh generator is challenging work. From the practical point of view such a mesh generator should *automatically* provide triangulation based on the CAD geometry for (arbitrary) complex geometric parts (typical injection moulded parts have many details).

Especially for parts with extremely thin-walled regions, the mesh generation results in a large number of elements. One needs a fine discretization in the thickness direction to describe the flow behaviour accurately. But, in contrast to the 2.5-D simulation, the mesh in the other directions is not independent of the division in the thickness direction.

In contrast to mould filling simulations in polymer processing, several commercial software programs based on a three-dimensional treatment have been developed specially for mould filling in metal castings. These programs have been used in industry for about ten years, see the proceedings [PIVK93], [CRCA95], [THBE98], [SAHC00] already cited in Section 6.2.1.2.

In our opinion the main reasons for this difference from mould filling in polymer processing are as follows. Usually, metal casting parts are not thin-walled, so the Hele–Shaw approximation is not appropriate. As already mentioned in Section 6.2.1.2, apart from the semi-solid processing, the rheology of molten alloys (over the liquidus temperature) is usually characterized by Newtonian behaviour and a much smaller viscosity (mostly constant viscosity is used). Therefore, for the numerical solution of the unsteady, incompressible Navier–Stokes equations (6.2) in (standard) metal casting, semi-implicit (projection) methods are often used. More precisely, all terms containing the velocity are treated explicitly in time and only the pressure is taken implicitly. This method is applied both in finite element (see [LUC95]) and also in finite volume context (see [LIPI96]). Using an explicit Euler method for time advancement in the momentum equations one gets the equations

$$\rho\, \vec{v}^{j+1} = \rho\, \vec{v}^{j} + \tau\, \vec{r}^{j} - \tau\, \nabla p^{j+1}, \qquad \operatorname{div} \vec{v}^{j+1} = 0$$

for the time instant t^{j+1}.

Here, $\vec{r}^{j}$ is a shorthand notation for the convective and viscous terms taken at the time t^j and $\tau = t^{j+1} - t^j$ is the time step. Inserting the relations from the momentum equations into the continuity equation, one obtains a Poisson equation for the pressure p^{j+1}. Afterwards the projection of the pressure gradient ∇p^{j+1} is added to the velocity, creating $\vec{v}^{j+1}$ as (discrete) divergence-free velocity. The main computational effort for each time step consists in the solution of a Poisson equation. But due to the explicit treatment of the convective and viscous terms, the overall time step is limited by the minimum of the convective (CFL number) and diffusive (viscous) time step limits. We note that the viscosity appears in the denominator in the viscous time step criterion. A discussion of such types of algorithms can be found e.g., in [PATA80], [PIRO89], [GDN98] and [FEPE99]. Because of the relatively high viscosity of plastic melts and also of elastomers, the just described algorithm is not appropriate for the simulation of injection and compression moulding or for semi-solid processing in casting.

Let us return to three-dimensional simulation in polymer processing. In what follows we will mention some special developments of the three-dimensional treatment in polymer processing. The abilities of several general-purpose, commercial and research-oriented CFD-codes to simulate the injection moulding process will not be discussed here.

In our opinion, the developments can be subdivided mainly in two directions. At first, a natural generalization of Hele–Shaw flows omitting the geometrical assumption (6.8), but assuming negligible inertia forces, leads to generalized Stokes flow (creeping flow). Over and above that, taking into account also the inertia effects which can play an important role in several geometric situations, one has to solve the unsteady Navier–Stokes equations (coupled with the energy equation).

For a general overview of finite element methods for Stokes flow problems we refer to the monographs [BRFO91], [BRAE92] and [BRSC96]. Many textbooks deal with the numerical solution of the non-stationary Navier–Stokes equations. We mention only [PATA80], [GDN98], [FEPE99] (finite volume methods) and [CUSS86], [PIRO89] (finite element methods).

In the mould filling problems which we have in mind, the free surface/boundary separating melt and escaping gas (air) represents an essential difficulty in connection with the solution of Stokes and Navier–Stokes flow problems. Due to this moving interface, both generalizations just mentioned are characterized by unsteady behaviour and a main problem consists in the treatment of the free surface. Usually, there exists no fixed domain formulation for generalized Stokes and Navier–Stokes flow (but see [STEL93] for a derivation of a fixed domain type formulation for an isothermal, Newtonian Stokes flow). Instead of this, front tracking (capturing) techniques are used. Typically, the position of the flow front (free boundary) is tracked using different variants of level-set type methods. The kinematic boundary (interface) condition for two (immiscible) fluids resulting from mass conservation forms the basis for such methods. On the other hand, the dynamic boundary (interface) conditions associated with momentum conservation express the fact that the forces acting on the fluids at the free surface are in equilibrium.

Let us mention three techniques of interface-capturing methods, i.e., the free surface is determined by cells of the triangulation which are partially filled. This can be done by following massless particles (Marker and Cell, MAC method) and by solving a hyperbolic transport equation for a scalar variable (volume fraction Volume of Fluid, VOF method) or level set function (level set method)). Here, we will only give a brief sketch. A more detailed discussion of the methodology of such volume tracking methods can be found e.g., in [RICO98].

The MAC method was originally described by Harlow and Welch in [HARW65]. The basic idea is to introduce a number of (massless) particles in each computational element and to use these particles to mark the different fluids (melt and gas) existing in the whole domain. The particles are transported with the melt velocity, such that the free surface is given by cells of the triangulation which contain particles but have also empty neighbour cells. Depending on the number of particles which have to be transported the computational effort can be large (especially in three dimensions). An implementation of the MAC method for the 2-D case is described in [GDN98], where also the results for a 2-D injection moulding example are presented. However, in contrast to Hele–Shaw flows in thin cavities, the cavity is assumed to be very thick (infinite) such that one dimension is neglected.

In the VOF method (see [HINI81]) the free boundary is tracked according to the solution of a hyperbolic transport equation for a scalar variable Ψ,

$$\frac{\partial \Psi}{\partial t} + \vec{v} \cdot \nabla \Psi = 0. \tag{6.24}$$

Using a triangulation the function Ψ represents the fractional volume of each cell occupied by the fluid (melt; volume ratio of melt per cell). One takes $\Psi = 1$ for full (filled by melt) cells and $\Psi = 0$ for empty (filled by air, gas) cells. Cells of the triangulation with values of Ψ between zero and one contain the free surface. If one is interested in a sharp interface, the shape of the free surface can be deduced from gradients of Ψ. As a further development of the original VOF method we refer to the SURFER scheme described by Lafaurie et al. in [LNSZZ94].

The third above mentioned technique is based on level set formulations (see the paper [OSSE88] and the monograph [SETH96]). This class of methods (sometimes also called pseudo-concentration method) can be considered as a variation of the VOF method. Similar to the VOF method, a hyperbolic equation (6.24) has to be solved, but now the level set function Ψ is usually defined as the signed distance from the free surface. More generally, this technique defines a function Ψ, which is greater than a critical value Ψ_c in the filled portion (with melt), whereas the air (gas) is characterized by $\Psi < \Psi_c$. The critical value Ψ_c defines the melt-air interface as iso-surface. Usually, the level set function Ψ should be reinitialized after each time step to guarantee mass conservation. In this connection we refer also to [ZZPH98] and, furthermore, to [THOM86], [SUSO94], [CHHO96] and [MEJA97].

Finally, we mention that there are several variants of these approaches. They involve not only the treatment of the free boundary but also the solution strategy of the field equations. One-fluid calculations can be performed for the melt applying the dynamic boundary conditions at the (sharp) interface. On the other hand, momentum and mass conservation can be solved separately in the two fluids (melt and gas) and again the dynamic boundary conditions are applied at the free boundary separating the two fluids. Alternatively, melt and air can be treated as a single fluid. The variation of the material properties (e.g., viscosity η) can be taken into account according to the volume fraction of each phase (VOF method) and the values of the level set function, respectively, i.e.,

$$\eta = \Psi\, \eta_l + (1 - \Psi)\, \eta_g, \quad \text{resp.} \quad \eta = \eta_g + (\eta_l - \eta_g)\, h(\Psi - \Psi_c),$$

where the subscripts l and g refer to liquid (melt) and gas (air), respectively. Here, h is the Heaviside function, where in the numerical simulation, this function is usually smoothed over a region of some finite thickness. In these cases, the dynamic boundary conditions are implicitly taken into account in the momentum equations (if necessary, surface tension effects can be transformed to volume terms, see e.g., [ZZPH98]).

Concerning the first generalization from Hele–Shaw flow to three-dimensional, creeping Stokes flow (small Reynold's number) we refer the readers to papers from the Industrial Materials Institute at the National Research Council Canada. In [HGGS98], Hétu et al. present a 3-D finite element model to predict velocity, pressure and temperature fields, which model consists in solving the Stokes and energy equation in the entire computational domain (cavity). The air in the empty region is assumed to be incompressible and has the role of a pseudo-fluid with small viscosity compared to the melt. The free boundary is treated by means of a level set method as mentioned above. Despite that the solution in the whole domain requires much more computational effort than a resolution only in the filled part (melt region), the authors prefer such a procedure because of some advantages (influence of trapped air, ability to solve multi-phase problems, better progression of flow front). Besides the usual material properties, the boundary conditions at the wall (melt/air – mould interface) are also determined in dependence on the values of the convected level set function. No-slip boundary conditions are imposed for the melt, whereas the conditions for the air have to allow that the air as pseudo-fluid is free to leave the cavity as pushed out by the liquid melt.
The Stokes flow as saddle point problem is solved by means of the Galerkin formulation using the Mini finite element (see [BRFO91], [BRAE92]).

Another reference concerning three-dimensional Stokes flow in injection moulding is the PhD thesis by Webelhaus (see [WEBE00]). In contrast to [HGGS98] the pseudo-fluid is not considered on the whole empty portion of the cavity but only in a part of the empty region close to the flow front. The boundary conditions for the computational domain (something larger than the filled portion) are given by an appropriate continuation of the boundary conditions of the melt region. This work contains also a comparison between different solution algorithms for Stokes flow problems as a saddle point formulation (variants of Uzawa algorithm, preconditioned conjugate residual algorithm, GMRES (generalized minimum residual method) with preconditioning, multigrid methods). Furthermore, based on the paper [FOMS93], the packing phase in injection moulding is also considered, such that a limited (weak) compressibility is included. The compressibility is given by an equation of state relating the pressure, specific volume (density) and temperature (pvT-diagram). Due to the limited compressibility (small Mach number) a similar solution strategy as in the incompressible case is applied.

The main difference is that in the weak compressible case the continuity equation as constraint in the saddle point problem contains also convective and unsteady terms. In terms of the above mentioned semi-implicit projection methods (pressure correction) this means that instead of a Poisson equation for the pressure as a result of mass conservation, now the equation for the pressure contains diffusive, convective and unsteady terms (see [FEPE99], Chapter 10).

In this connection we also refer to papers by authors from the Technical University Eindhoven. For instance, in [HAAG98] the filling phase for the gas-assisted injection moulding is simulated taking into account compressible material behaviour.

In the gas-assisted injection moulding process the first step consists in a partial injection of polymer melt (skin component) into the cavity. Afterwards, compressed gas in injected into the core of the melt. This process is used for producing hollow, light-weight parts.

Let us now review some developments concerning the simulation of injection moulding by means of the unsteady Navier–Stokes equations.
In [VERW98], [VTFG99] the non-stationary, non-Newtonian, incompressible 3-D Navier–Stokes equations are coupled with Hele–Shaw flow including the prediction of the fibre orientation pattern produced by the injection moulding process for relatively simple geometries (rectangular plates with different ribs). To reduce the computational demand, only selected features of the geometries are simulated with a 3-D model, whereas the entire part is simulated with Hele–Shaw flow. In other words, in this two-level approach the Hele–Shaw filling simulation provides boundary conditions for the three-dimensional finite element analysis which are limited to local geometrical features. The 3-D flow calculations are done with the built-in equations of the CFD-code FIDAP (see [ENGE93]) for velocity, pressure and temperature. A usual mixed formulation is used in which pressure is interpolated one order lower than velocity (note Ladyzhenskaya–Babuska–Brezzi (LBB) or inf-sup condition). A VOF method is used to treat the moving flow front. The coupled fibre orientation calculation is based on models developed in [ADTU87], [ADVA94] and is implemented on top of the software FIDAP.

Presently, the author of this monograph is a member of the staff of the software development group at the Magma Gießereitechnologie GmbH Aachen, Germany. This company has developed the commercial codes MAGMASOFT® and SIGMASOFT® for 3-D mould filling analysis (and related processes such as solidification and stress analysis) of several Newtonian casting processes as well as non-Newtonian semi-solid processing and injection moulding of plastics and elastomers. The software is capable of predicting the velocity, pressure and temperature as well as the position of the flow front, where especially for the filling process in injection moulding, the three-dimensional, non-Newtonian Navier–Stokes equations (including inertial terms) are solved by means of finite volume models. In contrast to a segregated solution technique, a fully coupled implicit algorithm is used for solving the momentum and continuity equations at each time step. The energy equation is solved after the new velocity and pressure have been obtained. We perform one-fluid calculations, where the flow front is treated by means of a VOF surface capturing technique (modified SURFER scheme as mentioned above). Besides this, reactive viscosity models are available in connection with simulation of the curing process for elastomers and thermosets. The fibre orientation can be simulated based on stochastic models (orientation tensors) as cited above (see [ADTU87], [ADVA94]). More details can be found in [LIPI96] (Newtonian metal casting) and [LISH00], [STKA00] (semi-solid processing and injection moulding).

Finally, we mention that other companies working on plastics simulation software also report development of three-dimensional software tools for injection moulding (see e.g., the annual technical conferences (ANTEC) of the Society of Plastics Engineers (SPE)).

6.3 Simulation results

In this section some results from numerical experiments will be presented. However, we do not give a complete survey of our simulation studies in injection and compression moulding based on the evolutionary variational inequality approach. Only the influence of some basic geometrical and operating conditions will be discussed for several real plastic products. Furthermore, we compare results obtained by the variational inequality approach to results obtained by the distance model (cf. Section 6.2.5) and by three-dimensional simulations (cf. Section 6.2.6).

6.3.1 Variation of gate location and thickness, non-isothermal effects and influence of narrow flow region

Variation of the gate location (gate design)

In Figure 6.5 we present simulation results for an injection-compression process (see Application 2.3) of a real plastic part. Its geometry is depicted in Figure 6.3 (bottom). The thickness ratio outside ($d1$) : inside ($d2$) of the dotted rectangle is given by $d1 : d2 = 2 : 1$. The circle and the other two rectangles (grey highlighted) are holes (inserts) in the plastic product.

Injection-compression moulding can be considered as an extension of the conventional injection process. After a pre-set amount of melt is fed into the open cavity, the two mould halves are brought together, such that the melt is compressed.

We have simulated the following injection-compression process including a variation of the gate location for this part. The initial charge for compression, consisting of the first two filling zones taken together, is obtained by the numerical simulation of injection of one third of the cavity volume (i.e., $\frac{1}{3}$ $\text{meas}_2(\Omega) \times d(x,0)$) through the corresponding gates (represented by dots ($\bullet$) in Figure 6.5). Then, we switch over to compression (the next four filling zones), reducing the gap thickness continuously to one third of $d(x,0)$. To detect the remaining air (white area) the last filling zone corresponds to a filling degree of about 96%.

Figure 6.5 illustrates the comparison of three different gate situations; left: two gates situated in the middle of the lower and upper side, middle: gates are located in the four corners and right: gates are at the interior hole (circle).

The main result of the simulations is the prediction of enclosed air (which cannot escape) for the cases illustrated in the left and middle part of Figure 6.5. The reason for the air traps occuring in this example is an unsuitable gate location. In

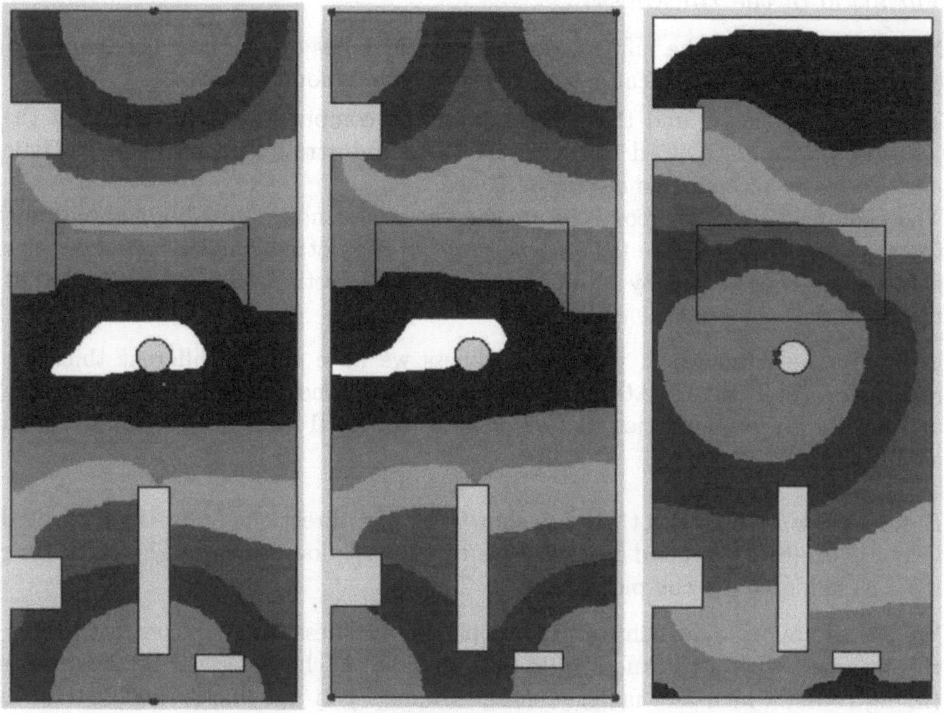

Figure 6.5: *Variation of gate location. Simulation results for an injection-compression process. Gates represented by •. Filling zones for six time instants. Degree of filling = 96 %, white area = air. Left and Middle: air traps. Right: suitable flow behaviour.*

contrast to this, the example with the gates at the interior hole (circle) shows a suitable flow behaviour.
Besides the appearance of an air trap, the situation with the gates located in the four corners is characterized by flow fronts merging together. This causes an unattractive, undesirable weld line in the middle of the plastic product reducing the part strength. Furthermore, the weld line is located in a good-visibility area.

Usually, multiple gates create problems of weld lines. But, on the other hand, they reduce the length of flow. In general, a single gate situation should be preferred.

At this point we remark that the 'jagged shape' of the flow fronts is caused by the underlying mesh (the flow fronts coincide with edges of the triangulation). In most of the figures presented here we have declined to 'smooth' the flow fronts (which can be done e.g., by interpolation).

Variation of the thickness

Now we will discuss the influence of varying thickness on the flow behaviour exemplified on a rather typical geometry for injection moulded parts.

The object geometry and the geometry for the computation (mid-plane) of the plastic product are plotted in Figure 6.3. This industrially manufactured plastic part is a cover plate for an electrical device.
The main features of its geometry are the two circled holes (grey highlighted) and the thickness ratio outside ($d1$) : inside ($d2$) of the dotted smaller rectangle. The gate region (represented by •) for the considered injection processes is located at the right wall.

To study the influence of varying thickness we have chosen different thickness ratios $d1 : d2$. In Figure 6.6 the results (six filling zones including the remaining air) are shown for the ratios $d1 : d2 = 1 : 1$ (top), $d1 : d2 = 2 : 1$ (middle) and $d1 : d2 = 4 : 1$ (bottom).

Besides these filling zones there are shown the flow directions (gapwise-integrated velocity, cf. (6.13)) related in each case to the last flow front (at the sixth time instant; filling nearly completed).

We have a uniform flow behaviour for constant thickness (top). In contrast to this, the situation with a thickness ratio of 2 (middle) is characterized by flow fronts (filling zones) which are 'faster' ('more advanced') in the thicker region than in the thinner one.
As already mentioned in the introductory Section 6.3.1, closely spaced filling zones (contours) indicate hesitation (here: in the thinner regions), whereas the widely spaced zones in the thicker regions (outside the rectangle) refer to an acceleration of the melt velocity. Obviously, this behaviour is also indicated in the left part of Figure 6.6 (higher velocities in the thicker region).

These tendencies are strengthened in the extreme case of a thickness ratio of 4, which leads to an air trap (bottom). Note that our simulation results show an air trap already for a thickness ratio of 3 (not contained in Figure 6.6).

Obviously, the entrapped air is caused by a large thickness ratio. This example confirms the tendency of the polymer melt to fill preferentially thicker areas with less resistance. As a result, for a thickness ratio of 4, a short shot is obtained, i.e., the polymer melt can not fill the entire cavity.

Non-isothermal effects

In Figures 6.7 and 6.8 we present some results of computations in which we consider especially thermal effects for the 'cover plate' (cf. also Figure 6.3).

We have determined an averaged temperature according to the first proposed approach discussed in Section 6.2.4.2. The mould temperature (cooling regime) is taken as constant and the relation between viscosity and temperature is given by the Arrhenius law (6.5).

The computation results for this plastic part, on the one hand, with a constant and, on the other hand, with a varying thickness (aspect ratio 2.5 : 1.5, which is the original ratio for this product) are shown in Figure 6.7 and 6.8, respectively. The flow fronts are depicted for selected time instants. Furthermore, the temperature and the pressure (for the last selected time instant) are plotted in each case. In Figure 6.8 we recognize the qualitative agreement between the computer-simulated and the experimental results.

Influence of a narrow flow region

The geometry for the example shown in Figure 6.9 is taken from the monograph [TUCK89], Section 5.9.4., cf. Figures 5.35–5.38. Important features of this geometry are the constant thickness and the different (converging and diverging, narrow and wide) flow regions near the neck area. Remembering Remark 6.3, the neck area represents a narrow passage where the geometrical limitation that one characteristic dimension (thickness) is much smaller than the others is not fully fulfilled. Usually, the Hele–Shaw flow predicts an overestimation of the flow through such narrow passages.

Figure 6.9 shows the predicted flow fronts at selected times obtained, on the one hand, by the distance model (top) and, on the other hand, by the variational inequality approach (bottom). The filling process is divided in 31 time steps (constant inflow). The depicted flow fronts computed by the distance model correspond to the time steps $1, \ldots, 30$, whereas the ten flow fronts shown in the middle part of this figure represent the time steps $3, 6, \ldots, 30$ (predicted by the variational inequality approach). The comparison between variational inequality approach and distance model will be discussed in Section 6.3.2.

Furthermore, in the right lower part of Figure 6.9 we have depicted the predicted pressure distribution (isobars) related to the last flow front (near completion of filling) according to the variational inequality approach. The spacing of the isobars indicates that high pressure gradients occur at the gate and the neck regions, where the velocity is greatest.

We mention that both the predicted flow fronts and the isobars according to our variational inequality model are in very good agreement with the simulation results presented in the above mentioned engineering monograph [TUCK89]. The results shown there by Güceri in Figures 5.35–5.38 are also based on finite element modelling.

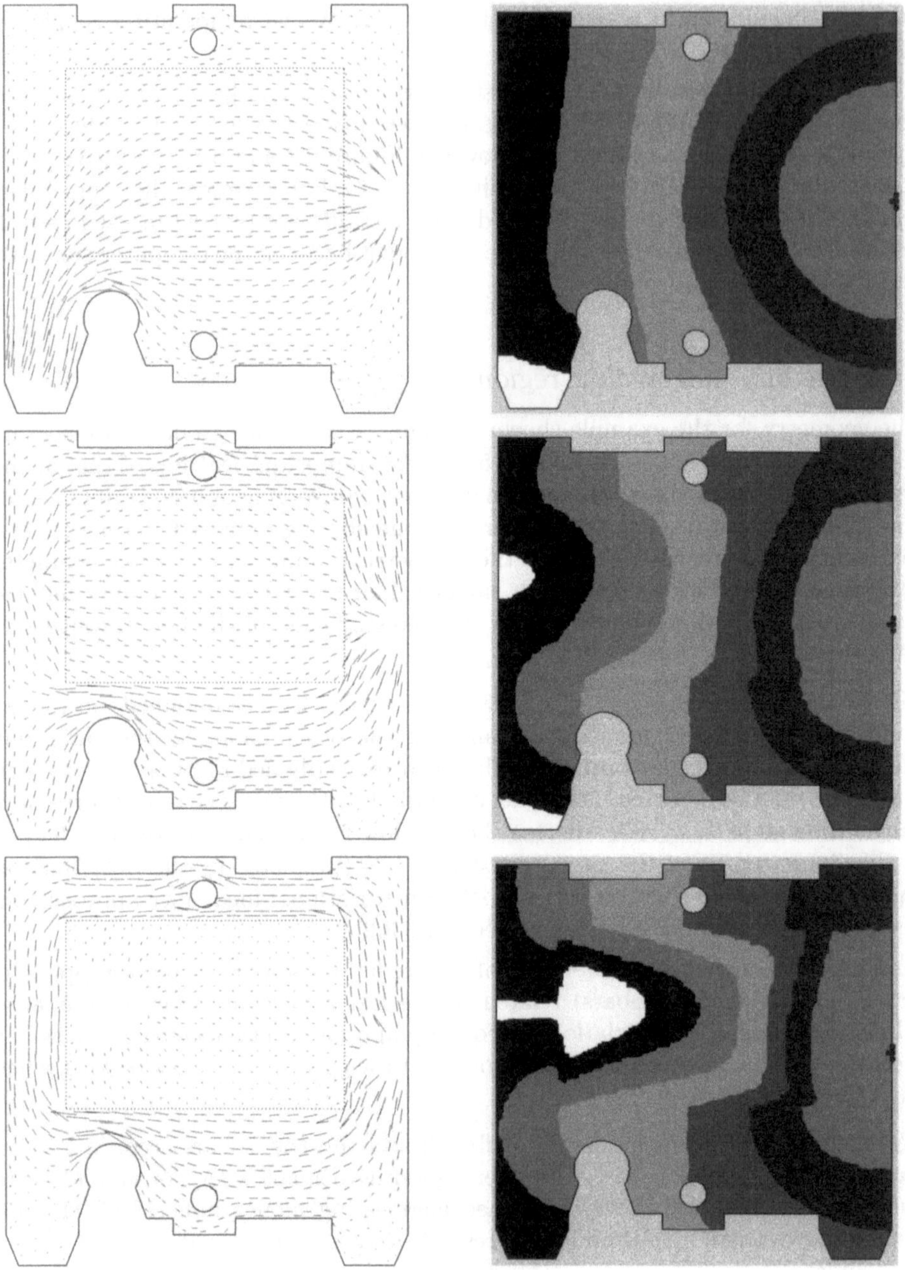

Figure 6.6: *Variation of thickness. Simulation results (including flow directions) for an injection process. Gates represented by • . Right: Filling zones for six time instants. Left: Corresponding flow direction (velocity) at the sixth time instant.*

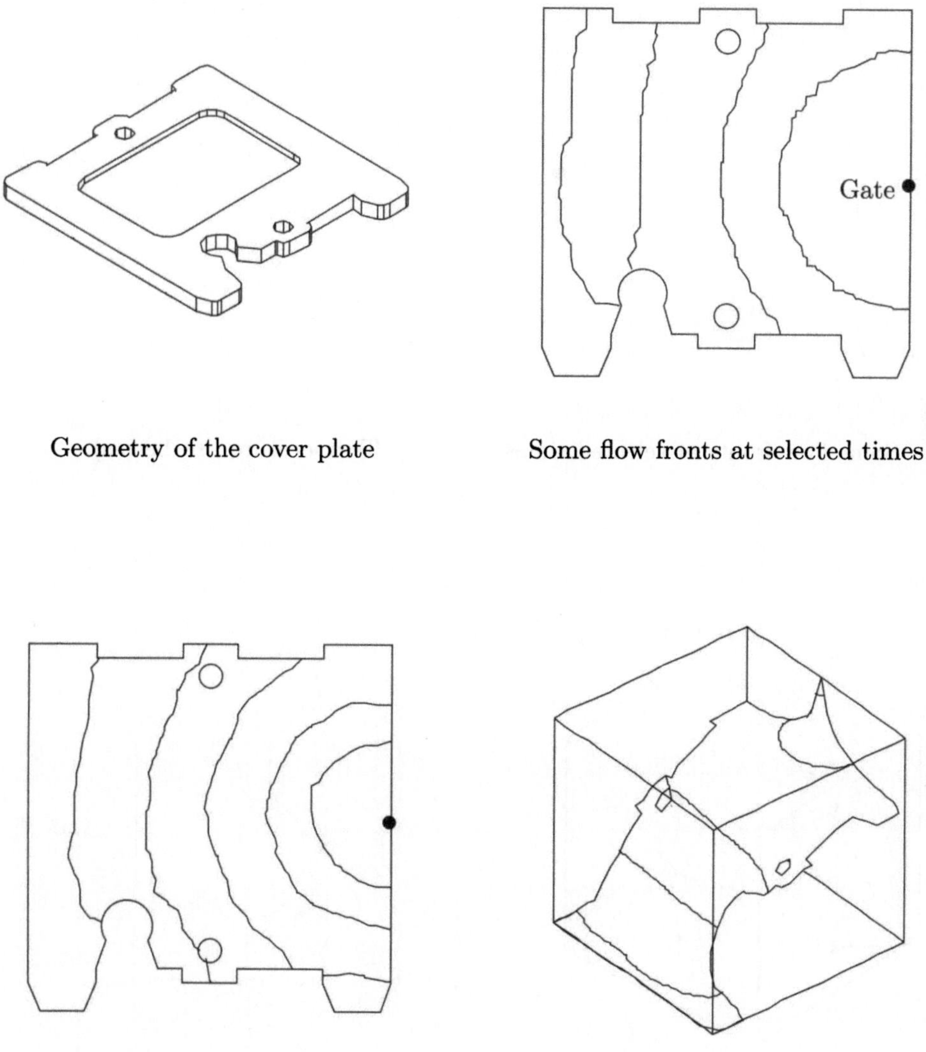

Figure 6.7: *Non-isothermal effects. Cover plate with constant thickness.*

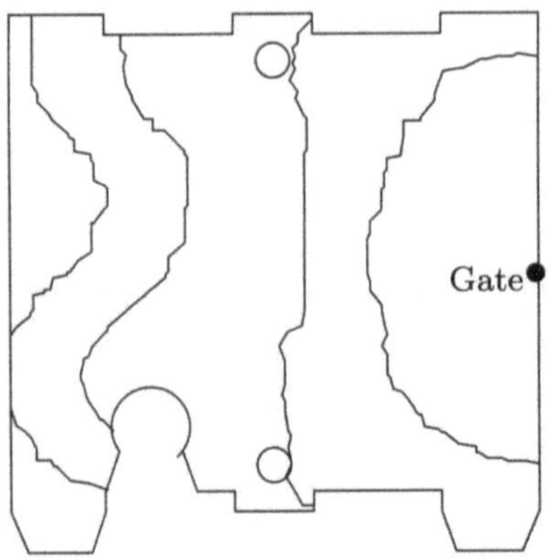

Some flow fronts at selected times

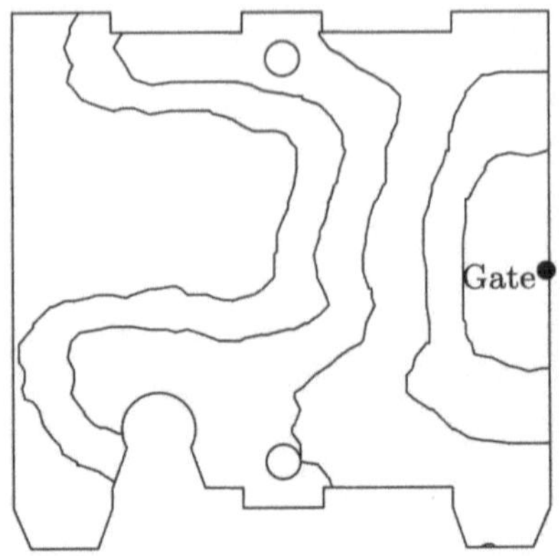

Temperature (isotherms)

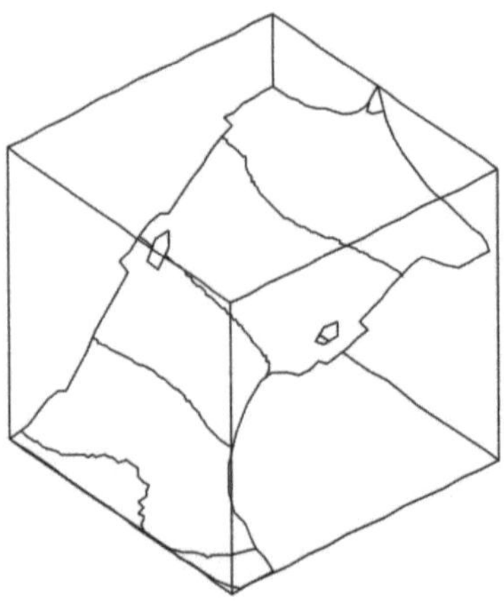

Solution surface of pressure

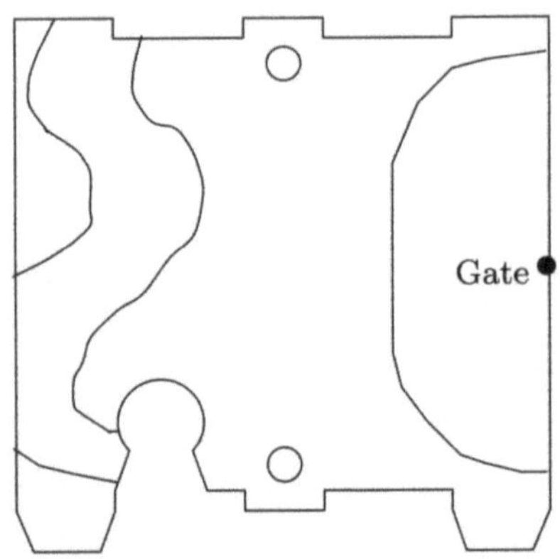

Filling pattern from experiment

Figure 6.8: *Non-isothermal effects. Cover plate with varying thickness ($d1 : d2 = 2.5 : 1.5$).*

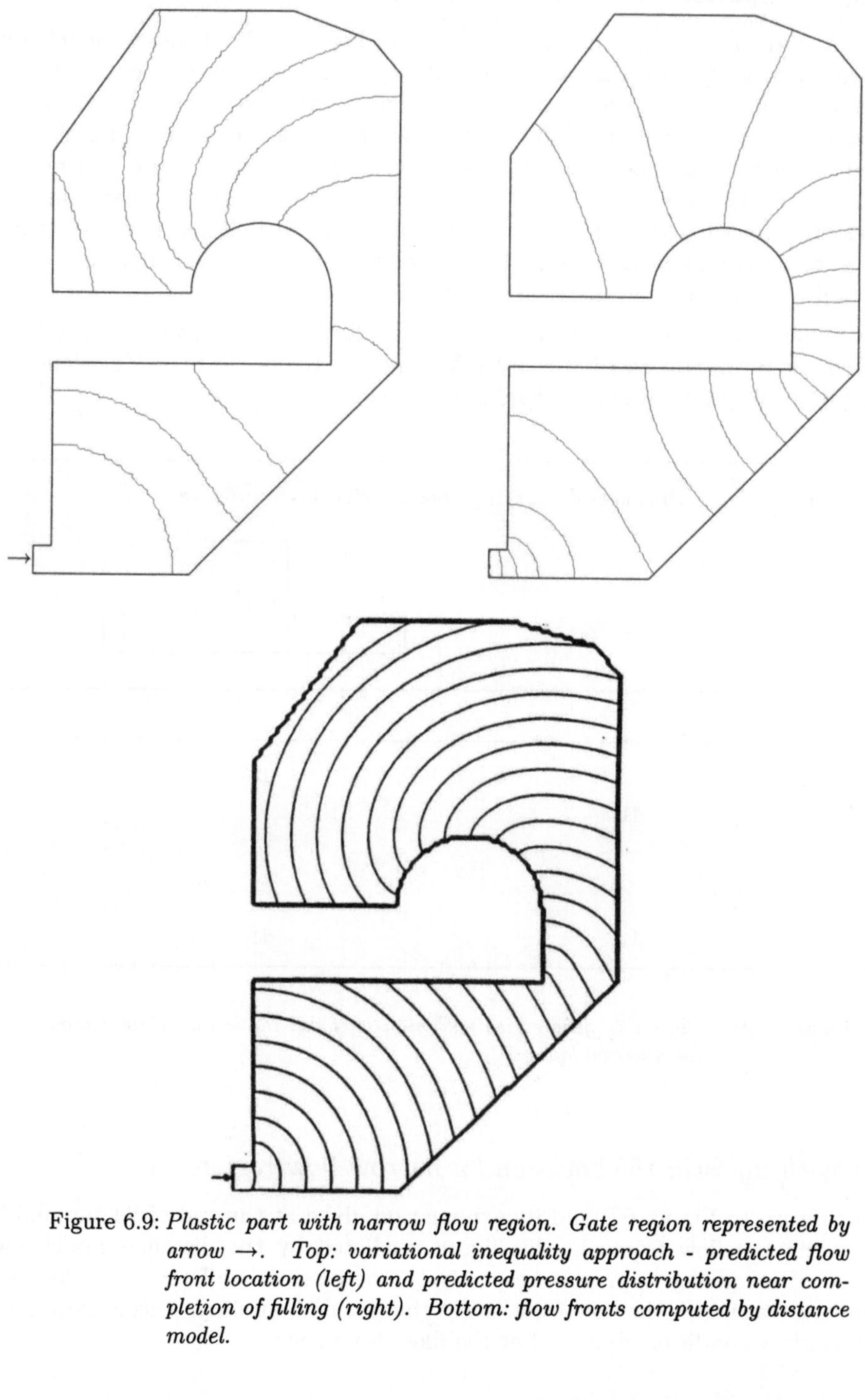

Figure 6.9: *Plastic part with narrow flow region. Gate region represented by arrow →. Top: variational inequality approach - predicted flow front location (left) and predicted pressure distribution near completion of filling (right). Bottom: flow fronts computed by distance model.*

6.3.2 Comparison with the distance model

Now we wish to find out some main differences between the distance model (see Section 6.2.5) and the variational inequality approach (see Section 6.2.4). Based on some numerical examples a comparison will be performed and, furthermore, remembering the derivation of the mathematical models, the reasons for the observed distinctions will be explained. Moreover, based on other numerical results and experiments (short shots) found in the literature, we want to examine our numerical simulation results. Due to these aims, we have chosen some relatively simple (geometrical) cavities from the engineering literature, which are not characterized by a complex geometrical shape.

The computations based on the distance approach were carried out and visualised by Frederik Berntsson and Peter Johansson at the Department of Mathematics, Linköping University, Sweden. Thanks to them.

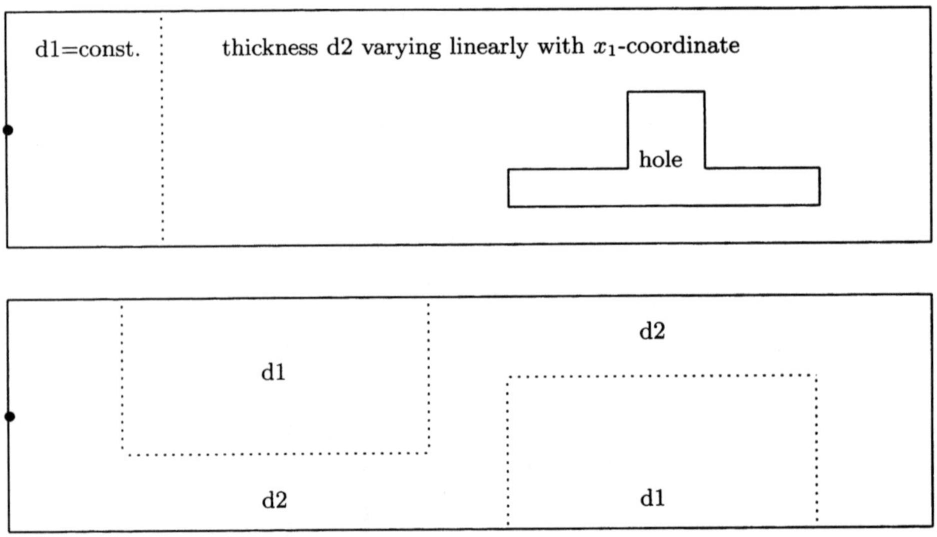

Figure 6.10: *Plate-like geometries with non-uniform thickness. Gate region represented by •.*

Part with uniform thickness and a narrow flow region

Let us return to Figure 6.9 and its explanation given at the end of Section 6.3.1. Comparing the flow fronts in this figure predicted by the distance model and variational inequality approach, respectively, the essential difference can be seen in the angle at which the flow fronts meet the (fixed) boundary. This is caused by the boundary conditions imposed at the fixed boundary.

Rectangular plates with varying thickness

The geometries for the next two examples (cf. Figures 6.10, 6.11, 6.12) come from the monograph [TAGO79]. To separate the influence of variable thickness on the distance model and variational inequality model, respectively, we have chosen two simple plates.

The first plate is 31 *mm* wide and 123 *mm* long. It is characterised by a constant thickness $d1 = 3.35$ *mm* and a thickness function $d2$ varying linearly in the x_1-direction from 1.8 *mm* to 1.68 *mm* (see Figure 6.10 (top)). The second plate contains subdomains of different thickness with a ratio $d1 : d2 = 2 : 3.5$ (see Figure 6.10 (bottom)).

In Figure 6.11 we can see the results from the distance model (flow fronts, weld lines) and the variational inequality approach (flow fronts and pressure distribution) for the first example. Moreover, in part (d) of this figure we have inserted some results from experiments and simulations taken from [TAGO79]. Solid lines represent short shots and 'x' stands for numerical simulation results based on an FEM formulation. Let us mention that the short shots show already at the beginning a tendency to advance further at the upper side than the lower. Perhaps, this is an indicator for some asymmetry in the cavity or in the operating conditions. The FEM-computations in part (d) of this figure were done on a relatively coarse grid (note the scattered distribution of points 'x' describing the flow fronts).
The comparison of the flow fronts predicted by the distance and variational inequality model (see (a) and (c)) shows a very good correspondence. We refer to the example in Figure 6.9 concerning an explanation for the slight difference in the angles at which the flow fronts meet the walls. Moreover, the agreement to the experiments (short shots) is very acceptable, taking into account the apparent asymmetry in the short shots as addressed above.

Summarizing the results presented in Figure 6.11, we emphasize that the rheology model used has only little effect on the filling pattern for cavities with a thickness varying mainly in the flow direction (here x_1-direction). Remembering the expression for the flow conductivity k (cf. (6.13)) this behaviour observed in Figure 6.11 is, not surprisingly, from the mathematical point of view.

The geometry considered in Figure 6.12 is again a plate with variable thickness. But now, compared to the geometry from Figure 6.11, the variation in thickness does not correspond to the main flow direction.
For purposes of comparison our results from the distance and variational inequality approach, Figure 6.12(d) contains short shots (solid lines) and simulation results of other finite element computations taken from [TAGO79]. Further experiments (short shots) and simulation results for this geometry can be found in [KRTA80] and [SYRJ91] (see also [JOHA97], Fig. 20).

Returning to the comparison of the distance and the variational inequality model illustrated in Figure 6.12 (see (a) and (c)), we observe that the distance model predicts a larger difference between the flow velocities in the thinner and thicker region than the variational inequality approach. For an explanation of this behaviour we refer to the next example in Figure 6.13.

Finally, we mention that the results obtained with the variational inequality model are in good agreement with the experimental results shown in Figure 6.12(d) and also finite element simulation results obtained by Syrjäla in [SYRJ91] (see also [JOHA97], Fig. 20).

As a further example we consider a simple rectangular plate with variable thickness, see Figure 6.13 (top). The plate is 60 mm wide and 100 mm long and contains two subdomains with different thickness. The thickness of the (dotted) smaller rectangle is one half of the thickness in the rest.

The filling zones according to the distance model are shown in Figure 6.13 (middle), whereas the filling zones predicted by the variational inequality approach are depicted in the lower part of this figure.
As already indicated in the previous example in Figure 6.12, the flow patterns show that the distance model predicts larger differences in the flow behaviour in the thinner and thicker region than do the variational inequality computations. In other words, the distance model tends to predict larger amounts of enclosed air for varying thickness than the variational inequality or other finite element/volume computations. This is also confirmed by comparisons between the distance model and the variational inequality approach performed for the cover plate (see Section 6.3.1, Figure 6.6 and [BERN96], Section 7.5). The distance model predicts an air entrapment already for a thickness ratio of 2, whereas the finite volume computations show entrapped air in the case of a thickness ratio of 3.

Looking for an explanation of the just mentioned behaviour, we recall that the distance model is based on 'small values' of n in the power-law model (6.6) (asymptotic solution for $n \to 0$). Combining the power-law model (6.6) with the flow conductivity k defined in (6.13), the latter exhibits bigger differences between thicker and thinner regions for small values of n than for n near 1. Therefore, the difference between distance and finite element/volume models depends also on the viscosity model used in a finite element/volume computation.

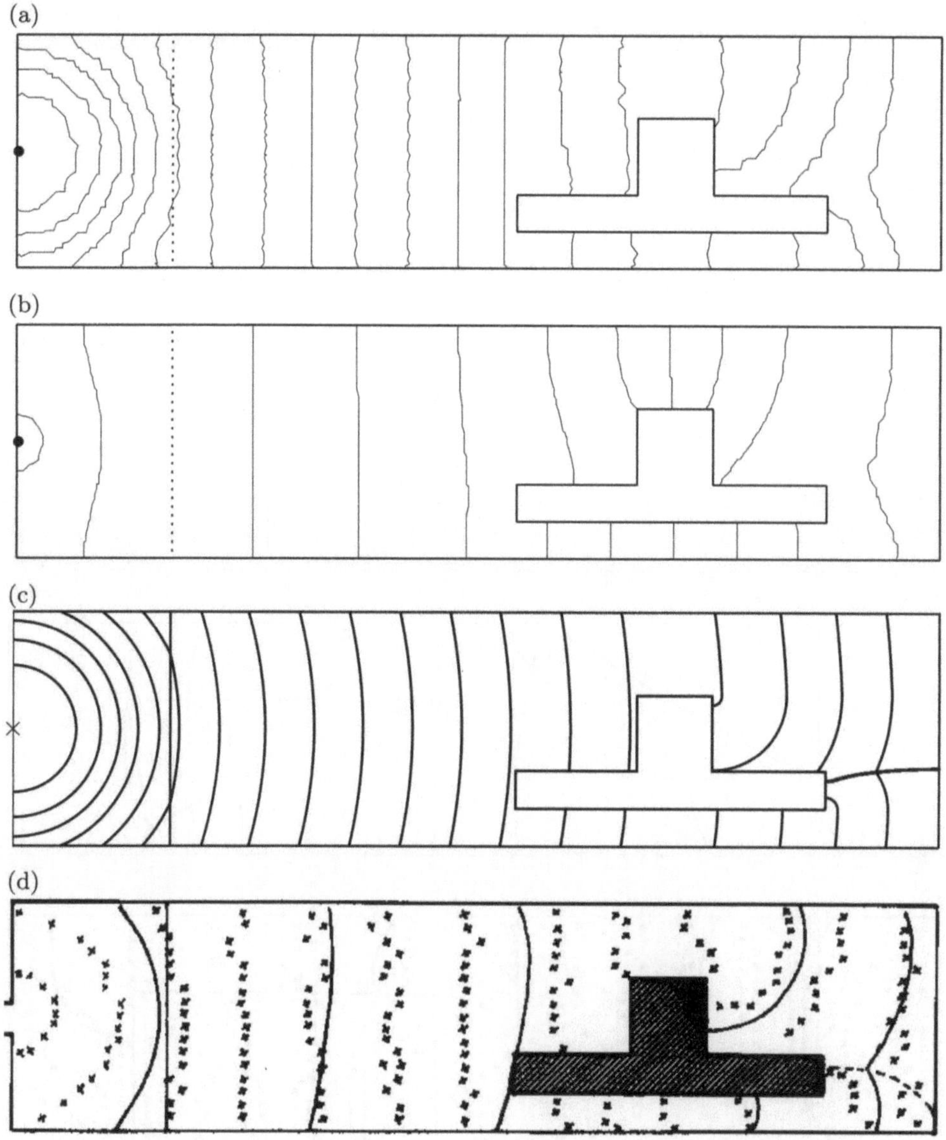

Figure 6.11: *Plate-like geometry I with non-uniform thickness – Comparison of distance, variational inequality (VI) approach and experiments. (a) Flow fronts according to VI model. (b) Predicted pressure distribution near completion of filling according to VI model. (c) Flow fronts and weld lines according to distance model. (d) from [TAGO79]: solid lines as short shots; 'x' stands for numerical solution.*

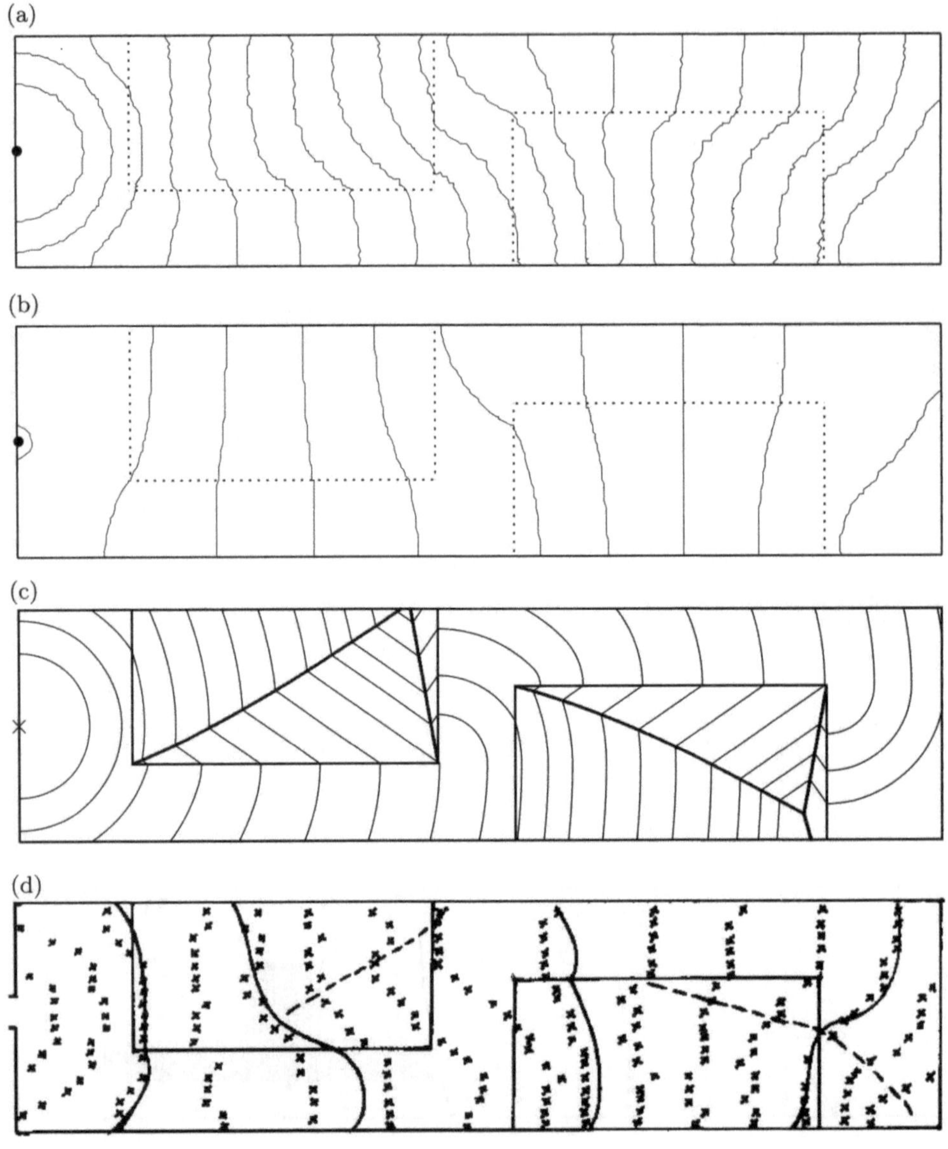

Figure 6.12: *Plate-like geometry II with non-uniform thickness – Comparison of distance, variational inequality (VI) approach and experiments. (a) Flow fronts according to VI model. (b) Predicted pressure distribution near completion of filling according to VI model. (c) Flow fronts and weld lines according to distance model. (d) from [TAGO79]: solid lines as short shots; 'x' stands for numerical solution.*

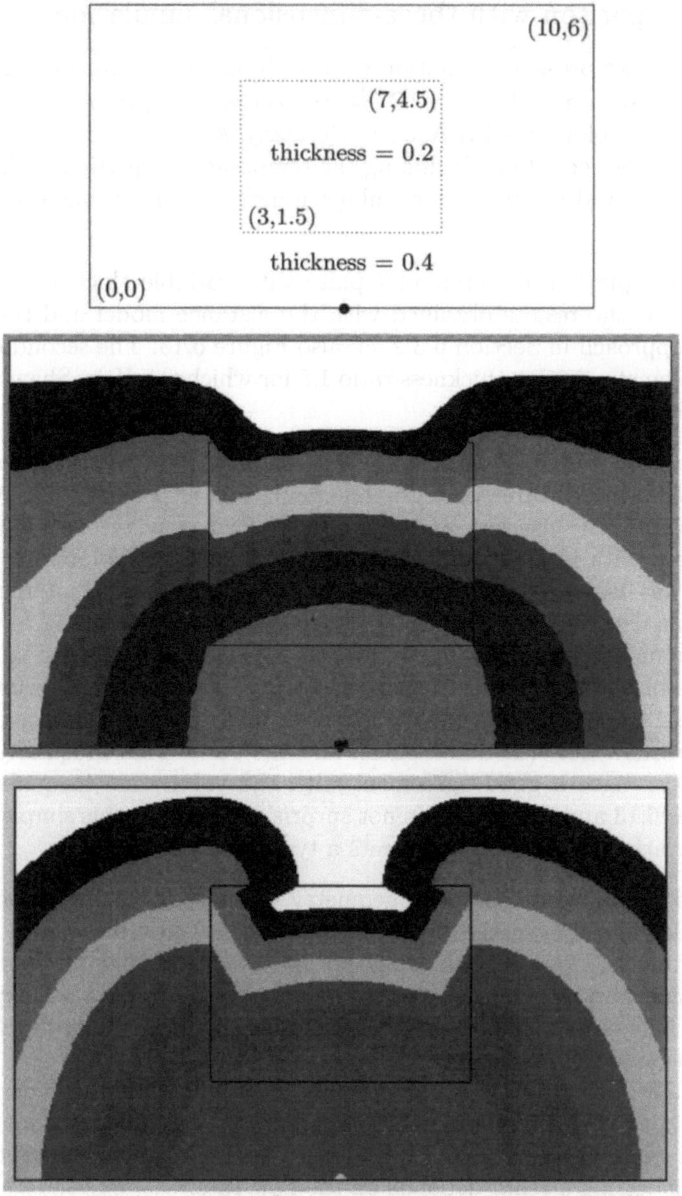

Figure 6.13: *Rectangular plate with non-uniform thickness – Comparison of distance and variational inequality approach. Gate region represented by •. Filling zones for seven time instants. Top: Geometry. Middle: Filling zones according to variational inequality model. Bottom: Filling zones according to the distance model.*

6.3.3 Comparison with three-dimensional simulations

In this section we present simulation results obtained by means of the 3-D mould filling analysis software SIGMASOFT® (cf. Section 6.2.6) from the company Magma Gießereitechnologie GmbH Aachen, Germany. As described in Section 6.2.6 the full Navier–Stokes equations including the temperature equation with viscous dissipation were solved in the a priori unknown melt domain (using a VOF method).

The first example is the rectangular plate with variable thickness for which we have discussed the results obtained with the distance model and the variational inequality approach in Section 6.3.2, cf. also Figure 6.13. The second example is a cascaded geometry with a thickness ratio 1:5 for which the Hele–Shaw assumptions are not fulfilled.

Using symmetry, only one-half of the rectangular plate with variable thickness was considered in the three-dimensional simulation, cf. Figure 6.14. We have chosen a Carreau–WLF viscosity model (cf. (6.7)) for a polystyrene 144 according to the measured data in [PGLA95], Chapter 10. A constant volumetric flow rate of 6.2 cm^3/s was assumed. In Figure 6.14 the filling stage and the temperature distribution are depicted for three time instants. The corresponding filling patterns should be compared with the fourth, fifth and sixth filling zone in Figure 6.13 (filling degrees of about 57.14%, 71.43%, 85.71%). The simulation results presented in Figure 6.14 are qualitatively the same as those obtained by means of generalized Hele–Shaw flow (distance model and the variational inequality approach). Especially, we have a good agreement with the variational inequality approach (cf. Figures 6.13 and 6.14). This is not surprising, because this simple rectangular plate with variable thickness represents a typical 2.5-D geometry.

Our second example is a cascaded geometry which is $L = 96\ mm$ long and $W = 20\ mm$ wide. The thin section is $L_1 = 38\ mm$ long and $4\ mm$ thick, whereas the thicker section is $L_2 = 58\ mm$ long and $20\ mm$ thick. The film-gate is located at the thinner part. Therefore, we have under consideration a 1:5 expansion flow problem. The length L_1 is chosen to ensure that the flow is fully developed before entering the expansion region.
Such a geometrical situation does not allow the application of a generalized Hele–Shaw flow in an accurate manner. The assumption (6.8) does not hold. Moreover, the sudden expansion in the cavity leads to a jetting phenomenon. Obviously, the shape of the jetting stream depends on the flow rate.

The part was moulded from a polycarbonate (Makrolen 2800, cf. [ZACH98]). Again, the viscosity of the material is described by a Carreau–WLF model. A constant volumetric flow rate is such that the total filling time is about 1.8 s.

In the PhD thesis [ZACH98] by Zachert (see also [IKV98], Chapter 13) a comparison for such simple geometries can be found between different software tools based on three-dimensional finite element/volume models, the 2.5-D software Cadmould (see also Section 6.2.1.3) and experimental results. The comparison between the 3-D software tools seems to us only of limited significance, because the flow algorithms on which the corresponding software tools are based are only partially appropriate for non-Newtonian flow of plastic melts.

Figure 6.15 shows the calculated location of the flow front and temperature distribution at selected times during the filling (in the cross-section through the part). More precisely, because of symmetry with respect to the width $W = 20\ mm$ (not with respect to the thickness, height), only one-half of the part is depicted (i.e., from $W = 0$ to $W = 10\ mm$). The upper part of this figure shows the filling pattern when the melt enters the expansion region. Afterwards, the melt does not immediately touch the lower and upper part of the mould as assumed in a Hele–Shaw flow (see the next two flow front locations in Figure 6.15). The lower part of this figure illustrates the flow front location at a filling degree of about 90%.

Finally, we mention that the simulation results shown in Figure 6.15 are in good agreement with experimental results presented in [ZACH98].

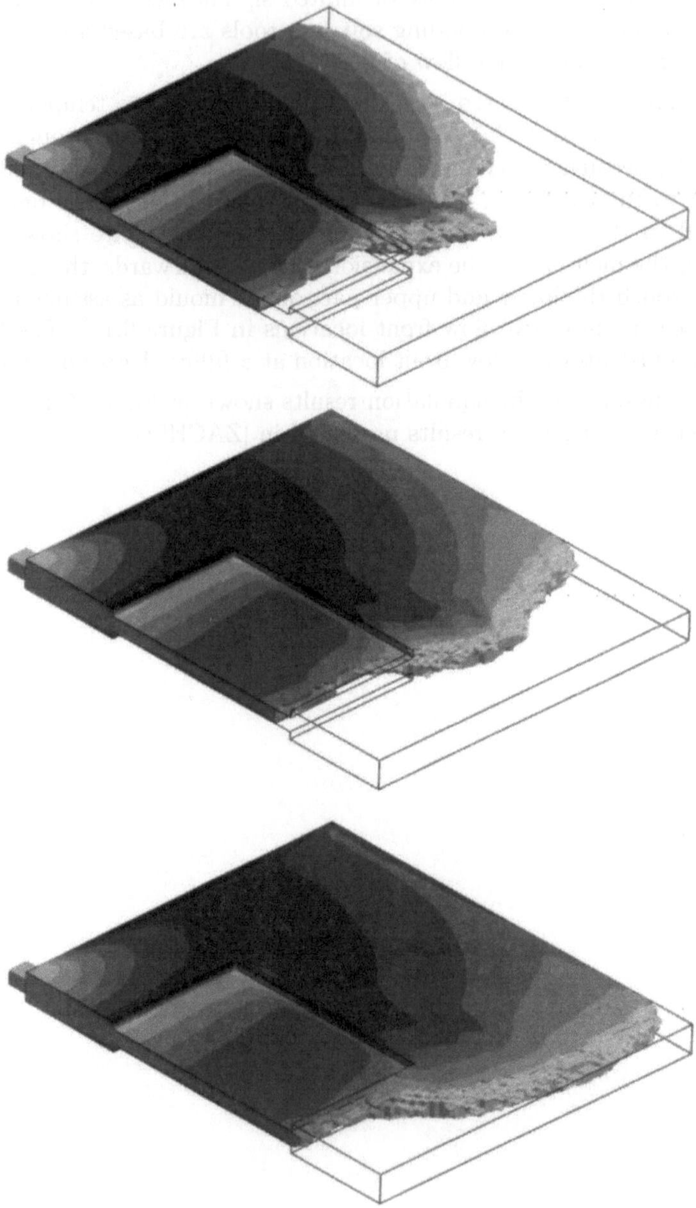

Figure 6.14: *Rectangular plate with non-uniform thickness – Three-dimensional simulation with mould filling analysis software SIGMASOFT®. Filling zones for three time instants (cf. with Figure 6.13).*

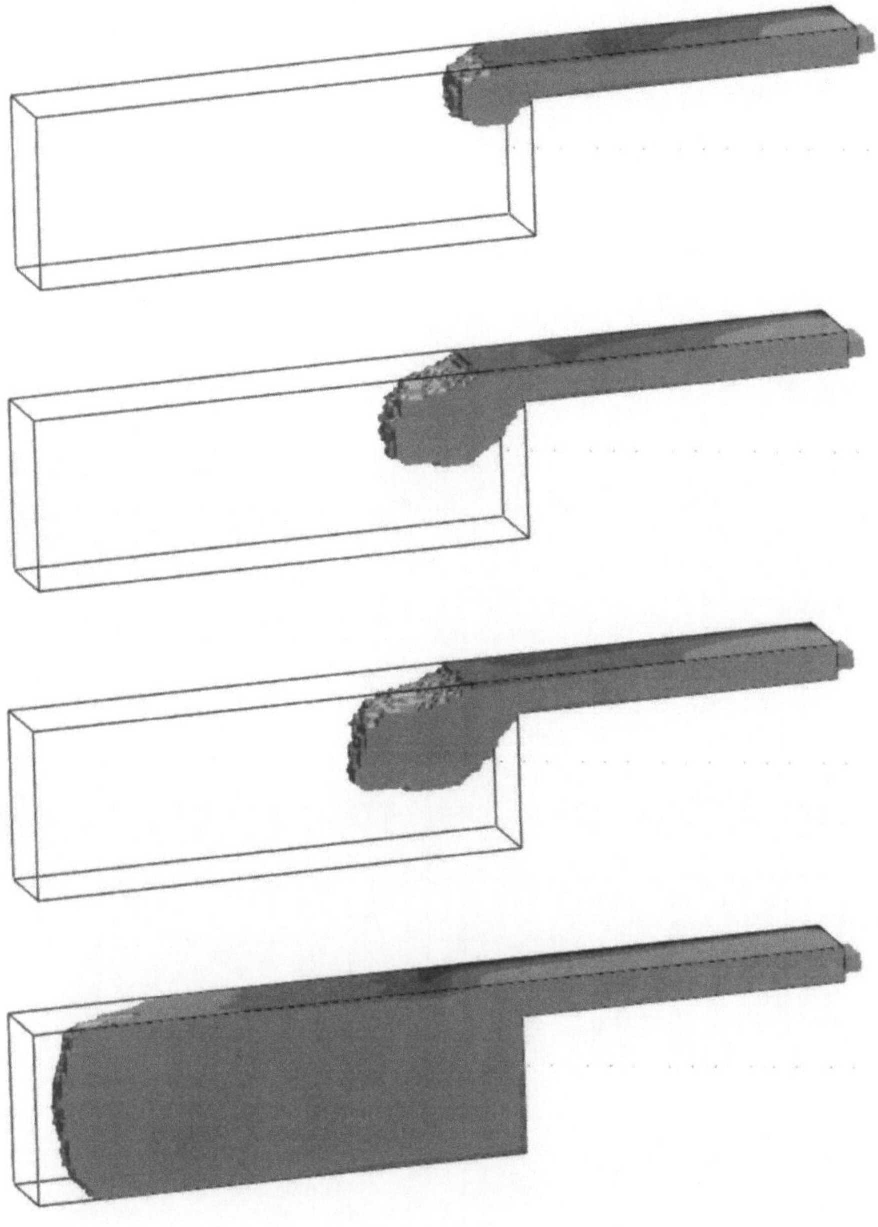

Figure 6.15: *Geometry with thickness ratio 1:5 (expansion flow problem) – Three-dimensional simulation with mould filling analysis software SIGMASOFT® . Filling zones for four time instants in a cross section through the part.*

International Series of Numerical Mathematics
Vol. 136, © 2002 Birkhäuser Verlag Basel/Switzerland

Chapter 7

Concluding Remarks

Let us conclude this book with a short discussion of the main results and some recommendations for further study.

In this work we have proposed an evolutionary variational inequality approach to a degenerate moving free boundary problem. In the mathematical analysis and the numerical treatment of this variational inequality formulation we have focused our attention on the following points.

- Mathematical modelling of application problems (injection and compression moulding, electro-chemical machining process).
- Derivation and comparison of different fixed domain formulations (weak and variational inequality formulations) for a general degenerate moving free boundary problem.
- Mathematical analysis of the evolutionary variational inequality formulation (existence, uniqueness and regularity results, statements with respect to the time evolution, maximum principle).
- Numerical treatment of elliptic and evolutionary variational inequalities (including mixed boundary conditions, two and three space dimensions) by means of finite element and finite volume approximations.
  The framework for our numerical study of these inequality problems includes among others the following topics.
  - Solvability and stability of the discrete problems.
  - Comparison of finite element and finite volume approximations.
  - Error estimates (convergence rate) for the finite element and finite volume solutions.
  - Discrete maximum principle as an important reason for the consideration of finite volume approximations.

  - Substitution of the finite volume inequality problems by penalization (regularization) equations. Iterative solution of the penalization problems.
  - Reducing the overall error between inequality and penalization solutions by means of an appropriate coupling of discretization and penalization parameters.
  - Numerical tests for model problems to confirm the theoretically derived convergence rates.
- Numerical simulation of the injection and compression moulding process. Discussion of the influence of basic geometrical and operating conditions. Comparison between different mathematical models.

Summarizing, our investigations cover mathematical modelling, mathematical and numerical analysis through to numerical simulation.

The main aim of Chapter 2 was the derivation of an evolutionary variational inequality as a fixed domain formulation for a general moving free boundary problem. We have discussed the connections to other fixed domain formulations and the advantages of the inequality formulation compared to the classical free boundary problem. We stress that the general advantages of fixed domain formulations with respect to the mathematical analysis and, in particular, to the numerical treatment of the underlying application problems are the motivation for derivation of the inequality formulation.

Mathematical-analytical aspects of the evolutionary inequality formulation were studied in Chapter 3. Special attention was given to treatment of the memory term, the time-dependent convex sets and different types of boundary conditions. We have proved the existence of a unique solution of the evolutionary inequality as a continuous mapping from the time interval $[0,T]$ to the Sobolev space $H^1(\Omega)$. Depending on the smoothness of the data we have shown that the evolutionary variational inequality solution $u = u(x,t)$ satifies the regularity

$$u \in \left(W^1_\infty(0,T;H^1(\Omega)) \cap L_\infty(0,T;W^2_p(\Omega))\right),$$
$$u(t) \in W^2_p(\Omega) \;\; \forall t \in [0,T], \quad 2 \le p < \infty.$$

Furthermore, the solution u fulfills a complementarity problem and Lewy–Stampacchia type inequalities. From the mathematical point of view, these properties are of interest for their own sake. On the other hand, this regularity of the solution and the complementarity problem were used in the numerical error analysis of the finite element and finite volume approximations.

Moreover, we have derived statements about the time evolution of the solution which show the well-posedness of the inequality formulation and reflect the physical behaviour of the underlying application problems in a correct way.

Based on a maximum principle we have discussed lower and, in particular, upper bounds for the inequality solution and its time-derivative. Finally, exploiting embedding theorems we have obtained that the solution is continuous (or even Lipschitz continuous) with respect to the space-time domain $\overline{\Omega} \times [0, T]$.

The derived mathematical results for this type of evolutionary inequalities are in qualitative agreement with known statements for parabolic variational inequalities related to one- and two-phase Stefan problems (see e.g., [RODR94]) as well as for the one-parameter (the time) family of elliptic inequalities asscociated with the classical Hele–Shaw problem (see [RODR87], Section 9.5 and [LORO86]).

In Chapter 4 we have applied the finite volume method to the numerical solution of interior and boundary obstacle problems of elliptic type (with mixed boundary conditions; two and three space dimensions). Depending on the choice of boxes, we have proved first and second order convergence between the finite volume and the finite element solution. Besides this, the convergence rate $O(h)$ (h as discretization parameter) was established for the finite volume solution in the energy norm.
Based on a discrete maximum principle we have discussed penalization (regularization) methods for solution of the finite volume inequalities. In particular, the penalization error is estimated in the energy and maximum norm. As a consequence, we have obtained a first order convergence statement for the total error by coupling discretization and penalization parameters. Furthermore, iterative methods for solution of the penalization problems have been proposed.
Summarizing, on the one hand, Chapter 4 contains new results especially for the convergence theory of the finite volume method applied to elliptic variational inequalities. On the other hand, it provides basic results for the numerical treatment of the evolutionary inequalities in Chapter 5.

The starting point for the just mentioned numerical analysis of the evolutionary inequalities in Chapter 5 consists in the discussion of an efficient and accurate discretization of the memory term. Our proposal does *not* require storing the (solution-) vectors of preceding time instants. The computational effort for the treatment of the memory term comprises only two matrix-vector multiplications at each time instant. These are two essential features for an efficient implementation.

The results of the numerical analysis for evolutionary inequalities can be summarized as follows. Convergence of first and second order has been proved between the finite element and finite volume solution depending on the choice of boxes. Based on the regularity of the variational inequality solution established in Chapter 3 and suitable a priori estimates, we have derived the convergence rate

$$O(h) + O(\tau)$$

in the norm of the space $L_\infty(0, T; H^1(\Omega))$ both for the finite element and finite volume solution. Here, h and τ denote the spatial and time discretization parameters, respectively.

Similar to Chapter 4, penalization and regularization methods were developed to solve the (discrete) evolutionary inequalities. Exploiting again a discrete maximum principle we have estimated the discrepancy between the finite volume (inequality) solution and the solutions of the penalization problems. Therefore, we were able to establish first order convergence for the overall error by coupling the discretization parameters (in space and time) with the penalization parameters.
Furthermore, we have discussed computational aspects of the proposed penalization problems and their iterative solution. Especially, in connection with injection and compression moulding as application problems, we use an extrapolation, which reduces the number of the necessary penalty-iterations (in general 2–5 at each time instant) to obtain the solution of the (discrete) penalization problem.

All the convergence rates theoretically predicted in Chapters 4 and 5 were confirmed by numerical experiments for test problems.

Finally, in Chapter 6 an overview was given of the mathematical modelling of injection and compression moulding by means of generalized Hele–Shaw flow and Navier–Stokes flow, including thermal and possible non-Newtonian effects. Mainly, we have discussed a mathematical model, including its numerical core problems, which is based on the considered evolutionary variational inequality. Simulation results both for simple geometries and for real plastic products demonstrate that the evolutionary variational inequality formulation as a fixed domain formulation can be successfully used for simulation of the flow process. With the help of numerical experiments we have pointed out and interpreted differences among various mathematical models.

Because of the wide range of problems considered in this work we could not give exhaustive answers and solutions to all problems. There are a lot of possible further research activities. Let us discuss only a few of them.

Due to the memory term (occuring also in the Neumann boundary conditions), the existence of a solution for the semi-coercive evolutionary inequality (cf. Remark 3.5[i]) is a more complicated problem than for semi-coercive elliptic inequalities (see e.g., [RODR87], Section 8.6 for the Signorini problem). However, it seems to be possible to find conditions which guarantee the existence of the solution in the semi-coercive case.
Moreover, other variational inequalities containing such a type of memory term should be investigated (e.g., with special nonlinearities in the differential operators A and B). In this context let us also mention that variational inequalities containing a different type of memory term (not resulting from a Baiocchi transformation) arise in application problems (viscoelasticity; for an introduction see [DULI76]).

In this work we have restricted ourselves to the numerical analysis and the use of penalization (regularization) methods for solution of the *discrete* elliptic and evolutionary inequalities. However, as already indicated at the beginning of Section 4.4 and in Remark 4.14, there is a wide variety of techniques for the numerical solution of such problems. In this context, we want to mention the modern way for the construction of fast solvers (adaptive multilevel algorithms) for nonlinear variational problems (characterized by a quadratic energy functional and an additional convex functional) which has been pointed out by Kornhuber in [KORN97] (see also Remark 2.10). Both the elliptic inequalities discretized by the finite volume method in Chapter 4 and the (finite element, finite volume) evolutionary inequalities (for a fixed time instant) belong to this class of problems.

On the basis of the simulation results presented in Section 6.3 a more detailed study of thermal and non-Newtonian effects should be performed. More numerical simulations and systematic comparisons with experiments seem to be necessary to get a deeper understanding of the *quantitative* influence and the complex *interaction* of these effects. This concerns both Hele–Shaw type flows (see Section 6.2.4) and 3-D Navier–Stokes flows (see Section 6.2.4). It should be emphasized that plastic melts and elastomers exhibit a complex rheological behaviour. Usually, the shear viscosity used in simulations described in Section 6.2.3 is experimentally measured by means of rheometers, assuming a steady laminar flow in a circular tube or capillary (Hagen–Poiseulle; steady simple shear experiment, see [TUCK89], Section 2.4.3). This is only an approach to the rheological behaviour occuring in complex flow situations of industrially manufactured parts.

Moreover, in the case of fibre suspensions, a coupled solution of the flow and the fibre orientation requires a modified constitutive relation between the viscous stress and shear rate tensor (a viscosity model with anisotropic properties caused by the fibre orientation). The flow kinematics can be deeply affected by the fibre orientation (see e.g., [ADVA94]).

From the practical point of view, not only the filling simulation is of interest, but also subsequent processes such as the packing phase and, in particular, a correct prediction of the shrinkage and warpage behaviour (mechanical properties of the part).

Finally, we mention the challenging task to develop accurate mathematical models and simulation software for other injection moulding techniques. As an example we mention the two- (or multi-) component injection moulding, i.e., for the case when the cavity is filled with more than one polymer. For instance, one uses recycled polymers as core component and 'new' polymers as skin component to get a suitable surface of the plastic product. Moreover, other application fields of plastic products require the use of gas as core component. From the practical point of view, the penetration behaviour of the core component into the skin component is essential for the application of these manufacturing techniques.

Bibliography

[ADGR95] S. Adam and C. Grossmann, Comparison principles applied to obstacle problems with penalties, *Numer. Func. Anal. and Optimiz.*, **16** (1995), 9&10, 1097–1110.

[ADAM75] R.A. Adams, *Sobolev spaces*, Academic Press, New York, 1975.

[ADVA94] S.G. Advani (ed.), *Flow and rheology in polymer composites manufacturing*, Elsevier, Amsterdam, 1994.

[ADTU87] S.G. Advani and C.L. Tucker III, The use of tensors to describe and predict fiber orientation in short fiber composites, *Journ. of Rheology*, **31** (1987), 8, 751–784.

[AASC91] J.F. Agassant, P. Avenas, J.Ph. Sergent and P.J. Carreau, *Polymer processing*, Hanser Publishers, Munich, Vienna, New York, 1991.

[ALTA90] M.C. Altan, A review of fiber-reinforced injection molding: flow kinematics and particle orientation, *Journ. of Thermoplastics Composite Materials*, **3** (1990), 275–313.

[ALLU83] H.W. Alt and St. Luckhaus, Quasilinear elliptic-parabolic differential equations, *Math. Zeitschrift*, **183** (1983), 311–341.

[ANG91] L. Angermann, Numerical solution of second order elliptic equations on plane domains, *R.A.I.R.O. Modél. Math. Numér. (M^2AN)*, **25** (1991), 2, 169–191.

[ANG95] L. Angermann, Error estimates for the finite element solution of an elliptic singularly perturbed problem, *IMA Journ. Numer. Anal.*, **15** (1995), 161–196.

[ARON95] G. Aronsson, *On some properties of a distance function and a connection to injection moulding*, Preprint LiTH-MAT-R-95-03, University of Linköping, 1995.

[ARON96] G. Aronsson, On p-harmonic functions, convex duality and an asymptotic formula for injection mould filling, *Euro. J. of Applied Mathematics*, **7** (1996), 417–437.

[ARJA92] G. Aronsson and U. Janfalk, On Hele–Shaw flows of power-law fluids, *Euro. J. of Applied Mathematics*, **3** (1992), 343–366.

[ATMR99] I. Athanasopoulos, G. Makrakis and J.-F. Rodrigues (eds.), *Free Boundary Problems: Applications and Theory*, Chapman & Hall/CRC Research Notes in Mathematics Series, Vol. 409, Boca Raton, Chapman & Hall/CRC, 1999.

[BATA81] K. Baba and M. Tabata, On a conservative finite element scheme for convective diffusion equations, *RAIRO Anal. Numer.*, **15** (1981), 3–25.

[BAIO77] C. Baiocchi, Estimations d' erreur dans L_∞ pour les inequations a obstacle, in *Mathematical Aspects of Finite Element Methods*, I. Galligani, E. Magenes (eds.), Lecture Notes in Math., Vol. 606, Springer, Berlin, Heidelberg, New York, 1977, 27–34.

[BACA84] C. Baiocchi and A. Capelo, *Variational and quasivariational inequalities – Applications to free boundary problems*, John Wiley & Sons, Chichester, 1984.

[BACO95] D.G. Baird and D.I. Collias, *Polymer processing, principles and design*, Butterworth-Heinemann, Newton, 1995.

[BARO87] R.E. Bank and J.R. Rose, Some error estimates for the box method, *SIAM J. Num. Anal.*, **24** (1987), 4, 777–787.

[BMYM98] A.K. Bashin, J.J. Moore, K.P. Young and S. Midson (eds.), *Semi-solid processing of alloys and composites*, Proc. of 5th conf. on semi-solid processing of alloys and composites, Colorado School of Mines, Golden, Colorado, 1998.

[BAUE99] H.J. Bauer, *Objektorientierter Entwurf und Implementierung multilevel-adaptiver Finite Element Methoden*, PhD Thesis, Technical University Munich, 1999.

[BBT95] G. Bayada, M. Boukrouche and M. El-A. Talibi, The transient lubrication problem as a generalized Hele–Shaw type problem, *Journ. for Anal. and its Appl.*, **14** (1995), 59–87.

[BEGI86] H. Begehr and R.P. Gilbert, Hele–Shaw type flow in $\mathbb{R}^n$, *Nonlinear Analysis, Theory, Methods & Appl.*, **10** (1986), 1, 65–85.

[BEGI87] H. Begehr and R.P. Gilbert, Non-Newtonian Hele–Shaw type flow in $n \geq 2$ dimensions, *Nonlinear Analysis, Theory, Methods & Appl.*, **11** (1987), 1, 17–47.

[BEVH99] F. Benkhaldoun, D. Hänel and R. Vilsmeier (eds.), *Finite Volumes for Complex Applications II*, Proceedings of the second international symposium on Finite Volumes for Complex Applications – Problems and Perspectives, Duisburg, Germany, 1999, Hermes, 1999.

[BENS82] A. Bensoussan, *Stochastic control by functional analysis methods*, North-Holland Publishing Company, Amsterdam, New York, Oxford, 1982.

[BELI82] A. Bensoussan and J.L. Lions, *Applications of variational inequalities in stochastic control*, North-Holland Publishing Company, Amsterdam, New York, Oxford, 1982.

[BELI84] A. Bensoussan and J.L. Lions, *Impulse control and quasi-variational inequalities*, Bordas, Gauthier-Villars, Paris, 1984.

[BEBR79] A.E. Berger, H. Brezis and J.C.W. Rogers, A numerical method for solving the problem $u_t - \triangle f(u) = 0$, *R.A.I.R.O. Analyse Numérique*, **13** (1979), 297–312.

[BERG98] A. Bergwall, *A geometric evolution problem arising in an asymptotic approach to compression moulding*, PhD Thesis, Linköping Studies in Science and Technology, Thesis 693, University of Linköping, 1998.

[BEEP92] M. Bern and D. Eppstein, Mesh generation and optimal triangulation, in *Computing in Euclidean Geometry*, D. Z. Du and F. H. Hwang (eds.), World Scientific, Singapore, New Jersey, London, Hongkong, 1992, 23–90.

[BERN83] E.C. Bernhardt (ed.), *Computer aided engineering for injection molding*, Hanser Publishers, Munich, Vienna, New York, 1983.

[BERN96] F. Berntsson, *Simulation tools for injection moulding*, Diploma Thesis, Report LiTH-MAT-EX-96-12, University of Linköping, 1996.

[BDF85] A. Bossavit, A. Damlamian and M. Frémond (eds.), *Free Boundary Problems: Applications and Theory*, Pitman Research Notes in Mathematics Series, Vol. 120, 121, 1983.

[BRAE92] D. Braess, *Finite Elemente*, Springer, Berlin, Heidelberg, New York, 1992.

[BRSC96] S.C. Brenner and L.R. Scott, *The mathematical theory of finite element methods*, second ed., Springer, Berlin, Heidelberg, New York, 1996.

[BREZ73] H. Brézis, *Opérators maximaux monotones et semi-groupes de contraction dans les espaces de Hilbert*, North-Holland Publishing Company, Amsterdam, New York, Oxford, 1973.

[BRSI71] H. Brézis and M. Sibony, Equivalence de deux inéquations variationelles et applications, *Arch. Rational Mech. Anal.*, **41** (1971), 254–265.

[BRFO91] F. Brezzi and M. Fortin, *Mixed and hybrid finite element methods*, Springer, Berlin, Heidelberg, New York, 1991.

[BHR77] F. Brezzi, W.W. Hager and P.A. Raviart, Error estimates for the finite element solution of variational inequalities, Part I – Primal Theory, *Numer. Math.*, **28** (1977), 431–443.

[CAGI89] G. Caginalp, Stefan and Hele–Shaw type problems as asymptotic limits of phase field equations, *Physical Review A*, **39** (1989), 11, 5887–5896.

[CAGI91] G. Caginalp, Phase field models and sharp interface limits: some differences in subtle situations, *Rocky Mountains Journ. of Math.*, **21** (1991), 603–616.

[CAI91] Z. Cai, On the finite volume element method, *Numer. Math.*, **58** (1991), 713–735.

[CMMC91] Z. Cai, J. Mandel and S. McCormick, The finite volume element method for diffusion equations on general triangulations, *SIAM J. Numer. Anal.*, **28** (1991), 2, 392–402.

[CMC90] Z. Cai and S. McCormick, On the accuracy of the finite volume element method for diffusion equations on composite grids, *SIAM J. Numer. Anal.*, **27** (1990), 636–655.

[CANN84] J.R. Cannon, *The one-dimensional heat equation*, Addison-Wesley Publishing Company, Menlo Park, California, 1984.

[CARR68] P.J. Carreau, *Rheological equations from molecular network theory*, PhD Thesis, University of Wisconsin, Madison, 1968.

[CHRA93] J.M. Chadam and H. Rasmussen (eds.), *Emerging applications in free boundary problems; Free boundary problems involving solids; Free boundary problems in fluid flow with applications*, Pitman Research Notes in Mathematics Series, Vol. 280, 281, 282, 1993.

[CHHO96] Y.C. Chang, T.Y. Hou and S. Osher, A level set formulation of Eulerian interface capturing methods for incompressible flows, *J. Comput. Phys.*, **124** (1996), 449–446.

[CHIP84] M. Chipot, *Variational inequalities and flow in porous media*, Springer, Berlin, Heidelberg, New York, 1984.

[CIAR78] P.G. Ciarlet, *The Finite Element Method for elliptic problems*, North-Holland Publishing Company, Amsterdam, New York, Oxford, 1978.

[CIJA81] P. Cizek and V. Janovsky, Hele–Shaw flow model of the injection by a point source, *Proc. of the Royal Soc. of Edinburgh*, **91A** (1981), 147–159.

[CORT85] Ph. Cortey-Dumont, On finite element approximation in the L_∞-norm of variational inequalities, *Numer. Math.*, **47** (1985), 45–57.

[CPS92] R.W. Cottle, J.-S. Pang and R.E. Stone, *The linear complementarity problem*, Volume in Computer science and scientific computing, Academic Press, 1992.

[CRAN84] J. Crank, *Free and moving boundary problems*, Clarendon Press, Oxford, 1984.

[CRDW84] M.J. Crochet, A.R. Davis and K. Walters, *Numerical Simulation of non-Newtonian flow*, Elsevier, Amsterdam, 1994.

[CRCA95] M. Cross and J. Campbell (eds.), *Modeling of casting, welding and advanced solidification processes VII*, Proc. of the seventh conference in a series on modeling casting and welding processes, London, England, 1995, TMS publication, 1995.

[CROW79] A.B. Crowley, On the weak solution of moving boundary problems, *Journ. Inst. Math. Appl. (IMA)*, **24** (1979), 43–57.

[CUSS86] C. Cuvelier, A. Segal and A.A. van Steenhoven, *Finite element methods and Navier–Stokes equations*, D. Reidel Publishing Company, Dordrecht, Boston, Lancaster, Tokyo, 1986.

[DALI90] R. Dautray and J.L. Lions, *Mathematical analysis and numerical methods for science and technology. Volume 1: Physical origins and classical methods*, Springer, Berlin, Heidelberg, New York, 1990.

[DEMA93] I. Demirdžić and D. Martinović, Finite volume method for thermo-elasto-plastic stress analysis, *Computer methods in Applied Mechanics and Engineering*, **109** (1993), 331–349.

[DEMU94] I. Demirdžić and S. Muzaferija, Finite volume method for stress analysis in complex domains, *Intern. Journ. for Numerical Methods in Engineering*, **37** (1994), 3751–3766.

[DEMU95] I. Demirdžić and S. Muzaferija, Numerical method for coupled fluid flow, heat transfer and stress analysis using unstructured moving meshes with cells of arbitrary topology, *Comput. Methods Appl. Mech. Engrg.*, **125** (1995), 235–255.

[DHLV95] J.I. Diaz, M.A. Herrero, A. Linan and J.L. Vazquez (eds.), *Free Boundary Problems: Applications and Theory*, Pitman Research Notes in Mathematics Series, Vol. 323, 1995.

[DIFR84] E. Di Benedetto and A. Friedman, The ill-posed Hele–Shaw model and the Stefan problem for supercooled water, *Transactions of the American Math. Society*, **282** (1984), 1, 183–204.

[DIJK59] E.W. Dijkstra, A note on two problems in connexion with graphs, *Numer. Math. 1*, 1959, 269–271.

[DORU82] J. Douglas Jr. and T.F. Russell, Numerical methods for convection-dominated diffusion problems based on combining the method of characteristics with finite element or finite difference procedures, *SIAM J. Numer. Anal.*, **19** (1982), 871–885.

[DULI76] G. Duvaut and J.L. Lions, *Inequalities in mechanics and physics*, Springer, Berlin, Heidelberg, New York, 1976.

[EKTE76] I. Ekeland and R. Temam, *Convex analysis and variational problems*, North-Holland Publishing Company, Amsterdam, New York, Oxford, 1976.

[ELLI80] C.M. Elliott, On a variational inequality formulation of an electrochemical machining moving boundary problem and its approximation by the finite element method, *Journ. Inst. Math. Appl. (IMA)*, **25** (1980), 121–131.

[ELLI81] C.M. Elliott, On the finite element approximation of an elliptic variational inequality arising from an implicit time discretization of the Stefan problem, *IMA Journ. Numer. Anal.*, **1** (1981), 115–125.

[ELLI87] C.M. Elliott, Error analysis of the enthalpy method for the Stefan problem, *IMA Journ. Numer. Anal.*, **7** (1987), 61–71.

[ELJA81] C.M. Elliott and V. Janovsky, A variational inequality approach to Hele–Shaw flow with a moving boundary, *Proc. of the Royal Soc. of Edinburgh*, **88A** (1981), 93–107.

[ELJA83] C.M. Elliott and V. Janovsky, An error estimate for a finite element approximation of an elliptic variational inequality formulation of a Hele–Shaw moving boundary problem, *IMA J. of Num. Anal.*, **3** (1983), 1–9.

[ELOC82] C.M. Elliott and J. Ockendon, *Weak and variational methods for moving boundary problems*, Pitman, London, 1982.

[ENGE93] M. Engelmann, *Fidap 7.0 theory manual*, Fluid Dynamics International Inc. 1993.

[EFHKW93] B. Erdmann, M. Frei, R.H.W. Hoppe, R. Kornhuber and U. Wiest, Adaptive finite element methods for variational inequalities, *East-West Journ. of Num. Math.*, **1** (1993), 4, 165–197.

[EWING85] R.E. Ewing, Finite element methods for nonlinear flows in porous media, *Comp. Methods in Appl. Mech. and Engin.*, **51** (1985), 421–439.

[FALK74] R.S. Falk, Error estimates for the approximation of a class of variational inequalities, *Math. Comp.*, **28** (1974), 963–971.

[FAPR83] A. Fasano and M. Primicerio (eds.), *Free Boundary Problems: Theory and Applications*, Pitman Research Notes in Mathematics Series, Vol. 78, 79, 1983.

[FFLM97] M. Feistauer, J. Felcman and M. Lukáčová-Medvid'ová, On the convergence of a combined finite volume-finite element method for nonlinear convection-diffusion problems, *Numer. Meth. for Partial Diff. Equ.*, **13** (1997), 163–190.

[FEPE99] J.H. Ferziger and M. Perić, *Computational methods for fluid dynamics*, Springer, Berlin, Heidelberg, New York, 1999.

[FICH64] G. Fichera, Problemi elastostatici con vincoli unilaterali: il problema di Signorini con ambigue condizioni al contorno, *Atti Accad. Naz. Lincei Mem. Cl. Sci. Fis. Mat. Natur Sez. Ia*, **7(8)** (1963-1964), 91–140.

[FOMS93] M. Fortin, H. Manouzi and A. Soulaimani, On finite element approximation and stabilization methods for compressible viscous flows, *Int. J. Numer. Meth. Fluids*, **17** (1993), 477–499.

[FORT92] S. Fortune, Voronoi diagrams and Delauney triangulations, in *Computing in Euclidean Geometry*, D. Z. Du and F. H. Hwang (eds.), World Scientific, Singapore, New Jersey, London, Hongkong, 1992, 193–233.

[FRAN95] C. Franz, *Numerische Behandlung von nichtlinearen Konvektions-Diffusionsgleichungen*, Diploma Thesis, TU Munich, 1995.

[FRIE88] A. Friedman, *Variational principles and free boundary problems*, Krieger, Malabar, Florida, 1988.

[FRSP93] A. Friedman and J. Spruck, *Variational and free boundary problems*, Springer, Berlin, Heidelberg, New York, 1993.

[GGZ74] H. Gajewski, K. Gröger and K. Zacharias, *Nichtlineare Operatorgleichungen und Operatordifferentialgleichungen*, Akademie-Verlag, Berlin, 1974.

[GERT95] H. Vogel, *Gerthsen, Physik*, Springer, Berlin, Heidelberg, New York, 1995.

[GISH89] R.P. Gilbert and P. Shi, Anisotropic Hele–Shaw flows, *Math. Methods in Appl. Sciences*, **11** (1989), 417–429.

[GLOW84] R. Glowinski, *Numerical methods for nonlinear variational problems*, Springer, Berlin, Heidelberg, New York, 1984.

[GLT81] R. Glowinski, J.L. Lions and R. Tremolieres, *Numerical analysis of variational inequalities*, North-Holland Publishing Company, Amsterdam, New York, Oxford, 1981.

[GDN98] M. Griebel, T. Dornseifer and T. Neunhoeffer, *Numerical simulation in fluid dynamics, a practical introduction*, SIAM, Philadalphia, 1998.

[GRIS85] P. Grisvard, *Elliptic problems in nonsmooth domains*, Pitman Advanced Publishing Program, Boston, London, Melbourne, 1985.

[GRRO94] Ch. Großmann and H.-G. Roos, *Numerik partieller Differentialgleichungen*, Teubner Studienbücher Mathematik, B.G. Teubner, Stuttgart, 1994.

[GUST85] B. Gustafsson, Application of variational inequalities to a moving boundary problem for Hele–Shaw flows, *SIAM J. Math. Anal.*, **16** (1985), 279–300.

[HAAG98] G. Haagh, *Simulation of gas-assisted injection moulding*, PhD Thesis, TU Eindhoven, 1998.

[HACK89] W. Hackbusch, On first and second order box schemes, *Computing*, **41** (1989), 277–296.

[HACK92] W. Hackbusch, *Elliptic differential equations, theory and numerical treatment*, Springer, Berlin, Heidelberg, New York, 1992.

[HACK94] W. Hackbusch, *Iterative solution of large sparse systems of equations*, Springer, Berlin, Heidelberg, New York, 1994.

[HAMI83] W. Hackbusch and H.D. Mittelmann, On multigrid methods for variational inequalities, *Numer. Math*, **42** (1983), 65–76.

[HARW65] F.H. Harlow and J.E. Welch, Numerical calculation of time dependent viscous incompressible flow of fluid with free surface, *The Physics of Fluids*, **8** (1965), 2182–2189.

[HAWE85] K. H. Hartwig and W. Weinelt, Numerische Lösung linearer elliptischer Variationsungleichungen mittels FDM, *WSR Technical University Karl-Marx-Stadt (Chemnitz)*, 8/1985.

[HANE96] J. Haslinger and P. Neittaanmäki, *Finite element approximation for optimal shape, material and topology design*, John Wiley & Sons, Chichester, 1996.

[HEIN87] B. Heinrich, *Finite Difference Methods on irregular networks*, ISNM Vol. 82, Birkhäuser, Basel, Boston, Berlin, 1987.

[HGGS98] J.-F. Hétu, D.M. Gao, A. Garcia-Rejon and G. Salloum, 3D Finite element method for the simulation of the filling stage in injection moulding, *Polymer Engineering and Science*, **38** (1998), 2, 223–236.

[HISH80] C.A. Hieber and S.F. Shen, A finite element/finite difference simulation of the injection molding filling process, *Journ. of Non-Newtonian Fluid Mechanics*, **7** (1980), 1–32.

[HINI81] C.W. Hirt and B.D. Nichols, Volume of fluid (VOF) method for the dynamics of free boundaries, *J. Comput. Phys.*, **39** (1981), 201–221.

[HHNL88] I. Hlaváček, J. Haslinger, J. Nečas and J. Lovišek, *Solution of variational inequalities in mechanics*, Springer, Berlin, Heidelberg, New York, 1988.

[HOZO92] K.-H. Hoffmann and Jun Zou, Parallel algorithms of Schwarz variant for variational inequalities, *Numer. Func. Anal. and Optimiz.*, **13** (1992), 5&6, 449–462.

[HOZO96] K.-H. Hoffmann and Jun Zou, Parallel solution of variational inequality problems with nonlinear source terms, *IMA J. Numer. Anal.*, **16** (1996), 31–45.

[HOSP90] K.-H. Hoffmann and J. Sprekels (eds.), *Free Boundary Problems: Theory and Applications*, Pitman Research Notes in Mathematics Series, Vol. 185, 186, 1990.

[HHHO94] Yu.E. Hohlov, S.D. Howison, C. Huntingford, J.R. Ockendon and A.A. Lacey, A model for non-smooth free boundaries in Hele–Shaw flows, *Quart. J. Mech. Appl. Math*, **47** (1994), 107–128.

[HOHL96] Yu.E. Hohlov, *Lectures on Hele–Shaw flows and related moving boundary problems*, Lecture notes of a seminar course at the Freiberg University of Mining and Technology, 1996, 86 pp.

[HOWI92] S.D. Howison, Complex variable methods in Hele–Shaw moving boundary problems, *Euro. J. of Applied Mathematics*, **3** (1992), 209–224.

[HOPP87] R.H.W. Hoppe, Multigrid algorithms for variational inequalities, *SIAM J. Numer. Anal.*, **24** (1987), 5, 1046–1065.

[HOKO94] R.H.W. Hoppe and R. Kornhuber, Adaptive multilevel methods for obstacle problems, *SIAM J. Numer. Anal.*, **31** (1994), 2, 301–323.

[IKED83] T. Ikeda, *Maximum principle in finite element models for convection-diffusion phenomena*, North-Holland Mathematics Studies 76, Lecture Notes in Numerical and Applied Analysis, Vol. 4, North-Holland Publishing Company, Amsterdam, New York, Oxford, 1983.

[IKV98] Proceedings 19-tes Kunststofftechnisches Kolloquium (in German), Institut für Kunststoffverarbeitung (IKV) der RWTH Aachen, Aachen, 1998.

[IKV00] Proceedings 20-stes Kunststofftechnisches Kolloquium (in German), Institut für Kunststoffverarbeitung (IKV) der RWTH Aachen, Aachen, 2000.

[ISAY87] A.I. Isayev, *Injection and compression molding fundamentals*, Marcel Dekker, New York, 1987.

[JAKA91] W. Jäger and J. Kacur, Solution of porous medium type systems by linear approximation schemes, *Numer. Math*, **60**, (1991), 407–427.

[JANF96] U. Janfalk, Behaviour in the limit as $p \to \infty$ of minimizers of functionals involving p-Dirichlet integrals, *SIAM J. Math. Anal.*, **27** (1996), 341–360.

[JERO83] J.W. Jerome, *Approximation of nonlinear evolution systems*, Academic Press, 1983.

[JERO82] J.W. Jerome and M.E. Rose, Error estimates for the multidimensional two-phase Stefan problem, *Math. Comput.*, **39** (1982), 377–414.

[JOHA97] P. Johansson, *On a weighted distance model for injection moulding*, PhD Thesis, Linköping Studies in Science and Technology, Thesis 604, University of Linköping, 1997.

[KACU85] J. Kačur, *Method of Rothe in evolution equations*, Teubner-Texte zur Mathematik, Vol. 80, Leipzig, 1985.

[KERK96] Th. Kerkhoven, Piecewise linear Petrov-Galerkin error estimates for the box method, *SIAM J. Num. Anal.*, **33** (1996), 5, 1864–1884.

[KENN95] P. Kennedy, *Flow analysis of injection molds*, Hanser Publishers, Munich, Vienna, New York, 1995.

[KIKU81] N. Kikuchi, Convergence of a penalty-finite element approximation for an obstacle problem, *Numer. Math.*, **37** (1981), 105–120.

[KIOD88] N. Kikuchi and J.T. Oden, *Contact problems in elasticity: a study of variational inequalities and finite element methods*, SIAM Studies in Applied Mathematics, Philadelpia, 1988.

[KIST80] D. Kinderlehrer and G. Stampacchia, *An introduction to variational inequalities and their application*, Academic Press, New York, 1980.

[KLV96] J.R. King, A.A. Lacey and J.L. Vazquez, Persistence of corners in free boundaries in Hele–Shaw flow, *Euro. J. Applied Math.*, **6** (1995), 455–490.

[KORN94] R. Kornhuber, Monotone multigrid methods for elliptic variational inequalities I, *Numer. Math.*, **69** (1994), 167–184.

[KORN96] R. Kornhuber, Monotone multigrid methods for elliptic variational inequalities II, *Numer. Math.*, **72** (1996), 481–499.

[KORN97] R. Kornhuber, *Adaptive monotone multigrid methods for nonlinear variational problems*, B.G. Teubner, Stuttgart, 1997.

[KRNE90] M. Krízek and P. Neittaanmäki, *Finite element approximation of variational problems and applications*, Harlow, Essex, Longman Scientific & Technical, 1990.

[KROE96] D. Kroehner, *Numerical schemes for conservation laws*, John Wiley and Teubner, 1996.

[KRTA80] W.L. Krueger and Z. Gogos, Injection molding into a rectangular cavity with inserts, *Polymer Engineering and Science*, **20** (1980), 426–431.

[KNT94] Y. Kuznetsov, P. Neittaanmäki and P. Tarvainen, Block relaxation methods for algebraic obstacle problems with M-matrices, *East-West J. Numer. Math.*, **2** (1994), 75–90.

[LMGS87] A.A. Lacey, J.A. Mc Geough and M. Shillor, Linear instability of the electroforming process, *Journ. of Engin. Math.*, **21** (1987), 149–154.

[LSU68] O.A. Ladyzhenskaya, V.A. Solonnikov and N.N. Ural'ceva, *Linear and quasilinear equations of parabolic type*, Trans. Math. of the American Math. Soc., Providence, 1968.

[LNSZZ94] B. Lafaurie, C. Nardone, R. Scardovelli, S. Zaleski and G. Zanetti, Modelling merging and fragmentation in multiphase flows with SURFER, *J. Comput. Phys.*, **113** (1994), 134–147.

[LAMB93] H. Lamb, *Hydrodynamics*, Cambridge Univerity Press paperback edition, 1993.

[LAWS77] C.L. Lawson, Software for C^1-surface interpolation, in *Mathematical Software III*, J.R. Rice (ed.), Academic Press, New York, 1977, 161–194.

[LMW84] R.D. Lazarov, V.L. Makarov and W. Weinelt, On the convergence of difference schemes for the approximation of solutions $u \in W_2^m$ $(m \geq 0.5)$ of elliptic equations with mixed derivatives, *Numer. Math.*, **44** (1984), 223–232.

[LETU87] C.C. Lee and C.L. Tucker III, Flow and heat transfer in compression mold filling, *Journ. of Non-Newtonian Fluid Mechanics*, **24** (1987), 245–264.

[LFTU84] C.C. Lee, F. Folgar and C.L. Tucker III, Simulation of compression molding for fiber-reinforced thermosetting polymers, *Journ. of Engineering for Industry, Transact. of ASME*, **106** (1984), 114–125.

[LIPI96] M.D. Lipinski, *Mold filling simulation for casting processes*, PhD Thesis, RWTH Aachen, 1996.

[LISH00] M.D. Lipinski, J. Steinbach and E. Hepp, Novel numerical algorithm for semisolid processing of complex castings, in [SAHC00], 2000, 365–372.

[LORO86] B. Louro and J.F. Rodrigues, Remarks on the quasi-steady one-phase Stefan problem, *Proc. of the Royal Soc. of Edinburgh*, **102A** (1986), 263–275.

[LUC95] R.W. Lewis, A.S. Usmani and J.T. Cross, Efficient mould filling analysis by an explicit finite element method, *J. Numer. Meth. Fluids*, **20** (1995), 493–506.

[MACO89] C.W. Macasko, *RIM fundamentals of reaction injection molding*, Hanser Publishers, Munich, Vienna, New York, 1989.

[MANV87] E. Magenes, R.H. Nochetto and C. Verdi, Energy error estimates for a linear scheme to approximate nonlinear parabolic problems, *R.A.I.R.O. Modél. Math. Numér. (M^2AN)*, **21** (1987), 4, 655–678.

[MANZ87] L.T. Manzione (ed.), *Applications of computer aided engineering in injection molding*, Hanser Publishers, Munich, Vienna, New York, 1987.

[MAMU97] S.R. Mathur and J.Y. Murthy, A pressure-based method for unstructured meshes, *Numerical Heat Transfer, Part B*, **31** (1997), 195–215.

[MAY94] I. May, *Untersuchungen zur Existenz und Eindeutigkeit von analytischen Lösungen für Hele–Shaw Strömungen mit kompakten Quellen und Senken*, Diploma Thesis, Freiberg University of Mining and Technology, 1994.

[MCGE74] J.A. McGeough, *Principles of electrochemical machining*, Chapman and Hall, London, 1974.

[MCGR74] J.A. McGeough and H. Rasmussen, On the derivation of a quasi-stationary model in electrochemical machining, *Journ. Inst. Math. Appl. (IMA)*, **13** (1974), 13–21.

[MCTH93] W. McLean and V. Thomée, Numerical simulation of an evolution equation with a positive-type memory term, *Journ. Australian Math. Soc., Ser. B*, **35** (1993), 23–70.

[MEJA97] M. Medale and M. Jaeger, Numerical simulation of incompressible flow with moving interfaces, *Intern. Journ. Numer. Meth in Fluids*, **24** (1997), 615–638.

[MENG80] G. Menges, Eine einfache Methode zur Bestimmung des Fließfrontverlaufs beim Spritzgießen von Thermoplasten, *Plastverarbeiter*, **31** (11), 1980, 671–676.

[MEIR92] A.M. Meirmanov, *The Stefan problem*, De Gruyter, Berlin, 1992.

[MEIS90] M. Meisel, *Integralbilanzmethode für Randwertaufgaben mit Lösungen im Sobolevraum* W_2^{1+r}, Ph. D. Thesis, TU Karl-Marx-Stadt/Chemnitz, 1990.

[MEYE90] G.H. Meyer, On computing free boundaries which are not level sets, in *Free Boundary Problems: Theory and Applications*, K.-H. Hoffmann and J. Sprekels (eds.), Pitman Research Notes in Mathematics Series, Vol. 185, 186, 1990, 138–154.

[MEYE93] G.H. Meyer, Front tracking for the unstable Hele–Shaw and Muskat problem, in *Flows in porous media*, J. Douglas Jr. and U. Hornung (eds.), Proceedings of the Oberwolfach Conference 1992, ISNM Vol. 114, Birkhäuser, 1993, 129–139.

[MIPA91] J.S.B. Mitchell and C.H. Papadimitriou, The weighted region problem: Finding shortest paths through a weighted planar subdivision, *J. Assoc. Comput. Mach.*, **38** (1991), 18–73.

[MPS88] K.W. Morton, A. Priestley and E. Süli, Stability of the Lagrange-Galerkin method with non-exact integration, *R.A.I.R.O. Modél. Math. Numér.* (M^2AN), **22** (1988), 4, 625–653.

[MOSC69] U. Mosco, Convergence of convex sets and of solutions of variational inequalities, *Advances in Mathematics*, **3** (1969), 510–585.

[MOSC77] U. Mosco, Error estimates for some variational inequalities, in *Mathematical Aspects of Finite Element Methods*, I. Galligani, E. Magenes (eds.), Lecture Notes in Math., Vol. 606, Springer, Berlin, Heidelberg, New York, 1977, 224–236.

[NAUM84] J. Naumann, *Einführung in die Theorie parabolischer Variationsungleichungen*, Teubner-Texte zur Mathematik, Vol. 64, Leipzig, 1984.

[NIEZ82] M. Niezgodka, Some aspects of approximation of optimal control problems for systems governed by parabolic equations involving free boundaries, in *Numerical treatment of free boundary value problems*, J. Albrecht, L. Collatz, K.-H. Hoffmann (eds.), ISNM Vol. 58, Birkhäuser, 1982, 211–223.

[NIST96] M. Niezgodka and P. Strzelecki (eds.), *Free Boundary Problems: Applications and Theory*, Pitman Research Notes in Mathematics Series, Vol. 363, 1996.

[NISH77] T. Nishida, A note on a theorem of Nirenberg, *J. Diff. Geom*, **12** (1977), 629–633.

[NIT77] J. Nitsche, L_∞-convergence of finite element approximation, in *Mathematical Aspects of Finite Element Methods*, I. Galligani, E. Magenes (eds.), Lecture Notes in Math., Vol. 606, Springer, Berlin, Heidelberg, New York, 1977, 261–274.

[NOVE88] R.H. Nochetto and C. Verdi, The combined use of a nonlinear Chernoff formula with a regularization procedure for two-phase Stefan problems, *Numer. Funct. Anal. and Optimiz.*, **9** (1987-88), 11-12, 1177–1192.

[OSSE88] S. Osher and J.A. Sethian, Fronts propagating with curvature-dependent speed: algorithms based on Hamilton-Jacobi formulations, *J. Comput. Phys.*, **79** (1988), 12–49.

[PGLA95] M. Pahl, W. Gleißle, H.M. Laun, *Praktische Rheologie der Kunststoffe und Elastomere*, VDI-Gesellschaft Kunststofftechnik, VDI Verlag Düsseldorf, 1995.

[PAGU97] T.D. Papathanasiou and D.C. Guell (eds.), *Flow-induced alignement in composite materials*, Woodhead Publishing Limited, Cambridge, England, 1997.

[PATA80] S.V. Patankar, *Numerical heat transfer and fluid flow*, McGraw-Hill, New York, 1980.

[PERI83] J.R.A. Pearson and S.M. Richardson, *Computational analysis of polymer processing*, Applied Science Publishers, London, 1983.

[PIRO89] O. Pironneau, *Finite elements methods for fluids*, John Wiley & Sons, Masson, 1989.

[PIVK93] T.S. Piwonka, V. Voller and L. Katgerman (eds.), *Modeling of casting, welding and advanced solidification processes VI*, Proc. of the sixth conference in a series on modeling casting and welding processes, Palm Coast, Florida, 1993, TMS publication, 1993.

[PRRO93] M. Primicerio and J.F. Rodrigues, The Hele–Shaw problem with nonlocal injection condition, *Gakuto Int. Ser., Math. Sci. Appl.*, **2** (1993), 375–390.

[RAJ94] V.T. Rajan, Optimality of the Delauney triangulation in $\mathbb{R}^d$, *Discrete Comput. Geom.*, **12** (1994), 189–202.

[REIS94] M. Reissig, The existence and uniqueness of analytic solutions for a moving boundary problem for Hele–Shaw flows in the plane, *Nonlinear Analysis, Theory, Methods & Appl.*, **23** (1994), 5, 565–576.

[RICH72] S. Richardson, Hele–Shaw flows with a free boundary produced by the injection of fluid into a narrow channel, *J. Fluid Mech.*, **56** (1972), 609–618.

[RICH81] S. Richardson, Some Hele–Shaw flows with time-dependent free boundaries, *J. Fluid Mech.*, **102** (1981), 263–279.

[RICH94] S. Richardson, Hele–Shaw flows with time-dependent free boundaries in which the fluid occupies a multiply-connected region, *Euro. J. of Applied Mathematics*, **5** (1994), 97–122.

[RICO98] W.J. Rider and D.B. Kothe, Reconstructing volume tracking, *Journ. of Comput. Physics*, **141** (1998), 112–152.

[RODR87] J.F. Rodrigues, *Obstacle problems in mathematical physics*, North-Holland Mathematics Studies 134, North-Holland Publishing Company, Amsterdam, New York, Oxford, 1987.

[RODR89] J.F. Rodrigues, The Stefan Problem revisited, in *Mathematical models for phase change problems*, Proceedings of the European Workshop held at Obidos, Portugal, October 1988, ISNM Vol. 88, Birkhäuser, 1989, 129–190.

[RODR94] J.F. Rodrigues, Variational methods in the Stefan problem, in *Phase transitions and hysteresis*, A. Visitin (ed.), Lecture Notes in Mathematics 1584, Springer, Berlin, Heidelberg, New York, 1994, 147–212.

[ROTH30] E. Rothe, Zweidimensionale parabolische Randwertaufgaben als Grenzfall eindimensionaler Randwertaufgaben, *Math. Ann.*, **102** (1930), 650–670.

[RUBI71] L.I. Rubinstein, *The Stefan Problem*, AMS, Providence, 1971.

[SAAD96] Y. Saad, *Iterative methods for sparse linear systems*, PWS Publishing Company, 1996.

[SATA58] P.G. Saffmann and G.I. Taylor, The penetration of a fluid into a porous medium or Hele–Shaw cell containing a more viscous fluid, *Proc. of the Royal Soc. of London*, **A 245** (1958), 312–329.

[SAHC00] P.R. Sahm, P.N. Hansen and J.G. Conley, *Modeling of casting, welding and advanced solidification processes IX*, Proc. of the ninth conference in a series on modeling casting and welding processes, (SIM 2000), Aachen, Germany, 2000, Shaker, 2000.

[SAMA84] A.A. Samarskij, *Theorie der Differenzenverfahren*, Geest & Portig, Leipzig, 1984.

[SLM87] A.A. Samarskij, R.D. Lazarov and V.L. Makarov, *Finite Difference Methods for differential equations with generalized solutions*, in Russian, Nauka, Moscow, 1987.

[SCHO84] R. Scholz, Numerical solution of the obstacle problem by the penalty method I, *Computing*, **32** (1984) , 297–306.

[SCHO86] R. Scholz, Numerical solution of the obstacle problem by the penalty method II, *Num. Math.*, **49** (1986), 255–268.

[SETH96] J.A. Sethian, *Level set methods*, Cambridge Monographs on Applied and Computational Mathematics, Cambridge University Press, 1996.

[STEF91] J. Stefan, Über die Theorie der Eisbildung, insbesondere über Eisbildung im Polarmeere, *Ann. Phys. u. Chem.*, **42** (1891),269–286.

[STEI91] J. Steinbach, Evolutionary variational inequalities with a Volterra term, *Numer. Func. Anal. and Optim.*, **12** (1991), 3&4, 395–411.

[STEI93] J. Steinbach, *Numerical solution of elliptic and evolutionary variational inequalities by means of the box method*, Report 413, 1993, Schwerpunktprogramm der DFG 'Anwendungsbezogene Optimierung und Steuerung', TU Munich, 89 pp.

[STEIa96] J. Steinbach, Comparison of finite element and finite volume schemes for variational inequalities, *East-West Journ. of Numerical Mathematics*, **4** (1996), 3, 207–235.

[STEIb96] J. Steinbach, Simulation of the mould filling process using a temperature dependent, viscous flow, in [NIST96], 1996, 344–356.

[STEIa97] J. Steinbach, A generalized temperature-dependent, non-Newtonian Hele–Shaw flow in injection and compression moulding, *Mathem. Methods in Appl. Sciences*, **20** (1997), 1199–1222.

[STEIb97] J. Steinbach, On box schemes for elliptic variational inequalities, *Numer. Func. Analysis and Optimiz.*, **18** (1997), 9&10, 1041–1066.

[STEIc97] J. Steinbach, An evolutionary variational inequality approach for the electrochemical machining process, in *Proceedings of the All-Russian Conference 'Advanced Electrotechnology in Machine Building Industry', June 1997*, Tula State University (Russia), 1997, 21–26.

[STEIa98] J. Steinbach, Evolutionary variational inequalities arising in non-isothermal Hele–Shaw flows, *Advances in Math. Sciences and Applications*, **8** (1998), 1, 357–371.

[STEIb98] On a variational inequality containing a memory term with an application in electro-chemical machining, *Journal of Convex Analysis*, **5** (1998), 1, 63–80.

[STKA00] J. Steinbach and D.M. Lipinski, Three-dimensional simulation of the injection moulding process including fibre orientation, in *Proc. of Materials Week 2000, International Congress on Advanced Materials, Munich, 2000.*

[STEL93] L.L. Stell, *A fixed domain method for injection governed by the Stokes equations*, PhD Thesis, Cornell University Ithaca, New York, 1993.

[STRE89] G. Strehlau, L_∞- error estimate for the numerical treatment of the obstacle problem by the penalty method, *Numer. Func. Anal. and Optimiz.*, **10** (1989), 1&2, 185–198.

[SUSO94] M. Sussman, P. Smereka and S. Osher, A level set approach for computing solutions to incompressible two-phase flow, *J. Comput. Phys.*, **114** (1994), 146–159.

[SYRJ91] S. Syrjäla, *Numerical modelling of injection mould filling and cooling*, Licentiate thesis, Tampere, 1991.

[TAGO79] Z. Tadmor and C.G. Gogos, *Principles of polymer processing*, John Wiley & Sons, New York, 1979.

[TANN00] R.I. Tanner, *Engineering rheology*, Oxford University Press, second ed., 2000.

[TARV94] P. Tarvainen, *Block relaxation methods for algebraic obstacle problems with M-matrices: theory and applications*, PhD Thesis, University of Jyväskylä 1994, Report 63 of the Department of Mathematics

[THBE98] B.G. Thomas and Ch. Beckermann (eds.), *Modeling of casting, welding and advanced solidification processes VIII*, Proc. of the eighth conference in a series on modeling casting and welding processes, San Diego, California 1998, TMS publication, 1998.

[THOM86] E. Thompson, Use of pseudo-concentration to follow creeping viscous flows during transient analysis, *Intern. J. Numer. Meth in Fluids*, **6** (1986), 749–761.

[TUCK89] C.L. Tucker III, (ed.), *Computer modeling for polymer processing, Fundamentals*, Hanser Publishers, Munich, Vienna, New York, 1989.

[VERD94] C. Verdi, Numerical aspects of parabolic free boundary and hysteresis problems, in *Phase transitions and hysteresis*, A. Visitin (ed.), Lecture Notes in Mathematics 1584, Springer, Berlin, Heidelberg, New York, 1994, 213–284.

[VERW98] B.E. VerWeyst, *Numerical predictions of flow-induced fiber orientation in 3-D geometries*, PhD Thesis, University of Illinois at Urbana-Champaign, 1998.

[VTFG99] B.E. VerWeyst, C.L. Tucker, P.H. Foss and J.F. O'Gara, Fiber orientation in 3-D injection molded features, *Inter. Polymer Processing XIV*, **4** (1999), 409–420.

[VOLG97] V.M. Volgin, Models for the evolution of the (shaping) boundary in the ECM process, in Russian, in *Proceedings of the All-Russian Conference 'Advanced Electrotechnology in Machine Building Industry', June 1997*, Tula State University (Russia), 1997, 27–41.

[WEBE00] F.-K. Webelhaus, *Numerische Simulation des Spritzgießprozesses unter Berücksichtigung dreidimensionaler Strömungseffekte und der Materialkompressibilität*, PhD Thesis, RWTH Aachen, 2000.

[WEOS97] E. Wegert and D. Oestreich, On a non-symmetric problem in electrochemical maching, *Math. Methods in the Applied Sciences*, **20** (1997), 841–854.

[WHCG85] A. Whelan, J.L. Craft and J.P. Goff, *Developments in injection moulding*, Elsevier, Amsterdam, 1985.

[WILF55] M.L. Williams, R.F. Landel and J.D. Ferry, The temperature dependence of relaxation mechanisms in amorphous polymers and other glass-forming liquids, *Journ. Am. Chem. Soc.* **77** 1955, 6, 3701–3707.

[ZACH98] J. Zachert, *Analyse und Simulation dreidimensionaler Strömungsvorgänge beim Spritzgießen*, PhD Thesis, RWTH Aachen, 1998.

[ZEIDI93] E. Zeidler, *Nonlinear functional analysis and its applications. Vol. I: Fixed-Point Theorems*, Springer, Berlin, Heidelberg, New York, 1993 (second corrected printing).

[ZEIDII90] E. Zeidler, *Nonlinear functional analysis and its applications. Vol. II/A: Linear monotone operators; Vol. II/B: Nonlinear monotone operators*, Springer, Berlin, Heidelberg, New York, 1990.

[ZEIDIII90] E. Zeidler, *Nonlinear functional analysis and its applications. Vol. III: Variational methods and optimization*, Springer, Berlin, Heidelberg, New York, 1990.

[ZZPH98] H. Zhang, L.L. Zheng, V. Prasad and T.Y. Hou, A curvilinear level set formulation for highly deformable free surface problems with application to solidification, *Numeric. Heat Transfer*, **34** (1998), 1–20.

List of Figures

2.1 Free boundary value problem: geometrical situation at a fixed time . . . 8
2.2 Schematic view of the injection moulding problem . . . 11
2.3 Schematic representation of the ground plan and the side view of a cavity . . . 12
2.4 Charge shape in compression moulding . . . 15
2.5 Cross section of an annular electro-chemical machining process . . . 17
2.6 Constitutive relations for the classical one-phase Stefan problem and the Hele–Shaw problem . . . 21

3.1 Illustration of the penalization function . . . 43

4.1 Construction of the boxes in the two-dimensional case . . . 82
4.2 One-dimensional obstacle problem . . . 108
4.3 Numerical results of the finite volume approximation for two obstacle problems . . . 133
4.4 Error behaviour between finite element and finite volume solution (elliptic inequalities) . . . 137
4.5 Error behaviour between exact and finite volume solution (elliptic inequalities) . . . 139

5.1 Error behaviour between exact and finite element solution (evolutionary inequalities) . . . 196
5.2 Error behaviour between finite element and finite volume solution (evolutionary inequalities) . . . 197
5.3 Error behaviour between exact and finite volume solution (evolutionary inequalities) . . . 198
5.4 Error behaviour between exact and finite volume (penalty) solution (evolutionary inequalities) . . . 200
5.5 Error behaviour between exact and finite volume (penalty) solution (evolutionary inequalities) . . . 201

6.1 Schematic representation of an injection-compression moulding machine 208
6.2 Dynamic viscosity as a function of shear rate and temperature 215
6.3 Object and computational geometry of plastic products 218
6.4 Pseudo-circle shaped flow front 229
6.5 Variation of gate location. Simulation results for an injection-compression process 239
6.6 Variation of thickness. Simulation results for an injection process 242
6.7 Non-isothermal effects – constant thickness 243
6.8 Non-isothermal effects – varying thickness 244
6.9 Plastic part with narrow flow region 245
6.10 Plate-like geometries with non-uniform thickness 246
6.11 Plate-like geometry I with non-uniform thickness – Comparison of distance, variational inequality (VI) approach and experiments 249
6.12 Plate-like geometry II with non-uniform thickness – Comparison of distance, variational inequality (VI) approach and experiments 250
6.13 Rectangular plate with non-uniform thickness – Comparison of distance and variational inequality approach 251
6.14 Rectangular plate with non-uniform thickness – Three-dimensional simulation 254
6.15 Geometry with thickness ratio 1:5 – Three-dimensional simulation 255

List of Tables

4.1 Error behaviour between finite element and finite volume solution (SOR with projection; elliptic inequalities) 135
4.2 Error behaviour between finite element and finite volume solution (penalty-iteration (4.32), (4.40); elliptic inequalities) . . 138
4.3 Error behaviour between exact and finite volume solution (elliptic inequalities) 138
4.4 Error behaviour between exact and finite volume solution for different penalty-iteration methods (elliptic inequalities) 140

5.1 Error behaviour between exact, finite element and finite volume solution (evolutionary inequality – Example 5.27) 195
5.2 Error behaviour between exact, finite element and finite volume solution (evolutionary inequality – Example 5.28) 195
5.3 Error behaviour between exact and finite volume solution for different penalty-iteration methods (evolutionary inequalities) . . 199

List of Symbols

Intervals, domains, regions, boundaries, constants

$\mathbb{R}^n$	n-dimensional Euclidean space
Ω	spatial problem domain (bounded connected open set in $\mathbb{R}^n$ with at least Lipschitz boundary)
$\mathrm{meas}_n(\Omega)$	n-dimensional Lebesgue measure of Ω
$\mathrm{diam}\,\Omega$	diameter of Ω
$\Gamma = \partial\Omega$	boundary of Ω
$\vec{n}$	outward unit normal to $\partial\Omega$
Γ_D, Γ_N	Dirichlet, Neumann/Newton part of the boundary Γ
Γ_0	points, where one type of the boundary condition changes into another
$(0,T)$, $[0,T]$	open, closed time interval
$Q = \Omega \times (0,T)$	space-time domain
$\Sigma = \partial\Omega \times (0,T)$	lateral boundary of Q
$\Sigma_D = \Gamma_D \times (0,T)$	Dirichlet part of Σ
$\Sigma_N = \Gamma_N \times (0,T)$	Neumann/Newton part of Σ
$\Omega(t) \subset \Omega$	unknown domain at the time instant $t \in [0,T]$
Γ_f, $\Gamma_f(t)$	interface, free boundary (which separate $\Omega(t)$ ('first phase') from the region $\Omega\backslash\Omega(t)$ ('second phase'))
$\mathcal{L} = \Omega(t) \times (0,T)$	('liquid') phase
C, c, C_i, c_i, $\bar{C}_i$, $\tilde{C}_i$, $i = 1,2,\ldots$	generic positive constants, independent of the related functions and discretization parameters

Notation for the general free boundary problem

x, t	space and time variable, respectively
$\theta = \theta(x,t)$	solution of the general free boundary problem (2.1)
$\eta = \eta(t)$	given function describing the solution in the second phase
$\lambda = \lambda(x,t)$	function describing the jump (of the quantity being conserved) at the interface
$\vec{w}$, $w_n = \vec{w} \cdot \vec{n}$	velocity and normal velocity of the interface, respectively
γ_D, κ, γ_N	coefficients in Dirichlet and Neumann/Newton boundary conditions
$\tilde{\theta} = \theta - \eta$	extension of θ into the whole space-time domain $\overline{Q}$
H	maximal monotone graph associated with the Heaviside function
$\chi \in H(\tilde{\theta})$	unknown function in the weak formulation
$\chi_{\Omega(t)}$	characteristic function of the set $\Omega(t)$
$u = u(x,t)$	new unknown function after the application of the Baiocchi transformation
F, g_D, g_N	transformed right-hand side and coefficients in the transformed boundary conditions, respectively
$A(x,t)$, $B(x,t)$	second order differential operators, coefficients of which depend on space and time
$a(t;v,w)$, $b(t;v,w)$	bilinear forms
$K = K_{\varphi,g_D}(t)$	time-dependent, non-empty, closed, convex subsets of the Sobolev space $H^1(\Omega)$ characterized by the obstacle φ and Dirichlet condition $g_D(t)$
J, ∂J	functional and its subdifferential
I_K	indicator functional with respect to the set K

Function spaces

$L_p(\Omega)$	Lebesgue space of measurable functions v defined on Ω for which $\|v\|_{L_p(\Omega)} = (\int_\Omega |v(x)|^p\, dx)^{1/p}$ is finite, $1 \le p < \infty$
$L_\infty(\Omega)$	Lebesgue space of measurable essentially bounded functions defined on Ω
(u, v)	scalar product on $L_2(\Omega)$, i.e., $(u, v) = \int_\Omega u(x)\, v(x)\, dx$
$W_p^k(\Omega)$	Sobolev space of functions whose generalized derivatives up to order k belong to $L_p(\Omega)$, $1 \le p \le \infty$, $k = 0, 1, \ldots$
$H^k(\Omega)$	Sobolev space of functions whose generalized derivatives up to order k belong to $L_2(\Omega)$, $k = 0, 1, \ldots$
$H_0^1(\Omega)$	space of functions from $H^1(\Omega)$ whose traces vanish on $\partial\Omega$ (closure of $C_0^\infty(\Omega)$ (space of infinitely differentiable functions with compact support in Ω) in $H^1(\Omega)$)
V	space of functions from $H^1(\Omega)$ whose traces vanish on Γ_D
V^*	dual space of V
$\langle .,. \rangle$	duality pairing between V^* and V
$L_p(\Gamma)$	Lebesgue space of measurable functions $v : \Gamma \longrightarrow \mathbb{R}$ for which $\|v\|_{L_p(\Gamma)} = (\int_\Gamma |v(x)|^p\, ds)^{1/p}$ is finite, $1 \le p < \infty$; analogous $L_p(\Gamma_N)$ associated with Neumann/Newton part $\Gamma_N \subset \Gamma$
$W_p^r(\partial\Omega)$	Sobolev space of functions $u : \partial\Omega \longrightarrow \mathbb{R}$ for real $r \ge 0$ (see [ZEIDII90], p. 1031)
γv	trace of v on Γ
ds	surface measure along Γ
$(u, v)_{\Gamma_N}$	scalar product on Γ_N, $(u, v)_{\Gamma_N} = \int_{\Gamma_N} u(x)\, v(x)\, ds$
w^+, w^-	positive and negative part of the function w, i.e., $w^+ = w \vee 0 = \max\{w, 0\}$ and $w^- = w \wedge 0 = \min\{w, 0\}$
$L_p(0, T; X)$	Lebesgue space of abstract (Bochner) measurable functions $v : (0, T) \longrightarrow X$ for which $\|v\|_{L_p(0,T;X)} = (\int_0^T \|v(t)\|_X^p\, dt)^{1/p}$ is finite; X as Banach space, $1 \le p < \infty$

$L_\infty(0,T;X)$	Lebesgue space of abstract (Bochner) measurable functions $v:(0,T)\longrightarrow X$ which are essentially bounded, $\|v\|_{L_\infty(0,T;X)} = \text{ess sup}_{t\in(0,T)}\|v(t)\|_X$; X as Banach space
$C([0,T];X)$	space of abstract continuous functions $v:[0,T]\longrightarrow X$, $\|v\|_{C([0,T];X)} = \max_{t\in[0,T]}\|v(t)\|_X$; X as Banach space
$W_p^1(0,T;X)$	space of functions $v\in L_p(0,T;X)$ for which $\partial_t v\in L_p(0,T;X)$ (i.e., time-derivative in generalized sense; see Section 3.1)
$W_r^{2,1}(Q)$	(Banach) space consisting of functions $z\in L_r(Q)$ having (generalized) derivatives $\partial_t z$, $\partial_{x_i} z$ and $\partial^2 z/\partial_{x_i}\partial_{x_j}$ in $L_r(Q)$ for $i,j=1,\ldots,n$

Spatial discretization

$\{\mathcal{S}_h\}_{h\in(0,h_0]}$	family of regular triangulations (consisting of closed simplices S) over a polygonal (polyhedral) domain $\overline{\Omega}$ in $\mathbb{R}^2$ ($\mathbb{R}^3$); cf. Section 4.1.2
$h_S = \operatorname{diam} S$	diameter of the simplex S
$h = \max_{S\in\mathcal{S}_h}$	spatial discretization parameter
ρ_S	$\sup\{\operatorname{diam} U : U$ is a n-ball contained in $S\}$
σ	(boundary) face of a simplex $S\in\mathcal{S}_h$
σ_h	set consisting of all (boundary) faces σ of any $S\in\mathcal{S}_h$ such that σ belongs to $\partial\Omega$
$\sigma_{h,2}$	set consisting of all (boundary) faces $\sigma\in\sigma_h$ such that $\sigma\subset\overline{\Gamma_N}$
ξ, η, ζ	notations for nodes/vertices of the triangulations
$\overline{\omega}$	set of nodes, i.e., $\overline{\omega}=\{\xi\in\overline{\Omega}:\xi$ is a vertex of at least one $S\in\mathcal{S}_h\}$
ω, γ	sets of interior and boundary nodes, i.e., $\omega=\{\xi\in\overline{\omega}:\xi\in\Omega\}$ and $\gamma=\{\xi\in\overline{\omega}:\xi\in\partial\Omega\}$, respectively
γ_D, γ_N	sets of Dirichlet and Neumann/Newton nodes, i.e., $\gamma_D=\{\xi\in\overline{\omega}:\xi\in\overline{\Gamma_D}\}$ and $\gamma_N=\{\xi\in\overline{\omega}:\xi\in\Gamma_N\}$, respectively
$\tilde{\omega}$	set of nodes related to the degrees of freedom, i.e., $\tilde{\omega}=\overline{\omega}\backslash\gamma_D$
Ω_ξ	union of those simplices $S\in\mathcal{S}_h$ which have ξ as a vertex

ω_ξ	difference star, i.e., $\omega_\xi = \{\eta \in (\overline{\omega}\backslash\{\xi\}) : \eta \in \Omega_\xi\}$ for $\xi \in \overline{\omega}$
$\mathcal{D}_h$	(dual) box partition of Ω associated to the (primary) triangulation $\mathcal{S}_h$
p^S	point of simplex $S \in \mathcal{S}_h$ which is used for construction of the boxes $D \in \mathcal{D}_h$, cf. Figure 4.1; 'well-known choices' for p^S: circumcentre or barycentre of corresponding simplex S
D_ξ	closed box (finite volume) associated to node $\xi \in \overline{\omega}$; $D_\xi \subset \Omega_\xi$, $\overline{\Omega} = \cup_{\xi\in\overline{\omega}} D_\xi$, cf. Figure 4.1
D_ξ^S	sub-box lying in simplex $S \in \mathcal{S}_h$; $D_\xi^S \subset D_\xi$, cf. Figure 4.1
$D_{\xi;\eta}^S$, $n=2$	triangle (sub-box) formed by the vertices ξ, $(\xi+\eta)/2$ (edge midpoint) and p^S, cf. Figure 4.1
$D_{\xi;\eta,\zeta}^S$, $n=3$	tedrahedron (sub-box) formed by the vertices ξ, $(\xi+\eta)/2$ (edge midpoint), $(\xi+\eta+\zeta)/3$ (barycenter of a face) and p^S
$\lambda_i = \lambda_i(x)$	barycentric coordinates of x with respect to S
$P_k(\Omega)$	space of polynomials of degree at most k defined on Ω
X_h^G	(piecewise linear) finite element subspace of $H^1(\Omega)$, i.e., $X_h^G = \{w \in C(\overline{\Omega}) : w_{\vert S} \in P_1(S) \text{ for all } S \in \mathcal{S}_h\}$
V_h^G	(piecewise linear) finite element subspace of V, i.e., $V_h^G = \{w \in X_h^G : w = 0 \text{ on } \Gamma_D\}$
Φ_ξ^G	nodal basis functions of V_h^G, X_h^G with $\Phi_\xi^G(\eta) = \delta_{\xi\eta}$ (Kronecker symbol), $\xi, \eta \in \overline{\omega}$
V_h^C	space of discontinuous piecewise constant functions with respect to $\mathcal{S}_h$, i.e., $V_h^C = \{w \in L_\infty(\Omega) : w_{\vert S} = \text{const} \quad \forall S \in \mathcal{S}_h\}$
X_h^B, V_h^B	spaces of discontinuous piecewise constant functions with respect to $\mathcal{D}_h$, i.e., $X_h^B = \{w \in L_\infty(\Omega) : w_{\vert D} = \text{const } \forall D \in \mathcal{D}_h\}$ and $V_h^B = \{w \in X_h^B : w = 0 \text{ on } \Gamma_D\}$
Φ_ξ^B	nodal basis functions of V_h^B, X_h^B
v_h, w_h	grid function vectors $(v_h)_{\xi\in\tilde{\omega}}$, $(w_h)_{\xi\in\overline{\omega}}$ defined on $\tilde{\omega}$ (resp. $\overline{\omega}$)
$v_{h,\xi}$	component of the grid function vector v_h associated to the node $\xi \in \overline{\omega}$

$v_h^+,\ v_h^-$	positive and negative part (componentwise) of the grid function vector v_h, i.e., $v_h^+ = \max\{v_h, 0\}$ and $v_h^- = \min\{v_h, 0\}$
e_h^ξ	unit grid function vector associated with a node $\xi \in \overline{\omega}$
$L_2(\tilde{\omega}),\ L_2(\overline{\omega})$	linear spaces formed by the grid function vectors; isomorph to $V_h^G,\ X_h^G$; $L_2(\tilde{\omega}) := \mathbb{R}^{\dim V_h^G}$, $L_2(\overline{\omega}) := \mathbb{R}^{\dim X_h^G}$
$\langle v_h, w_h \rangle$	Euclidean scalar product for grid functions vectors
P^G	(Galerkin) prolongation $P^G : L_2(\tilde{\omega}) \longrightarrow V_h^G$
P^B	(Box) prolongation $P^B : L_2(\tilde{\omega}) \longrightarrow V_h^B$
$\|v_h\|_{0,\overline{\omega}},\ \|v_h\|_{0,\gamma}$	norms for grid function vectors, given by $\|v_h\|_{0,\overline{\omega}} := \left\|P^B v_h\right\|_{L_2(\Omega)}$ and $\|v_h\|_{0,\gamma} := \left\|P^B v_h\right\|_{L_2(\partial\Omega)}$
$\lvert w_h\rvert_{1,\overline{\omega}},\ \|w_h\|_{1,\overline{\omega}}$	semi-norm and norm for grid function vectors, given by $\lvert w_h\rvert_{1,\overline{\omega}} := \left\lvert P^G w_h\right\rvert_{H^1(\Omega)}$ and $\|w_h\|_{1,\overline{\omega}} := \left(\lvert w_h\rvert_{1,\overline{\omega}}^2 + \|w_h\|_{0,\overline{\omega}}^2\right)^{1/2}$
$L_G\ \ M_G,\ R_G$	Galerkin stiffness, mass and boundary matrices
$L_B,\ M_B,\ R_B$	box (finite volume) stiffness, mass and boundary matrices
$f_{Gh},\ g_{Gh},\ f_{Bh},\ g_{Bh}$	Galerkin and box right-hand side vectors
Λ_B	box matrix defined by $\Lambda_B = L_B + M_B + R_B$
$a_B(P^G v_h, P^B w_h)$	box bilinear form, i.e., $a_B(P^G v_h, P^B w_h) = \langle \Lambda_B v_h, w_h \rangle$
K_h	(discrete) constraint set
$v_h/\operatorname{meas}_n(D)$	grid function vector (right-hand side) divided (componentwise) by the area of the boxes
$w_h/\operatorname{meas}_{n-1}(\partial D)$	grid function vector divided (componentwise) by the area of the part of the box boundaries lying on Γ
Π_h	(piecewise linear) Lagrange interpolation operator, i.e., $\Pi_h : C(\overline{\Omega}) \longrightarrow X_h^G$; $\Pi_h v$ as X_h^G-interpolant
$\mathcal{L}_h$	lumping operator with respect to a box partition $\mathcal{D}_h$ (cf. Remark 4.8)

Time discretization

$[0,T],\ \tau = T/N$	time interval, time step
$t^j = j\ \tau$	time instants for $j = 0,\dots,N$
v_h^j	grid function vector at time instant t^j
$L_G^j\ M_G^j,\ R_G$	Galerkin stiffness, mass and boundary matrices at time instant t^j
$L_B^j,\ M_B^j,\ R_B$	box (finite volume) stiffness, mass and boundary matrices at time instant t^j
$f_{Gh}^j,\ g_{Gh}^j,\ f_{Bh}^j,\ g_{Bh}^j$	Galerkin and box right-hand side vectors at time instant t^j
$\Lambda_G^j,\ \Lambda_B^j$	Galerkin and box matrix at time instant t^j (weighted with parameter $\alpha \in [0,1]$, i.e., $\Lambda_G^j = (1-\alpha)\ (L_G^j + M_G^j) + \alpha\ (L_G^{j-1} + M_G^{j-1}) + R_G$; analogous Λ_B^j
K_h^j	(discrete) constraint set at time instant t^j
$I_\tau^0(P^G v_h)$	piecewise constant in time function of $P^G v_h$ (cf. Remark 5.7)

Notation for the application problems (Chapters 2, 6)

Applications 2.2, 2.3: injection and compression moulding

$\Omega,\ \Omega(t)$	mid-surface of the cavity and plastic melt region, respectively
$\tilde{d}(x)$	(gap) thickness of the cavity
$d(x,t)$	effective thickness for the flow
$\partial_t d(x,t)$	closing speed of the mould in compression moulding
$k = k(x,t)$	flow conductivity of the plastic melt
T	filling time
Γ_E	fixed boundary of the cavity (lateral container wall)
Γ_G	boundary gate in injection
Ω_{G_i}	gates
γ_G	known flow rate through the boundary gate
ρ	density of the fluid

p	pressure
$\vec{v} = (v_1, v_2, v_3)$	velocity vector
$\nabla\vec{v}$	velocity gradient tensor
$\bar{\vec{v}}$	gapwise-averaged velocity vector
$\vec{g}$	gravitational force vector
η	shear (dynamic) viscosity
$\underline{\sigma}$, $\underline{\tau}$, $\underline{\dot{\gamma}}$	stress tensor, viscous stress tensor, rate of strain (shear rate) tensor
$\dot{\gamma}$	shear rate (scalar measure for the rate of deformation)
D/Dt	material derivative
e, ϑ, ϑ_{SO}	(specific) internal energy, temperature, no-flow temperature
$\vec{q}$, $\tilde{\lambda}$, c_p	heat flux vector, thermal conductivity, specific heat
$\mathcal{L}^z$	(three-dimensional) melt region (for temperature)
n	power law exponent
δ	Delta distribution
$d(x_a, x_b)$	weighted interior distance between points x_a and x_b (in distance concept)
$d_h(x_a, x_b)$	approximation of $d(x_a, x_b)$
Ψ	VOF or level set function

Application 2.4: electrochemical machining process

$A(t)$, $\Omega(t)$, $\partial\Omega$	anode, electrolyte, cathode surface
T	machining time
$k = k(x,t)$	conductivity of the electrolyte
$\vec{E}$, $\vec{j}$	electric field, electric current
Φ	potential
γ_A	given value of the poential Φ at the anode
λ_E	electro-chemical equivalent

Index

Air trap, 211, 238, 240
Application problem
compression moulding, *see* Compression moulding
ECM problem, *see* Electro-chemical machining process
elasto-plastic torsion, *see* Elasto-plastic torsion problem
injection moulding, *see* Injection moulding
porous media flow, *see* Porous media flow
quasi-steady Stefan problem, *see* Stefan type problem

Baiocchi transformation, 2, 3, 23, 24, 27, 191, 205
Barycentric boxes, 83, 132
Barycentric coordinates, 83, 110
Bilinear form, 25, 32
Boundary element method
application in polymer processing, 18
Boundary value problem, 41, 55, 56, 63, 75
Box method, *see* Finite volume method
Bramble–Hilbert Lemma, 93

Characterics
method of, 224
Chernoff's formula, 22
Circumcentric boxes, 84
Coincidence set, 107, 124, 190
Complementarity problem, 25, 72, 76, 171
finite-dimensional, 76, 110, 155
Complex variable theory, 204
Compression moulding, 2, 3, 5
distance concept, 225
evolutionary variational inequality, 26, 46, 56
fibre-reinforced, 207
free boundary problem, 16, 217, 221
mathematical modelling, 14–16, 217–221
simulation aims, 210
well-posed problem, 16, 19
Conservation law, 212–213
energy, 213, 222
mass, 212
momentum, 212, 219
Constitutive law, 212, 213
Convex set, 28, 76
convergence (Mosco), 36, 37, 79
time-dependent, 3, 26, 31, 32, 46–55
time-independent, 40–46
Cure reaction, 208, 209, 213

Dijkstra's algorithm, 230
Discretization
spatial, general notation, 76–77, 81–83
time, general notation, 144
Distance concept, 6, 225–231
convex duality, 228
Dijkstra's algorithm, 230

interior distance, 227
pseudo-circle principle, 229
shortest path, 229–231
weighted interior distance, 227
weighted region problem, 230
Duality technique, 228

Elasto-plastic torsion problem, 134
Elastomer, 11, 207
rheology, 213–217
Electro-chemical machining process, 2, 3
evolutionary variational inequality, 26, 46, 48, 50, 53–54, 56
free boundary problem, 17, 205
mathematical modelling, 16–18
well-posed problem, 18
Elliptic regularity
estimates, 55, 60
theory, 55, 60, 71, 76
Elliptic variational inequality, 74–76
comparison of FEM and FVM, 102
ECM problem, 26
elasto-plastic torsion, 134
finite element error analysis, 79
finite element formulation, 4, 78–79
finite volume error analysis, 103–110
finite volume formulation, 4, 86–88
Hele–Shaw problem, 26, 205
numerical solution methods, 110
penalization (discrete), 111–126
regularity statements, 76
Signorini problem, 126
with time as parameter, 26
Enthalpy formulation, 19, 20, 205
Evolutionary variational inequality, 2, 3, 6
abstract, 40
boundedness of FEM/FVM solution, 151–153
comparison of FEM and FVM, 158–162
complementarity problem, *see* Complementarity problem
continuity of solution, 69–70
derivation, 23–29
finite element a priori estimate, 162–163
finite element error analysis, 170–177
finite element formulation, 4, 145–147
finite volume a priori estimate, 166–167
finite volume error analysis, 170, 177–179
finite volume formulation, 4, 147–149
general formulation, 32–34
Lipschitz continuity of solution, 69–70
maximum principle, 50–52
memory term, 2, 3, 26
penalization (discrete), 179–187
regularity with respect to time, 40–55
semi-coercive, 39
solvability of FEM/FVM formulation, 151–152
spatial regularity, *see* Spatial regularity
stability of FEM/FVM solution, 153–155
time evolution of FVM solution, 156–157

Filling pattern, 210, 225, 227, 230
Finite element method (FEM)
comparison with FVM, 93–103, 158–162
general notation, 76–79
mixed, 224

Finite volume method (FVM)
as generalized finite difference, 80, 109
cell-centered, 81
comparison with FEM, 93–103, 158–162
general notation, 81–86
vertex-centered, 81
Fixed domain formulation, 3, 6
variational inequality, 2, 3, 23–29
weak (enthalpy) formulation, 3, 19
Fixed-point Theorem, 36
Flow conductance, *see* Fluidity
Flow front, 18, 19, 210, 227, 230, 232, 234, 236, 237, 239, 240
Fluid
(weak) compressible, 212, 236
generalized Newtonian, 213, 214
incompressible, 204, 212
Newtonian, 213, 214
non-Newtonian, 213
viscous, 212
Fluidity, 14, 220
Fountain flow, 211, 232
Free boundary, 9, 19, 24, 107, 109, 211, 212, 217, 226, 232, 234–236
Free boundary problem, 1–3, 204, 227
application of Baiocchi transformation, 2, 3, 23, 24, 27, 205
compression moulding, 16, 217, 221
connection to variational inequality, 1, 2
definition (classical formulation), 8–10
distributional formulation, 23
electro-chemical machining process, 17
electro-forming process, 205
fixed domain formulation, 18–29, 234
flow in porous media, 22, 23, 29, 205, 224
injection moulding, 13, 217, 220, 226
maximum principle, 9
numerical example, 132, 134, 135, 191, 192, 194
Stefan type, 3, 9, 10, 19, 20, 23, 24, 26, 27, 205, 224
variational inequality formulation, 23–29
weak (enthalpy) formulation, 19–23
well-posed problem, 9, 16, 18, 19, 204
Free surface, *see* Free boundary
Front capturing, *see* Front tracking
Front fixing
application in polymer processing, 18
Front tracking
application in ECM process, 16
application in Hele–Shaw flow, 205
application in polymer processing, 18
level set method, 234–235
Marker and Cell method, 234–235
Volume of fluid method, 234–235, 252

Gagliardo–Nirenberg inequality, 61, 68
Gronwall inequality
continuous, 35
discrete, 35

Heaviside function, 20
maximal monotone graph of, 20, 27, 42
Hele–Shaw flow
classical, 3, 13, 26, 204–205, 217, 224
constitutive relations, 22

generalized, 6, 13, 211, 217–221, 231
simplifications, 219, 221, 222, 231
well-posed problem, 16, 204

Indicator functional, 28, 42
convex, l.s.c, proper, 28
Injection moulding, 2, 3, 5
generalized Stokes flow, 236–237
distance concept, 225–231
evolutionary variational inequality, 26, 46, 56
fibre-reinforced, 207, 237
free boundary problem, 13, 217, 220, 226
gas-assisted, 237
mathematical modelling, 11–14, 217–221
Navier–Stokes flow, 237–238
production cycle, 208–209
simulation aims, 210
three-dimensional simulation, 231–238
well-posed problem, 16, 19
Integral transformation
Baiocchi, 2, 3, 23, 24, 27, 205
Interpolation operator
piecewise linear, 79, 88, 92, 98, 104, 106, 120, 137, 152, 172, 195

Jetting effect, 211, 232

Level set methods, 16, 18, 234–235
Lewy–Stampacchia inequalities, 4, 42, 70–72, 76, 129
Lions-Stampacchia Theorem, 36, 76, 91, 152
Lubrication theory, 13, 14, 209, 211, 219, 221, 231
Lumping operator, 86, 97

Mach number, 236
Marker and Cell method, 234–235
Maximum principle
discrete, 4, 110–111, 122, 156, 179
evolutionary variational inequality, 50–52
free boundary problem, 9
weak, 44
Metal casting, 209–210, 237
rheology, 233
semi-solid processing, 210, 213, 233, 237
three dimensional simulation, 233

Navier–Stokes equations, 212, 233, 234, 237–238, 252
numerical treatment, 233–234, 237–238
Navier–Stokes flow
generalized, 6
No-slip condition, 219, 221, 236

Obstacle problem, *see* variational inequality
elliptic, *see* Elliptic variational inequality
evolutionary, *see* Evolutionary variational inequality

Parabolic variational inequality, 3, 27
Peclet number, 223
Penalization problem, 4, 41
discrete, 112, 113, 131, 179, 180
Lagrange multiplier, 116
Lewy–Stampacchia type, 4, 41–43, 46, 53, 113, 180
solution method for, 122–126, 187–191
Phase field problem, 205
Piecewise constant in time function, 153, 159, 170, 176, 186
Polymer
rheology, 213–217
thermoplastics, 11, 207
thermosets, 11, 207

Porous media flow, 22, 23, 29, 205, 224
Pseudo-circle principle, 6, 229
Pseudo-concentration method, *see* Level set methods

Relaxation algorithm
  SOR method, 29
  SOR method with projection, 110, 132, 136, 191, 194
Reynold's equation, 14
Reynold's number, 219, 236
Rothe method, 4, 29, 55

Saddle point, 116, 236
Semi-discretization in time, *see* Rothe method
Semigroup
  nonlinear, 22
Shear rate, 212, 214, 222, 226
  tensor, 212
Shortest path, 229–231
Signorini problem, 39, 126
  (discrete) penalization, 131
  comparison of FEM and FVM, 127
  connection to interior obstacle problem, 129–131
  finite element error analysis, 127
  finite element formulation, 127
  finite volume error analysis, 128
  finite volume formulation, 127
  regularity statements, 126
Sobolev space
  definition, 34
  embedding, 39, 44, 61, 69, 70, 152, 227
  interpolation properties, 79, 92, 106, 120, 152, 173
SOR method with projection, 110, 132, 136, 139, 191, 194
Spatial regularity
  Dirichlet conditions, 56–63
  Neumann/Newton conditions, 63–69
Stefan type problem
  classical one- and two-phase, 3, 10, 20, 23, 24
  constitutive relations, 22
  enthalpy method, 19, 20, 205
  for super-cooled water, 205
  one-phase, zero-specific heat, 2, 3, 9, 204
  quasi-steady, one-phase, 10
  variational inequality for, 24, 27
Stokes flow
  generalized, 6, 234, 236–237
  saddle point formulation, 236
Stress tensor, 212
Subdifferential, 27, 42
Suction problem
  ill-posed problem, 9, 204

Thermoplastics
  amorphous, 207
  liquid crystalline, 207
  main properties, 11, 207–208
  semi-crystalline, 207
Thermosets
  cure reaction, 208, 209, 213
  main properties, 11, 207–208
Thixotropic effects, 213
Three-dimensional flow simulation, 6
  difficulties, 232
  effects, 232
  injection moulding, 231–238
  metal casting, 233
Trace
  concept, 34
  Theorem, 34, 65, 120
Triangulation
  admissible, 77, 144
  global inverse assumption, 117, 122, 183
  local inverse relation, 117, 184
  regular, 77, 144

weakly acute, 103, 109, 110, 156, 179

Variational inequality, 1
connection to free boundary problem, 1, 2
evolutionary, *see* Evolutionary variational inequality
of first kind, 26
of obstacle type, 26
of second kind, 27
parabolic, 3, 27
Vector-valued function
definition, 34
embedding, 35
Viscoelastic effects, 213, 214
Viscosity
dynamic, 212
shear, 212, 214, 221, 222, 233
Viscosity model
Arrhenius, 214, 216, 222
Carreau, 216, 222
Carreau–WLF, 216, 252
modified Cross, 216, 222
power law, 215, 216, 222, 225, 226
reactive, 237
WLF, 214
Volume of fluid method, 234–235, 237, 252

Weighted region problem, 230
Weld line, 211, 231, 239

MIX
Papier aus verantwortungsvollen Quellen
Paper from responsible sources
FSC® C105338

If you have any concerns about our products,
you can contact us on
ProductSafety@springernature.com

In case Publisher is established outside the EU,
the EU authorized representative is:
Springer Nature Customer Service Center GmbH
Europaplatz 3, 69115 Heidelberg, Germany

Printed by Libri Plureos GmbH
in Hamburg, Germany